Water Science & Technology
Diffuse Pollution VIII

Diffuse Pollution VIII

Selected Proceedings of the 8th IWA International Conference on Diffuse/Nonpoint Pollution, held in Kyoto, Japan, 25–29 October 2004

Issue editors: S Fujii[*], K Yamada[**] and T Kunimatsu[***]

[*] Research Center for Environmental Management, Kyoto University, Otsu City, Shiga, Japan
[**] Dept. of Environmental Systems Engineering, Ritsumeikan University, Kusatsu City, Shiga, Japan
[***] School of Environmental Science, The University of Shiga Prefecture, Hikone City, Shiga, Japan

Publishing

Programme Committee
Michael Burkart USA
Vladimir Chour Czech Republic
Brian D'Arcy UK
Ray Earle Ireland
Shigeo Fujii Japan
Hiroaki Furumai Japan
Sung Ryong Ha Korea
Ralph Heath South Africa
Jiri Holas Czech Republic
Robert Humphries Australia
Lyn Kirshner USA
Takao Kunimatsu Japan
Habib Muhammetoglu Turkey
Vladimir Novotny USA
Eduardo von Sperling Brazil
Nitayaporn Tonmanee Thailand
Govert Verstappen The Netherlands
Kiyoshi Yamada Japan

National Organising Committee
Shigeo Fujii Kyoto University, Japan
Hiroaki Furumai University of Tokyo, Japan
Takanobu Inoue Toyohashi University of Technology, Japan
Yukio Komai Hyogo Prefectural Institute of Public Health and Environmental Sciences, Japan
Takao Kunimatsu University of Shiga Prefecture, Japan
Victor Muhandiki Ritsumeikan University, Japan
Kiyoshi Yamada Ritsumeikan University, Japan

Organised by
IWA Specialist Group on Diffuse Pollution
Japan Society on Water Environment, Research Committee on Nonpoint Pollution

Sponsored by
Ritsumeikan University

Organised by
Ministry of Environment
Ministry of Land, Infrastructure and Transport
Ministry of Agriculture, Forestry and Fisheries
Prefecture Government of Shiga
Research Centre for Environment, Ritsumeikan University
Research Centre for Environment Quality Control, Kyoto University
The University of Shiga Prefecture

British Library Cataloguing in Publication Data
A CIP catalogue record for this book is available from the British Library
ISBN: 1 84339 553 3
ISBN-13: 978 1 84339 553 9

Contents

- 1   **Management of agricultural nonpoint source pollution in China: current status and challenges** X. Wang
- 11   **Distribution of dioxins in surface soils and river-mouth sediments and their relevance to watershed properties** M. Kanematsu, Y. Shimizu, K. Sato, S. Kim, T. Suzuki, B. Park, K. Hattori, M. Nakamura, H. Yabushita and K. Yokota
- 23   **Study of water quality distribution in Lake Biwa in consideration of runoff pollutant loads from its catchment basin** A. Ichiki, A. Sasaki, N. Sakata, K. Nakakura and H. Yamate
- 33   **Estimation of annual pollutant loadings in two small catchments and examination of their differences caused by regional properties** S. Fujii, M. Moriya, P. Songprasert and H. Ihara
- 45   **Effects of soil erosion on water quality and water uses in the upper Phong watershed** S. Sthiannopkao, S. Takizawa and W. Wirojanagud
- 53   **Long-term fluctuation and regional variation of nutrient loads from the atmosphere to lakes** T. Kunimatsu and M. Sudo
- 63   **Water chemistry gradient in a degraded bog area** R. Iqbal, S. Akimoto, K. Tokutake, T. Inoue and H. Tachibana
- 73   **Characteristics of sediments in a newly constructed reservoir in Japan** Y. Sakurai and S. Haruta
- 79   **Evaluation of nutrient loads from a mountain forest including storm runoff loads** T. Kunimatsu, T. Otomori, K. Osaka, E. Hamabata and Y. Komai
- 93   **Evaluation of hydrological processes in a mountainous small basin using a quinone biomarker** M. Fujita, H. Haga, K. Nishida and Y. Sakamoto
- 101   **Nitrogen removal function of recycling irrigation system** T. Hitomi, I. Yoshinaga, Y.W. Feng and E. Shiratani
- 111   **Recycling mineral nutrients to farmland via compost application** Y.Y. Liu, M. Ukita, T. Imai and T. Higuchi
- 119   **Study on the potential of farmland soils as non-point sources of nitrogen and phosphorus in Japan** M. Ukita, X. Shi, T. Higuchi, Y. Arkin and M. Fukada
- 131   **Decrease in herbicide concentrations and affected factors in lagoons located around Lake Biwa** M. Sudo, M. Nishino and T. Okubo
- 139   **Estimation of pesticide runoff from paddy fields to rural rivers** A. Numabe and S. Nagahora
- 147   **Economic valuation of reduction in nitrogen outflow from a paddy field area equipped with a recycling irrigation facility** Y.W. Feng, E. Shiratani, I. Yoshinaga, and T. Hitomi
- 155   **On-site treatment of turbid river water using chitosan, a natural organic polymer coagulant** M. Sekine, A. Takeshita, N. Oda, M. Ukita, T. Imai and T. Higuchi
- 163   **Study on purification mechanism in soil penetration facility for effluents from urban area and control strategies** K. Yamada, D. Ujiie and K. Nishikawa
- 175   **Trace metal levels in sediments deposited in urban stormwater management facilities** J. Marsalek, W.E. Watt and B.C. Anderson
- 185   **Lead isotope ratios in urban road runoff** M. Shinya, K. Funasaka, K. Katahira, M. Ishikawa and S. Matsui
- 193   **The characteristics and measuring technique of refractory dissolved organic substances in urban runoff** K. Wada, S. Yamanaka, M. Yamamoto and K. Toyooka

- 203 **Runoff and loads of nutrients and heavy metals from an urbanized area** H. Shirasuna, T. Fukushima, K. Matsushige, A. Imai and N. Ozaki
- 215 **Dispersion and dry and wet deposition of PAHs in an atmospheric environment** N. Ozaki, K. Nitta and T. Fukushima
- 225 **Characteristics of litter waste in highway storm runoff** L.-H. Kim, J. Kang, M. Kayhanian, K.-I. Gil, M.K. Stenstrom and K.-D. Zoh
- 235 **Correlation analysis among highway stormwater pollutants and characteristics** Y.H. Han, S.L. Lau, M. Kayhanian and M.K. Stenstrom
- 245 **Characteristics of particle-associated PAHs in a first flush of a highway runoff** R.K. Aryal, H. Furumai, F. Nakajima and M. Boller
- 253 **Water quality modeling to evaluate BMPs in rice paddies** J.H. Jeon, C.G. Yoon, H.S. Hwang and K.W. Jung
- 263 **Evaluation of AnnAGNPS in cold and temperate regions** S. Das, R.P. Rudra, P.K. Goel, B. Gharabaghi and N. Gupta
- 271 **Nonlinear regression approach to evaluate nutrient delivery coefficient** M.S. Bae and S.R. Ha
- 281 **Comparative study of two watershed scale models to calculate diffuse phosphorus pollution** A. Kovacs
- 289 **Integrating principles of nitrogen dynamics in a method to estimate leachable nitrogen under agricultural systems** M. Burkart, D. James, M. Liebman and E. van Ouwerkerk
- 303 **Indicator of risk of water contamination by phosphorus from Canadian agricultural land** E. van Bochove, G. Thériault, F. Dechmi, A.N. Rousseau, R. Quilbé, M.-L. Leclerc and N. Goussard
- 311 **Estimation of particulate nutrient load using turbidity meter** K. Yamamoto and T. Suetsugi
- 321 **Application of monitored natural attenuation to remediate a petroleum-hydrocarbon spill site** C.M. Kao, W.Y. Huang, L.J. Chang, T.Y. Chen, H.Y. Chien and F. Hou
- 329 **Comparison of several methods for BAP measurement** J. Nakajima, Y. Murata and M. Sakamoto
- 337 **Evaluation of atmospheric deposition of nitrogen to the Feitsui Reservoir in Taipei** S.L. Lo and H.A. Chu

Management of agricultural nonpoint source pollution in China: current status and challenges

Xiaoyan Wang*,**

*College of Resource, Environment & Tourism, Capital Normal University, Beijing 100037, P.R. China
**Key Laboratory of Resource, Environment & GIS, Beijing City, Beijing 100037, P.R. China

Abstract Water quality in China shows an overall trend of deterioration in recent years. Nonpoint source pollution from agricultural and rural regions is the leading source of water pollution. The agricultural nonpoint source pollutants are mainly from fertilization of cropland, excessive livestock and poultry breeding and undefined disposal of daily living wastes in rural areas. Agricultural nonpoint sources contribute the main source of pollution to most watersheds in China, but they are ignored in management strategy and policy. Due to the lack of full understanding of water pollution control and management and the lack of perfect water quality standard systems and practical legislative regulations, agricultural nonpoint source pollution will become one of the biggest challenges to the sustainable development of rural areas and to society as a whole. The system for agricultural nonpoint source pollution control in China should include an appropriate legislation and policy framework, financing mechanisms, monitoring system, and technical guidelines and standards. The management of agricultural nonpoint source pollution requires multidisciplinary approaches that will involve a range of government departments, institutions and the public.
Keywords Agricultural nonpoint source pollution; challenge; China; management; Miyun Reservoir

Introduction

After effective control of point sources pollution, nonpoint source pollution has become a major problem causing degradation of water quality; therefore, nonpoint source control has been increasingly discussed in some developed countries. Since the 1960s, the USA and other countries have been aware of nonpoint source pollution. Some legislation such as the Clean Water Act of 1972 and the Coastal Zone Act Reauthorization Amendments of 1990 in the USA encourage states to enact regulation of nonpoint sources (Malik et al., 1994). The European Union issued the Water Framework Directive in 2000. The Directive mandates the introduction of measures to control diffuse pollution (Reeves et al., 2003). Mandates or incentives have been applied to so-called "Best Management Practices" (BMP) to reduce soil erosion and other runoff, and to restrict use of fertilizer and pesticide by taxes (Helfand, 1995).

Unlike point sources, emissions from nonpoint sources are influenced by stochastic events such as temperature and precipitation. The loads are not measured with certainty but in fact represent a probability distribution around the actual discharged loads. This lack of ability to connect, scientifically, the source of effluent with its entry into a water body has made the traditional, effluent-based regulations impossible to implement (Parker, 2000). These difficulties may explain the new focus on environmental management approaches at the international and national levels (Reeves et al., 2003).

China is one of the largest producers and consumers of chemical fertilizers in the world, and the excessive nutrient discharge from agricultural watersheds is considered to be an important source of nonpoint source pollution (Yan et al., 1999). However, there are few researches related to these aspects in China. Research on nonpoint source pollution (NPSP) started in China in the 1980s. There has been more and more progress on

doi: 10.2166/wst.2006.033

load distribution. Compared with point sources, however, few abatement efforts have been implemented on nonpoint sources, and studies aiming at nonpoint source regulation were also rare. This paper mainly discusses the situation and difficulties in management of agricultural NPSP in China.

Current situation of NPSP in China

To support its large population, China's agriculture has one of the highest fertilizer and pesticide application rates in the world. Large amounts of fertilizer and pesticide enter surface waters with runoff, which mainly occurs in the summer months due to more runoff-producing rainstorm events in that season. Surface water inputs from direct drift, leaching and erosion are additional important sources contributing to nonpoint source pollution.

Due to the limitations of the nation's conditions, the government policies seldom involve agricultural pollution. Hence, agriculture contributes more and more pollutants to the water environment in China.

Since the 1970s, the eutrophication of major lakes and water systems in China has been getting worse rapidly. According to the monitoring data of water quality in large major rivers, lakes and reservoirs in the country in recent decades, the water quality shows an overall tendency of deterioration, and such a tendency has been kept at a relatively steady level in recent years. More than 85% of the nation's lakes are at serious eutrophication stages. Over 50% nonpoint source pollution discharges to the lakes in eastern China (Jin *et al.*, 1990).

Nearly 82% of 532 main rivers in China are contaminated by excessive nitrogen at various levels. The higher the stream order, the heavier the pollution is. The water quality in rainy seasons is worse than that in dry seasons (SEPA, 2002).

Figure 1 illustrates the water quality in typical rivers, with length over 121,000 kilometres, in the year of 2002. The proportion of grades I, II and III, which are suitable for domestic use, to the whole length is 61.4%. About 38% of the total length of rivers are so degraded (falling in the category of ultra-V) that they have lost many functions, including use for irrigation, and themselves become sources of pollution.

It is remarkable that nonpoint source pollution from agricultural and domestic wastewater aggravates rapidly, and this is the main pollution source in some regions. In Lake Taihu and Lake Dianchi, the ratios of their nonpoint sources (mainly nitrogen) to their total sources are 75% and 70%, respectively (Table 1).

Investigation has revealed that nonpoint source pollution from agricultural and rural regions is the leading source of water pollution. Due to different physical characters and social productions, there are various types of nonpoint source pollution. In general,

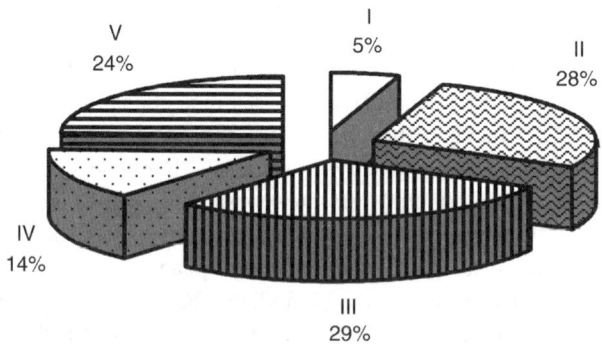

Figure 1 Water quality of typical rivers of China in the year of 2002

Table 1 Load contributions from point and nonpoint source to rivers in 2000

Pollutant	Point source		Nonpoint source		Contribution of point source (%)	Contribution of nonpoint source (%)
	discharge (Million ton)	into stream (Million ton)	discharge (Million ton)	into stream (Million ton)		
COD	18.94	12.45	75.74	8.31	60	40
NH_3-N	1.82	1.10	7.07	0.71	61	39
TN	3.81	2.10	28.64	4.24	33	67
TP	0.67	0.36	6.12	0.62	37	63

Data source from SEPA, 2002

nonpoint source pollution in China results from human activities. The lack of best management practices in agriculture and the conventional lifestyle in rural areas are two important factors contributing to the agriculture pollution. For example, improper farming practices, discharge of livestock and human wastes, and the lack of management policies on the agricultural environment are important reasons. Table 2 shows the contribution of nonpoint sources to the watershed of Lake Dianchi.

Chemicals overuse

China has the highest population in the world, but has limited arable land. In order to meet the food demands of such a huge, growing population, the agriculture has been more and more extremely intensive in using inorganic fertilizers to increase crop productivity. Rapid growth in China's per hectare chemical fertilizer application, from less than 10 kg in 1960 to more than 800 kg in 2000, is required to increase grain production, but has caused many environment problems such as groundwater pollution (Li and Zhang, 1999; Editorial Committee of Agricultural Yearbook in China, 2003). In 17 provinces, per hectare chemical fertilizer application is higher than the recommended level (225 kg) in the world.

Due to high profit from growing vegetables, fruits and flowers, farmers commonly use high rates of N and P fertilizers, about 10 times higher (average fertilizer application rate is 569–2,000 kg per hectare) than for grain crops. Increasing vegetable area with high fertilizer input is one of the biggest potential problems for eutrophication of water bodies in watersheds (Zhang et al., 2004).

In addition, the average ratio of N:P:K in fertilizers in 1993 was 1:0.31:0.11, with much higher N- and P-rates than the world average (1:0.5:0.5). Large amount of N-fertilizer and poor utilization rates may lead to nitrate leaching and hence polluting groundwater. These effects were investigated in high-application regions at levels over 500 kg/hm^2 (Zhang et al., 1996). The increase of P fertilizer application causes long-term accumulation of phosphorus in soils, resulting in greater overall losses of phosphorus to aquatic ecosystems.

Table 2 Nutrient load of croplands from different sources in Lake Dianchi (kg/ha) (Modified from Zhang et al., 2004)

Source	1960s		1980s		Present	
	N	P_2O_5	N	P_2O_5	N	P_2O_5
Fertilizer	5	1	98	16	401	248
Manure	19	11	160	85	223	122
Village	29	8	56	15	66	18
Total	53	20	314	116	690	387

Livestock breeding

According to the State Environmental Protection Administration (SEPA) survey on pollution conditions relating to nationwide livestock and poultry breeding, in general the discharge of animal waste had approached an alarming level, while less than 10 per cent of the total breeding farms had experienced environmental impact evaluations. Animals on farms produced 1.9 billion tons of excrement during 1999, almost 2.5 times greater than the amount of solid waste discharged by China's industrial sector (Zhang, 2000). The estimated annual losses of COD, BOD and NH_3-N from manure in 2001 were 7.28 million tons, 4.99 million tons and 1.32 million tons, respectively. The COD loss from manure is near to the sum of COD from industrial and municipal wastewater. The manure loss from mid-small-size livestock farms is nearly 70% of the nation's total manure loss.

At the same time, animal breeding farms in the rural region had tried to develop into highly animal-concentrated farms. The N and P losses from such a concentrated region have reached very high levels, as much as 1,000 kg N and 600 kg P_2O_5 per hectare agricultural land, far surpassing the carrying capacity of soil to these organic nutrients (Zhang et al., 2004).

The investigation of the feed, excreta, soil, surface and subsurface waters in the livestock farms in Jiangsu Province, using the case region survey and nutrient analysis approaches, showed that the discharge of liquid manure and the run-off loss of solid manure disposed on the ground were the main environment problems in livestock production. The surface run-off from the wastes of livestock production is the main route causing environmental problems (Yang et al., 2001).

Rural residence

Due to the lack of sewage treatment and excessive waste accumulation in rural areas in China, especially in the transitional areas with fast urbanization, waste from daily life is discharged randomly even into rivers. TN and TP concentrations in the village runoff are nearly 10 times those from farm runoff (Wang et al., 2003b; Wan et al., 2000). The nonpoint source pollution from agricultural and rural areas will become one of the biggest challenges to sustainable development in China.

In short, agricultural nonpoint source pollution in China has become the predominant source to the aquatic system, especially on the seasonal climate zones in eastern and southeastern China with higher crop production and high population density. Due to diversity in physical conditions and farming practices, the pollution contribution varies (see Table 3).

The most direct and important impact of agricultural NPS on the environment is the degradation of China's drinking water resource. It is estimated that the economic loss by NPS-induced water pollution in China is 0.5 ~ 1% of GDP. The impacts of NPS on soil are cumulative, persistent, and deteriorating, and consequently influence groundwater quality and food security greatly. NPS has significantly negative impacts on rural social development and upon society as a whole.

Table 3 Nonpoint source pollution from different sources in some regions in China

Region	Farm runoff (%)	Livestock (%)	Rural residence (%)	Data source
Shanghai	12.25	18.67	18.09	Zhang et al., 1997
Dianchi Lake	53			Wang et al., 2003a
Miyun Reservoir	30	61		Wang et al., 2003b
Hangjiahu	23.62	43.8	16.07	Qian et al., 2002
Taihu Lake	47.6	33.4	19	Wan et al., 2000

Concerned policies and regulations in China

The maintenance of water quality has long been recognized as important for environmental and economic purposes and has prompted the introduction of statutory measures to protect the water environment. China has implemented some environmental regulations such as industrial air pollution and waste dumping. But much less attention has been paid to the environmental and health effects of farming chemicals. The environmental regulations designed for the industrial sector should be modified and enforced for the rural sector that has developed rapidly since the early 1980s. Several environmental protection laws and regulations include "Environmental Protection Law of the People's Republic of China", "Law of the People's Republic of China on Prevention and Control of Water Pollution", "Water Law of the People's Republic of China" and "Law of the People's Republic of China on Water and Soil Conservation".

At present, no legal and operational framework is established in China to control agricultural management practices, leading to high nonpoint source pollutions. Most environmental laws and regulations are constituted around point source pollution, though a few involve nonpoint source pollution but lack pertinence, feasibility and maneuverability. The direct cause is the deficiency of financial support from both the central and local governments. The total financial input to the environmental protection is less than 1% of the nation's GNP.

Another fundamental problem that remains is that many of the existing agricultural development policies are inconsistent with environmental protection policies: for the central government, it is drawing the frame of policy and legislation; whereas, the local government has no technical standards and corresponding management practice; for supervisor branches, there is lack of technical standard and system for source monitoring; for farmers, they have no technical standard to obey in fertilizer applications and cultivation practices, even in drinking water protection areas.

Other options to address nonpoint source pollution

In general, there are two main policy approaches for aquatic environmental pollution control, i.e. mandatory means and economic incentive. Many developed countries and developing countries had taken the mandatory means as the basic instrument through issuing permit licenses and implementing environmental rules or standards. Due to the inflexibility of mandatory means on environmental impact and economic efficiency, the economic means becomes a supplement, or a substitution, or combined measure for aquatic environmental pollution control. On the other hand, managers have paid more attention to public participation and education. Research and practice on agricultural pollution control have been accomplished in many western countries for a few decades. A series of effective control measurements have been designed and implemented from different viewpoints and more incentive policies are applied. Quantitative frames and evaluation policies have become an approach to get optimal schemes by integrated evaluation of environmental-economic benefit, for example, allocation of pollution load on the basis of gross load control using economic measures such as tax, subsidy and pollution trading to eliminate nonpoint source pollution (Wang, 2003).

In recent years, nonpoint source contributions have remained and have increased as a main source causing surface water pollution in China. It has been noted that the conventional command-and-control regulations are ineffective in controlling agricultural nonpoint source pollution, whereas the watershed abatement trading between point and nonpoint sources may be more cost-effective. The feasibility of point-nonpoint effluent trading in China has been discussed in detail by Zhang and Wang (2002).

Due to the vast area in China, there are huge discrepancies in regional natural conditions, agricultural production, and social and economic development. Different policies can be set for different watersheds. In a watershed with advanced development, such as Lake Taihu, the incentive approach with tax, such as tax on pollutant and chemical charges to restrain improper activities of farmers, can be practicable. In the watershed with the least advanced development, such as Lake Dianchi, the incentive approach becomes dominant, where the point source pollution has considerable portions and a tradeable market is necessary.

The hallmark advantage of economic incentive programs is, of course, their potential to minimize compliance costs by optimizing cost-effectiveness. The magnitude of cost savings depends upon factors such as differences in dischargers' marginal costs and transaction costs. In spite of limited reliable data, a preliminary analysis indicated that the proposed incentive program could provide significant cost savings compared to mandatory BMPs. At least two additional factors make incentive programs attractive. First, they are based on decentralized decision-making and preserve the flexibility of individual farmers to respond to changes in economic, environmental and technological conditions. Second, the programs encourage innovation by providing direct financial rewards for creating better and cheaper pollution control methods.

As discussed above, nonpoint emissions and abatement abilities in different watersheds may display different features, and the variance of nonpoint emissions may increase or decrease with the abatement levels. In practice, a case-by-case study is compelling and a specific model should be developed at local levels (Young & Karkoski, 2000).

In summary, in the case of diffuse sources of pollution, appropriate regulation must take account of the unobservability and unverifiability of individual emissions. Which of the methods of pollution control is appropriate depends on other factors: the information availability, type of resources to be regulated, degrees of uncertainty, social cost of damage, the number of polluters to be controlled, monitoring costs, and transaction costs. Each case must be decided on its own merit. Much remains to be done in the area of diffuse sources.

Apart from the management practice regulations and limited trading programs, the treatment of diffuse sources has remained mostly theoretical. The main problem is that the controlling prosecution has been mainly concentrated on end-control engineering projects such as establishment of sewage farm and artificially-built or reclaimed wetland at bayou or watercourse, even in the watersheds where agriculture is the leading contributor to eutrophication of water systems. To solve the problems, it was suggested that proper regulations to confine the environmentally unfriendly agricultural activities should be established and implemented in different watersheds as soon as possible. Other complex issues, such as regulation, monitoring and administration, must also be considered with care (O'Shea, 2002). Education and technical and financial assistance are effective to encourage adoptions of less polluting farm practices. Research findings indicate that the adoption of an improved management practice is most strongly influenced by producer perceptions of its effect on profitability (Feather and Cooper, 1995).

Future challenges for this problem in China

The control of diffuse pollution is an important aspect of maintaining water quality to be tackled by the regulators. No country has successfully finished this tough task. Due to the lack of full understanding of water pollution control and the lack of perfect water quality standard systems and practical legislative regulations, it is much more difficult in China because of diversity in the vast areas, inefficient management on water and land and other harassments.

- Difficulty in the technology

Non-point source emissions rely partly on some random variables such as rainfall. It is hard to identify and measure nonpoint source emissions at the source level. It is impossible to get complete information on nonpoint sources, including discharges and variations by other factors. An alternative solution is to find the effect–result relation between pollution damage and economic behavior, which is the basis of management policy (Russell and Shogren, 1993; He and Wang, 1999).

- Limitation in the national situation

Traditionally, farmers in China are not responsible for the control of pollution caused by agricultural activities. The conventional command-and-control regulation is therefore ineffective for agricultural nonpoint sources, but the economic incentive-based policies can be applied (Wang, 2003). However, there is institutional limitation to prevent environmental quality objectives from being brought into policy-making.

- Attitude of farmers

Farmers possess land but not water resources. This separation of property right leads to less initiative for farmers to reduce water pollution, whereas more care is taken of their benefits from farming (He and Wang, 1999).

- Difficulty in substitution of pesticide

Reliance on pesticides is the primary or sole means of pest control; however, it may have a number of harmful side effects. Since the mid-1970s, researchers have described unintended consequences from pesticide use in agriculture, particularly in developing countries. Several economic studies have questioned whether the current patterns of pesticide use are economically and socially efficient. Policy makers in many countries have begun to regulate pesticide use, and have taken an interest in alternative methods for controlling agricultural pests. However, it is impossible for alternative methods to replace pesticide completely (Widawsky et al., 1998).

- Number of pollutants and spatial discrepancy

Numerous pollutants cause difficulty in controlling nonpoint source pollution from acquired information and reduce the level of cooperation in pollution alleviation.

Although nonpoint source pollution varies spatially and temporally a rather uniform policy and standards are performed. These controls are distinguished between the characteristics of pollutants and the characteristics of receiving waters; therefore, the impact on the local environment can be directly assessed and control can be exercised (DETR, 1997).

In short, the agricultural pollution problems were typical: a pressing pollution problem existed, the technology to control the problem was available, but the concern about costs and the administrative difficulty of creating a compliance program were stalling progress. The lack of aggressive enforcement was not simply due to a lack of legal authority, it reflected the common notion that nonpoint sources cannot be held individually accountable for specific pollution-control requirements because the sources are not only diffuse but also too many to administer individually. The same reasoning has led policymakers nationwide to an apparent impasse in the search to control nonpoint sources. The crux of this problem is finding a way to: (1) make individual farmers accountable for the pollution they generate; (2) maximize cost-effectiveness for farmers and minimize transaction costs for farmers and regulators; (3) make the control program flexible and practical for farmers to implement; and (4) simplify the administration by regulatory agencies.

Conclusions

Nonpoint source pollution from agricultural and rural regions is thought to be the leading source of water pollution and will become one of the biggest challenges to sustainable development of rural areas and upon society as a whole. The main agricultural nonpoint source pollutants are from fertilization of cropland, excessive livestock and poultry breeding and undefined disposal of daily living waste in rural areas.

There are no perfect water quality standard systems and practical legislative regulations in China to control agricultural management practices, leading to high nonpoint source pollution inputs because of the diversity of the vast areas, inefficient management of waters and lands and other harassments. The system for agricultural nonpoint source pollution control in China should include an appropriate legislative and policy framework, financing mechanisms, monitoring system, and technical guidelines and standards.

The management of agricultural nonpoint source pollution requires a multidisciplinary approach that will involve a range of government departments and institutions. The integration of water quality management on surface water and groundwater for both point and nonpoint source pollutions at the scale of watershed should be based on the full knowledge of the functions and interactions between water and the related ecological factors in the watershed.

Economic approaches will be the effective practice of nonpoint source pollution control, public participation and education, as well as an information sharing policy.

Acknowledgements

The authors are most grateful to Prof. Fujii from Kyoto University, Japan and two anonymous referees for their helpful insights, which contributed to an obvious improvement of the manuscript. Special thanks to Dr. P. Wang from UMCES, USA for correcting English writing and valuable comments. This work was jointly supported by the National Ministry of Education and Beijing Municipal Education Commission. Thanks to the Water soil Protecting Station of Miyun County for help in the fieldwork.

References

DETR (1997). *Economic Instruments for Water Pollution*. Department of the Environment Transport and the Regions, pp. 71.

Editorial Committee of Agricultural Yearbook in China (2003). *Agricultural Yearbook in China (1986–2002)*, Beijing: Chinese Agriculture Press (in Chinese).

He, P. and Wang, J.J. (1999). Current situation, puzzle and challenge of control and management on Nonpoint source pollution. *Agro-environmental Protection*, **18**(5), 235–236 (in Chinese).

Helfand, G.E. (1995). Alternative pollution standards for regulating nonpoint source pollution. *Journal of Environmental Management*, **45**(3), 231–241.

Jin, X.C., Liu, H.L., Tu, Q.Y., Zhang, Z.S. and Zhu, X. (1990). *Lake Eutrophication in China*, China Environment Science Press, Beijing, China (in Chinese).

Li, Y. and Zhang, J.B. (1999). Agricultural diffuse pollution from fertilizers and pesticides in China. *Water Science and Technology*, **39**(3), 25–32.

Malik, A.S., Larson, B.A. and Ribaudo, M. (1994). Economic incentives for agricultural nonpoint source pollution control. *Water Resources Bulletin*, **30**(3), 471–480.

O'Shea, L. (2002). An Economic Approach to Reducing Water Pollution: point and diffuse sources. *The Science of the Total Environment*, **282–283**, 49–63.

Parker, D. (2000). Controlling agricultural nonpoint water pollution: cost of implementing the Maryland Water Quality Improvement Act of 1998. *Agricultural Economics*, **24**(1), 23–31.

Feather, P.M. and Cooper, J. (1995). Voluntary Incentives for Reducing Agricultural Nonpoint Source Water Pollution. In *Agriculture Economic Research Service Agriculture Information Bulletin*, pp. 716.

Qian, X.H., Xu, J.M. and Shi, J.Ch. (2002). Comprehensive Survey and Evaluation of Agricultural Nonpoint Source Pollution in Hang-Jia-Hu Waternet Plain. *Journal of Zhejiang University*, **28**(2), 147–150 (in Chinese).

Reeves, A.D., Kirk, E.A. and Sherlock, K.L. (2003). Institutional challenges faced when tackling the issue of diffuse pollution. In *Proceedings of 7th Diffuse Pollution Conference*, Dublin, Ireland, 12, pp. 39–43.

Russell, C.S. and Shogren, J.F. (1993). *Theory, Modeling and Experience in the Management of Nonpoint-source Pollution*, Kluwer Academic Publishers, Boston.

SEPA (State Environmental Protection Administration of China) (2002). *Report on the State of the Environment in China in 2001*. Available online www.zhb.gov.cn /649368273124589568 /index.shtml (in Chinese).

Yang, J.S., Chen, D.M., Liu, G.M. and Li, H.F. (2001). Nutrient cycling and environmental effect of livestock production in case regions of Jiangsu Province. *China Environmental Science*, **21**(5), 468–471.

Young, T.F. and Karkoski, J. (2000). Green evolution: are economic incentives the next step in nonpoint source pollution control? *Water Policy*, **2**(3), 151–173.

Wan, X.H., Qiu, D. and Zhao, X.M. (2000). Analysis of Pollution Characters of the Big Animal Farms in the Taihu Area. *Agricultural Environment and Development*, **64**(2), 35–38 (in Chinese).

Wang, J.N., Ge, Ch.Zh. and Zhang, Y. (2003a). *Institution and Policy of Control of Water Pollution in China*, China Environment Science Press, Beijing, China (in Chinese).

Wang, X.Y. (2003). *Nonpoint source pollution and its management*, Ocean Press, Beijing, China (in Chinese).

Wang, X.Y., Guo, F. and Hu, Q.J. (2003b). Nonpoint source pollution loading from the watersheds of Chaohe River and Baihe River around Miyun Reservoir. *Urban Environment and Urban Ecology*, **16**(1), 31–33 (in Chinese).

Widawsky, D., Rozelle, S., Jin, S.Q. and Huang, J.K. (1998). Pesticide productivity, host-plant resistance and productivity in China. *Agricultural Economics*, **19**(1–2), 203–217.

Yan, W.J., Yin, C.Q. and Zhang, S. (1999). Nutrient budgets and biogeochemistry in an experimental agricultural watershed in Southeastern China. *Biogeochemistry*, **45**, 1–19.

Zhang, C.G. (2000). Review of the livestock revolution in China. In *Implication of Asian economic crisis for the livestock industry*, pp. 203–213.

Zhang, W. and Wang, X.J. (2002). Modeling for point-non-point source effluent trading: perspective of non-point sources regulation in China. *Sci. Total Environ.*, **292**(1–3), 167–176.

Zhang, W.L., Tian, X.L., Zhang, N.X. and Li, X.Q. (1996). Nitrate pollution of groundwater in Northern China. *Agriculture, Ecosystems and Environment*, **59**(3), 223–231.

Zhang, D.D., Zhang, J.Q., Wang, Y.G. and Zhang, X.H. (1997). The Main Non-point Source Pollution in Shanghai Suburbs and Harness Countermeasure. *Environmental Science in Shanghai*, **16**(3), 1–3 (in Chinese).

Zhang, W.L., Wu, S.X., Ji, H.J. and Kolbe, H. (2004). Estimation of Agricultural Non-Point Source Pollution in China and the Alleviating Strategies I. Estimation of Agricultural Non-Point Source Pollution in China in Early 21 Century. *Scientia Agricultura Sinica*, **37**(7), 1008–1017.

Distribution of dioxins in surface soils and river-mouth sediments and their relevance to watershed properties

M. Kanematsu*, Y. Shimizu*, K. Sato*, S. Kim*, T. Suzuki*, B. Park*, K. Hattori*, M. Nakamura**, H. Yabushita** and K. Yokota***

*Research Center for Environmental Quality Control, Kyoto University, 1-2 Yumihama, Otsu, Shiga 520-0811, Japan (E-mail: kanematsu@biwa.eqc.kyoto-u.ac.jp)
**Hiyoshi Ecological Services Co., 908 Kitanosyo, Omihachiman, Shiga 523-0806, Japan
***Lake Biwa Research Institute (LBRI), 1-10 Uchidehama, Otsu, Shiga, 520-0806, Japan

Abstract The dioxins toxic equivalent (TEQ) concentration in surface soils, river sediments and river-mouth sediments was measured by the CALUX assay in the Yasu and Ado River basins around Lake Biwa, Japan. In order to examine the distribution of dioxins in each watershed, we evaluated and compared the correlation between the dioxins TEQ concentration and the solid characteristics (i.e. organic carbon content and particle size distribution) of all samples. In both basins, the dioxins TEQ concentration in forest soil correspondingly showed a very good linear relationship to organic carbon content. On the other hand, the dioxins TEQ concentration in paddy field was significantly high, although organic carbon content was relatively low. Generally, the smaller particles have the higher dioxins TEQ concentrations in surface soils, and river sediments were composed of very coarse particles and had relatively low dioxins TEQ concentration. Therefore, we expected high dioxins TEQ concentration in river-mouth sediment, which was, however, not the case. Although the dioxins TEQ concentration in river-mouth sediments is low, the degree of dioxins pollution was different in each basin. The difference was considered to come from the difference of watershed properties including land use, river-slope, dam construction as well as the surface soil pollution.
Keywords Dioxins; CALUX assay; river sediment; river-mouth sediment; surface soil; watershed property

Introduction

In recent years, the contamination of the micro-organic pollutants (MOPs) has been of considerable concern and their risk not only to humans but also to the ecosystem is a worry. Because of persistency, bioconcentration potential, ubiquitous distribution and strong toxicity, dioxins are one of the most dangerous pollutants among the various MOPs. Dioxins are released into the atmospheric environment from incinerators of municipal/industrial wastes, and the deposition of them on the ground can occur via various pathways: dry gaseous, dry particulate and wet forms (Lohmann and Jones, 1998). Additionally, some chlorinated organic pesticides which contain dioxins as byproducts, such as pentachlorophenol (PCP) and chlorinitrofen (CNP), were applied in the paddy fields in Japan, especially during 1960–1980s (Masunaga *et al.*, 2001). Due to their high hydrophobicity, it is said that dioxins are mainly sorbed onto the organic component of surface soils. They are gradually washed away with the soil erosion caused by rain and/or snow, and flow into the water environment with particulate matter (i.e. SS and sediment). Then, highly-chlorinated dioxins exist and move with particulates in a higher ratio into the water environment. Therefore, the movement and fate of dioxins should be considered with particulate matter in watersheds.

In this research, the Yasu and River basins around Lake Biwa (the largest freshwater lake in Shiga prefecture, Japan, in Figure 1) were chosen, and a number of surface soils,

doi: 10.2166/wst.2006.034

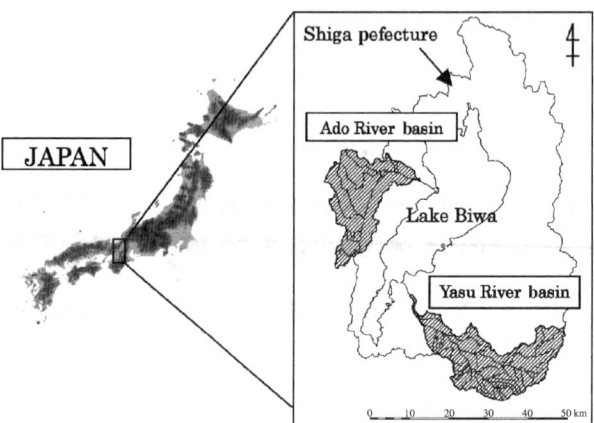

Figure 1 The location of the Lake Biwa, Yasu River basin and Ado River basin

river sediments and river-mouth sediments are collected in these basins. We measured the dioxins TEQ concentration in all of them by the chemically activated luciferase expression (CALUX) assay. We also measured the characteristics of the soil and sediment (i.e. organic carbon content and particle size distribution), and compared the relationship between the dioxins TEQ concentration and the characteristics of all samples to grasp the characteristics of dioxins distribution and movement in each watershed.

Sampling

Surface soils and river sediments

In this research, sampling points of surface soils and river sediments were determined considering land use and tributaries in each watershed. Sampling points of surface soils and river sediments are shown in Figure 2. Both surface soils and river sediments were collected from approximately the top 5 cm by a stainless shovel after removal of crude materials such as fallen leaves and stones. After sampling, they were freeze-dried immediately by freeze-drier (FDU-830, EYELA), and then kept in refrigerator until further analysis.

River-mouth sediments

Sampling points of river-mouth sediments were determined considering the flow regime and water depth at the river-mouth. At the Yasu River-mouth (Figure 2c), we mainly collected them along three directions. On the other hand, at the Ado River-mouth (not shown), we collected them along the shore because water depth suddenly increased. By the way, the Yasu River had flowed into Lake Biwa from different points (north and south river mouths) until 1979. Therefore, we also sampled sediments near these old river-mouths. We expected that sediments collected near the present river-mouth and the old Yasu River-mouth had accumulated from mainly 1980 and until mainly 1979, respectively. River-mouth sediments were collected from about the top 10 cm by core sampler (diameter: 5 cm). After sampling, they were also freeze-dried and stored in the refrigerator.

Methods

Organic carbon content of all samples was analyzed by high temperature combustion (900 °C) using TOC analyzer (TOC-5000A, Shimadzu Co.) with a solid sample module (SSM-5000A, Shimadzu Co.). The measured value of the inorganic carbon content was

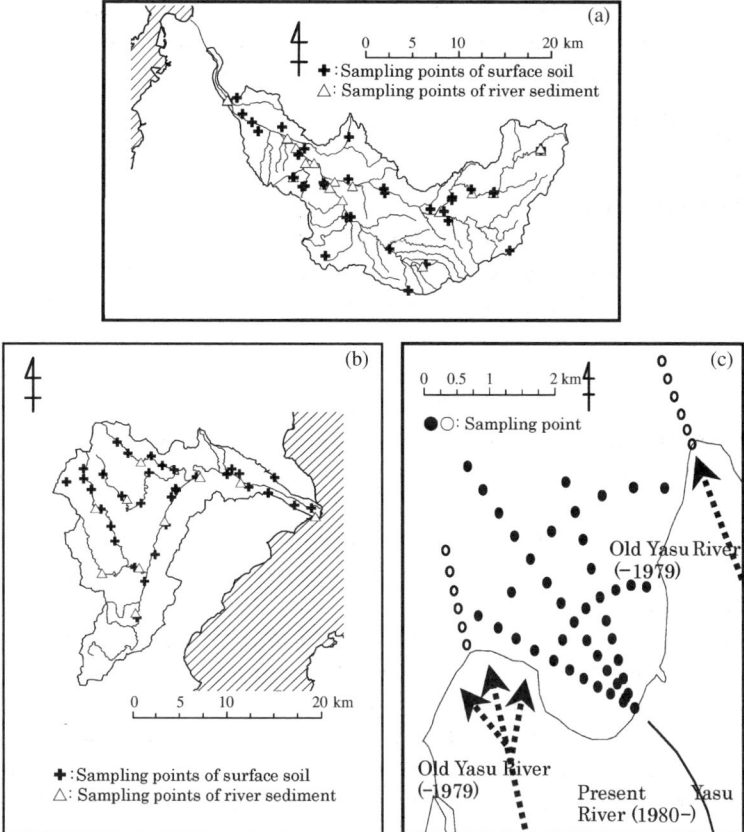

Figure 2 (a) Sampling Points in the Yasu River Basin (b) Sampling Points in the Ado River Basin (c) Sampling Points of Yasu River-mouth sediments (● and ○ collected near the present Yasu river-mouth and the old river-mouth, respectively)

relatively low and negligible. Therefore, we used the value of the total carbon as the value of organic carbon.

Surface soil and river sediment were fractionated by using the stainless sieves (JIS-Z8801, open diameter: 2,000, 500, 250 and 106 μm) to know the particle size distribution. Since river-mouth sediment consisted of very small particles, we analyzed its particle size distribution by using the particle size analyzer (SALD-2100, Shimadzu Co.). High Resolution Gas Chromatography with High Resolution Mass Spectrometry (HRGC/HRMS) is officially used for the measurement of the dioxins concentration in Japan. The HRGC/HRMS method has many strong points, but it takes labor and costs to measure the dioxins concentration. Therefore, this method was not fit to use in this research, because there were a number of samples which must be measured. The CALUX assay is relatively rapid (only five days) and requires only approx. 2–10 g of the solid samples (Denison et al. (1998)).

Additionally, a good correlation has already been confirmed between the results from the CALUX assay and TEQ values from HRGC/HRMS. Therefore, in this research, the CALUX assay was used to measure the dioxins TEQ concentration. The CALUX assay method is based on the cellular response mechanism to dioxin toxicity for estimating TEQ concentration. The dioxins bind to a cytosolic protein, the aryl-hydrocarbon (Ah) receptor; then the receptor-chemical complex migrates to the cell's nucleus and binds with the AH-receptor nuclear translocator protein (Arnt). This Ah receptor–Arnt complex interacts with specific DNA sequences and swiches the gene expression of regulated genes on. The CALUX assay uses a genetically modified mammalian cell line (mouse hepatoma H1L6.1) that contains the firefly luciferase under transactivational control of the Ah-receptor.

Results and discussion
Distribution of dioxins in surface soils and river sediments

The relationship between organic carbon content and the dioxins TEQ concentration in surface soils (forest and paddy soils) and river sediments of the Yasu and Ado River basins are shown in Figure 3. In both basins, the dioxins TEQ concentration in forest soil showed a very good linear relationship to organic carbon content. However, the regression lines of forest soil in both basins were different from each other. The location of incinerators around Shiga prefecture is shown in Figure 4. It is found that the number of incinerators around the Yasu River basin is more than that of the Ado River basin. Therefore, we can estimate that, generally, the dioxins concentration in the atmosphere around the Yasu River basin is higher than that around the Ado River basin. The difference of the inclination of the regression lines could come from the difference of dioxins concentration in the atmosphere.

In paddy fields, although organic carbon content is relatively low, the dioxins TEQ concentration is quite high. It was indicated that the dioxins in pesticides (e.g., PCP and CNP) as impurities still remain even now (Masunaga *et al.*, 2001). Moreover, as a result of the fractionation of surface soils of both river basins by sieves, it is found that smaller particles have higher dioxins TEQ concentration per unit mass (data not shown). This result comes from the fact that smaller particles have larger surface area per unit mass. On the other hand, river sediments consist of relatively coarser particles than surface soils, and have substantially lower organic carbon content and dioxins TEQ concentration value. Therefore, we considered that the smaller particles, likely to have the higher dioxins TEQ concentration, flow more easily downstream and/or to the river mouth. Then, is the river-mouth the "haunt" of hydrophobic organic pollutants?

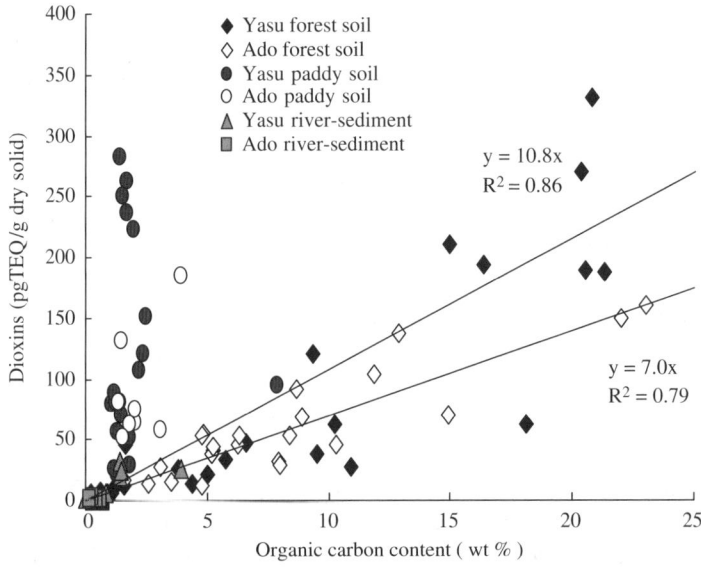

Figure 3 Correlations of dioxins TEQ concentration and organic carbon content in surface soil and river sediment

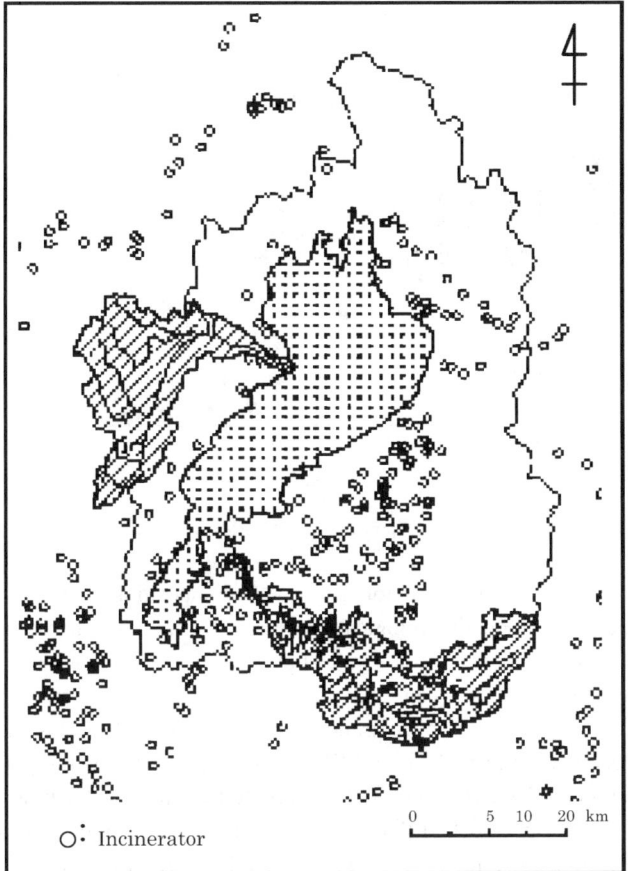

Figure 4 Location of incinerators discharging dioxins around Shiga prefecture

Distribution of dioxins in river-mouth sediment

Particle size distribution and organic carbon content in river-mouth sediment. The relationship between the median particle size (chosen as a representative particle size) and organic carbon content are shown in Figure 5. It indicates that the values of organic carbon content in river-mouth sediments are relatively lower than those of surface soils. Furthermore, we can find that the particle size of river-mouth sediments is very fine, that is, they are almost all composed of silt and clay. Therefore, it could be estimated that river-mouth sediments are mainly derived from smaller surface soil particles and that the organic matter of surface soils is torn off and/or decomposed by various physical, chemical and biological reactions through various hydrological processes. From Figure 5, there is almost no difference in the relationship between the median particle size and organic carbon content between the Yasu and Ado River-mouth sediments.

Dioxins TEQ concentration in river-mouth sediment. The relationship between the dioxins TEQ concentration value and organic carbon content in both river-mouth sediments is shown in Figure 6, and the relationship between the dioxins TEQ concentration and the median particle size is shown in Figure 7. Generally, in both river-mouth sediments, the dioxins TEQ concentration increases with increase in organic carbon content and with a reduction in the median particle size. However, the dioxins TEQ concentration in river-mouth sediments is unexpectedly not so high in either river mouth, although the median particle size is quite a bit smaller than that of surface soils.

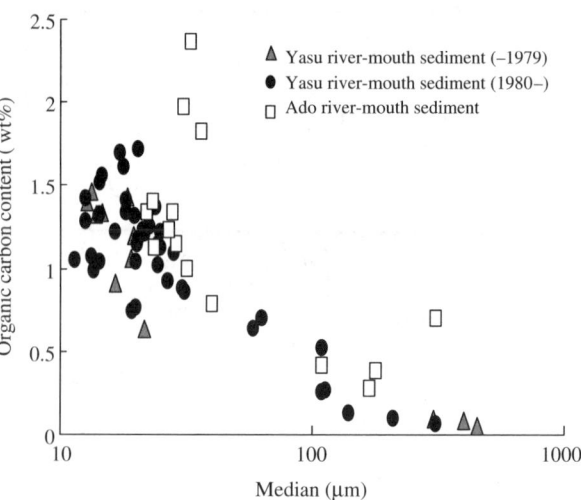

Figure 5 Relationship between the median particle size and organic carbon content in river-mouth sediment

These results are attributed to the exfoliation and decomposition of the organic matter in surface soils through various hydrological processes. On the other hand, if we focus our attention on the river-mouth sediments collected from the present and past (−1979) river-mouths of the Yasu river, we can find that the past Yasu River-mouth sediment has relatively higher dioxins TEQ concentration than the present. It seems to be caused by the usage of pesticides such as PCP and CNP mainly in 1960–1980. Figure 8 shows the organic carbon normalized dioxins TEQ concentration in the Yasu River-mouth. In Figures 6 and 7, we can also find that the dioxins TEQ concentration in the Yasu River mouth is higher than that in the Ado River mouth in general. The difference of the dioxins TEQ concentration per unit organic carbon content is not only due to the

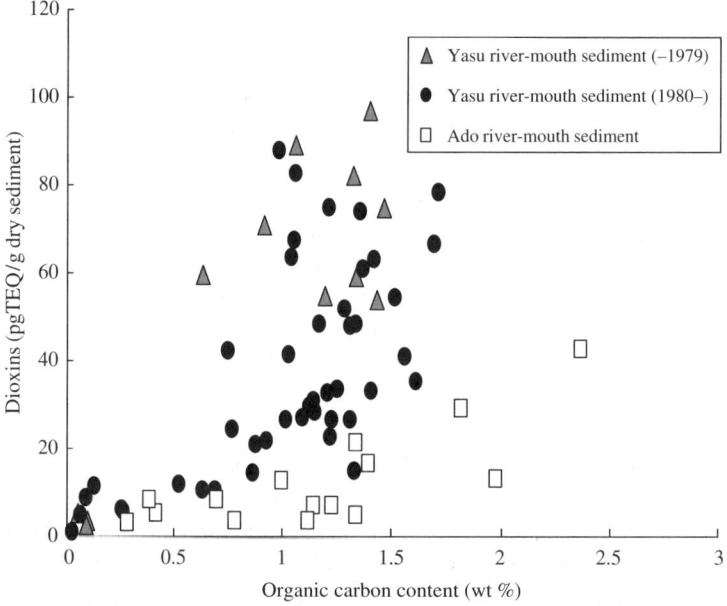

Figure 6 Relationship between the dioxins TEQ concentration and organic carbon content in river-mouth sediment

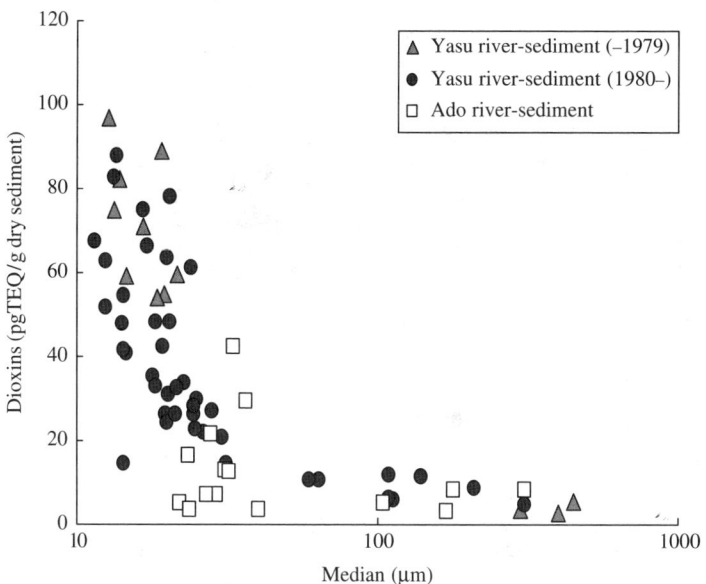

Figure 7 Relationship between the dioxins TEQ concentration and particle size in river-mouth sediment

different degree of pollution by dioxins in both basins (see Figure 4) but also due to some watershed properties.

Comparison of the dioxins TEQ concentration in surface soils, river sediments and river-mouth sediments

In order to examine the distribution of dioxins in both basins in detail, the dioxins TEQ concentrations in surface soils, river sediments and river-mouth sediments are plotted together against organic carbon content in Figure 9 (the Yasu River basin) and Figure 10 (the Ado River basin), respectively. In Figure 9 (the Yasu River basins), the plots of river-mouth sediments are close to paddy soils. On the other hand, in Figure 10

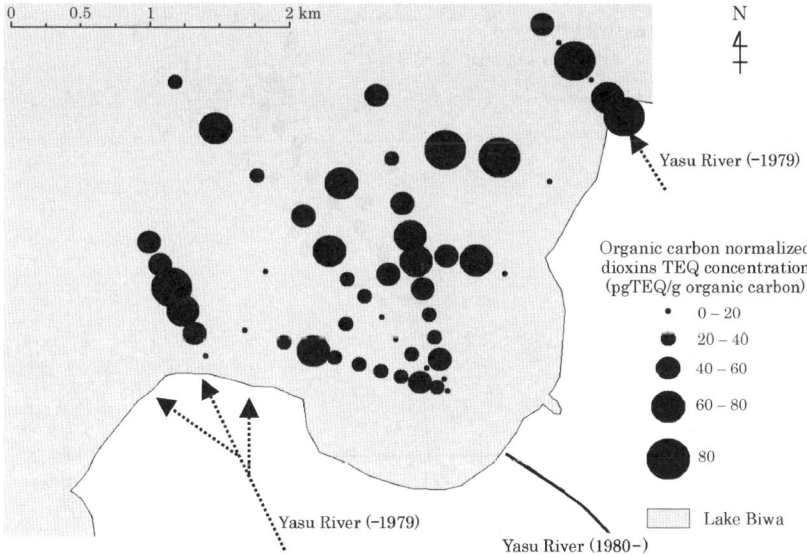

Figure 8 Organic carbon normalized dioxins TEQ concentration in the Yasu River-mouth

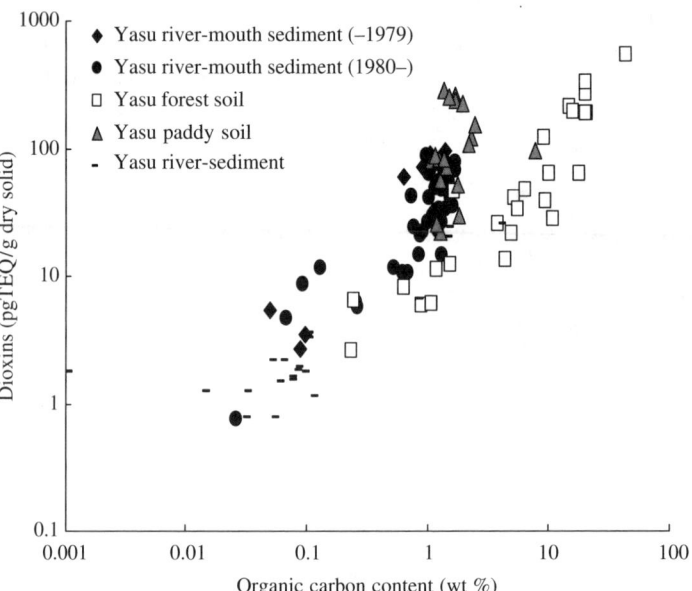

Figure 9 Comparative correlation of dioxins TEQ concentration and organic carbon content in all samples in Yasu river basin

(the Ado River basins), those of river-mouth sediments are widely extended between forest soil and river sediment. It may be concluded that the Yasu River-mouth sediments are more influenced by paddy soils than forest soil, and that Ado River-mouth sediments are more influenced by forest soils than paddy soils. These results may be explained by the difference of watershed properties, therefore we focus on the watershed properties in the next section (i.e. land use, river slope, dam construction, and rain intensity).

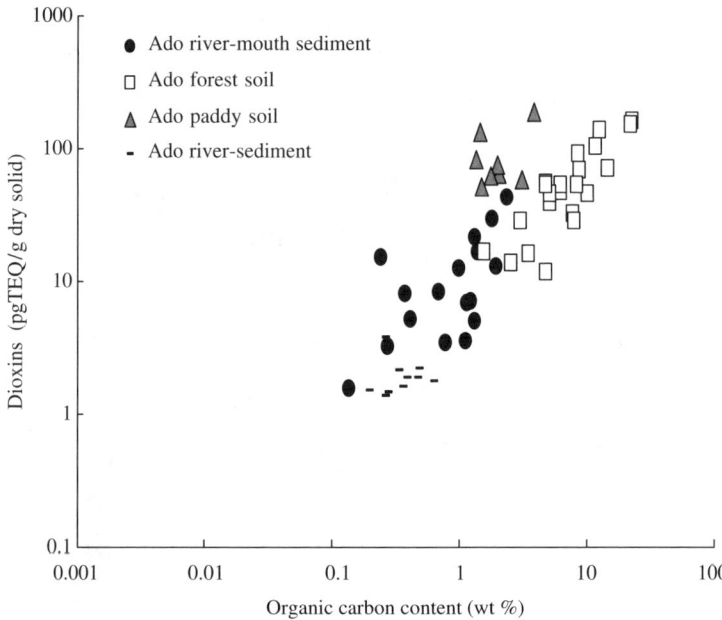

Figure 10 Comparative correlation of dioxins TEQ concentration and organic carbon content in all samples in Ado river basin

Watershed properties

Land use. The land use of the Yasu and Ado River basins is shown in Figure 11. Moreover, the ratio of each land use for the Yasu and Ado River basins is shown in Table 1. These two basins have different land use ratios. Especially important is the different ratio of paddy fields. One of the most notable factors of the different degree of the dioxins TEQ concentration in river-mouth sediment may be the difference in ratio of paddy field, polluted by dioxins contained in pesticides as byproducts.

River slope and dam construction. The slope of the Yasu and Ado River mainstream and the locations of dam and sluice gate are schematized in Figure 12. As shown in Figure 12, the slope of the Ado River mainstream is steeper than that that of the Yasu River mainstream. Furthermore, although there are two dams and two sluice gates in the Yasu River mainstream, the Ado River mainstream has no dam and/or sluice gates at all. Consequently, the arrival ratio of soil derived from forest to river-mouth is probably influenced by these situations (i.e. the downward movement of SS is disturbed by the dam).

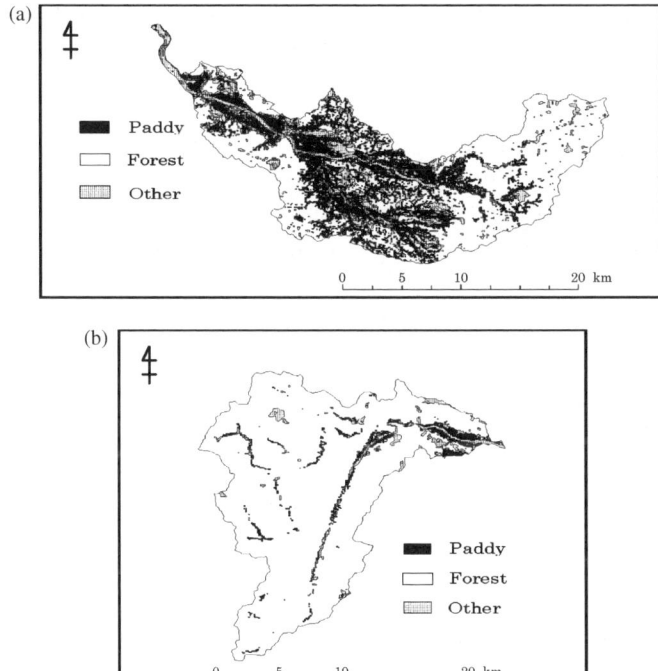

Figure 11 Land use of (a) the Yasu River basin and (b) the Ado River basin

Table 1 Land use ratio of the Yasu and Ado River basins

Land usage ratio	Yasu River basin	Ado River basin
Area (km^2)	395.1	309.3
Forest	61.1% (26)	91.5% (24)
Paddy	19.3% (30)	3.9% (10)
Others	19.6% (8)	4.7% (0)

*The values in parenthesis indicate the number of samples collected in this research

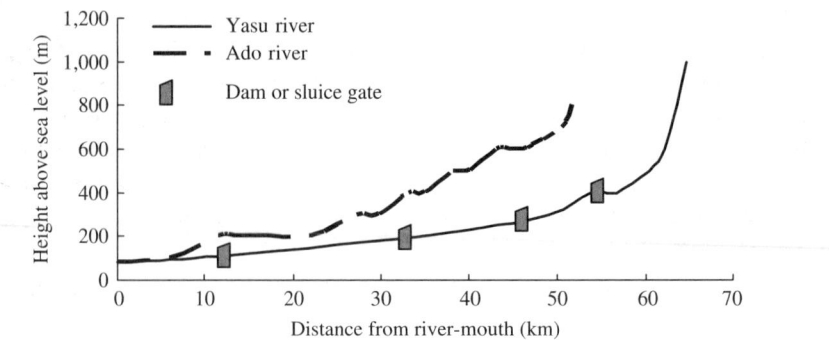

Figure 12 Slope of mainstream and location of dam and sluice gate

Rain intensity. The amount of rainfall is considerably different between the Yasu River basin and the Ado River basin; the amounts of rainfall in Yasu River basin and in the Ado River basin are about 1,600–1,800 mm and about 2,000–2,600, respectively (http://www.longlife.pref.shiga.jp/). This difference comes from the different climate in winter of both basins. That is to say, in winter, the rain and/or snow falls more in the Ado River basin than in Yasu River basin. Therefore, the amount of forest soil erosion by rain and/or snow in the Ado River basin may be more than that in Yasu River basin. This may be one of the reasons why the Ado River-mouth sediments are more influenced by forest soils than Yasu River-mouth sediments.

Mass balance of dioxins in watersheds

As mentioned above, the dioxins TEQ concentration in river-mouth sediments is relatively low. Therefore, considering the mass balance of dioxins in a watershed, we thought that the bottom of the lake near river-mouth is not the "sink" of dioxins and/or dioxins in surface soils may not move significantly. In order to know more about the movement and fate of dioxins in a watershed, future research including the analyses of lake sediment and river and lake waters must be carried out.

Conclusions

In this research, the dioxins TEQ concentration in surface soils, river sediments and river-mouth sediments was measured by the CALUX assay in the Yasu and Ado river basins around Lake Biwa, Japan. We evaluated and compared the correlation between the dioxins TEQ concentration and the solid characteristics (i.e. organic carbon content and particle size distribution) of all samples. The following conclusions were derived from the results and discussion:

1. In both the Yasu and Ado River basins, the dioxins TEQ concentration in forest soils showed a good linear relationship to organic carbon content. However, the regression lines of forest soils in both basins were different from each other. This result may be attributed to the different dioxins concentration in atmosphere in each basin.
2. It was found that river-mouth sediments were mainly composed of very fine particles such as silt and clay derived from surface soils. However, the dioxins TEQ concentration and organic carbon content were relatively low. This result is due to the exfoliation and decomposition of the organic matter in surface soils through various hydrological processes.
3. As a whole, the value of the dioxins TEQ concentration in river-mouth sediments collected from the old Yasu River-mouth (–1979) were relatively higher than that collected from the present Yasu River-mouth (1980–). This finding leads to a

conclusion that the influence of the dioxins contained in pesticides ever used in the past remains even now.
4. As a noticeable characteristic common to all, the dioxins TEQ concentration increases with increase in organic carbon content and with reduction in the particle size in river-mouth sediments. However, it was revealed that the degree of dioxins pollution varied between the Yasu and Ado River-mouth sediments. This may be attributed to watershed properties such as land use, river slope, dam/sluice gate construction and precipitation as well as the different degree of dioxins pollution in surface soils.

Acknowledgements

This research was financially supported by the Ministry of the Environment, Japan. We thank Mr H. Murata, M. Nakamura, and other members of Hiyoshi Ecological Services Co. for offering helpful cooperation and comments on the CALUX assay for dioxins analysis. Thanks are offered to Lake Biwa Research Institute for helping with the sampling of river-mouth sediments. Assistance by several professionals and graduate students in our laboratories is also acknowledged.

References

Denison, M., Brouwer, A. and Clark, G. (1998) US patent # 5, 854, 010.

http://www.longlife.pref.shiga.jp/.

Lohmann, R. and Jones, K. (1998). Dioxins and furans in air and deposition: A review of levels, behaviors and processes. *Science of the Total Environment*, **219**(1), 53–86.

Masunaga, S., Takasuga, T. and Nakanishi, J. (2001). Dioxins and dioxins-like PCB impurities in some Japanese agricultural formulations. *Chemosphere*, **44**, 873–885.

Study of water quality distribution in Lake Biwa in consideration of runoff pollutant loads from its catchment basin

A. Ichiki*, A. Sasaki**, N. Sakata*, K. Nakakura*** and H. Yamate****

*Dept. of Civil and Environmental Systems Engineering, Ritsumeikan University, 1-1-1 Nojihigashi, Kusatsu, Shiga 525-8577, Japan (E-mail: *a-ichiki@se.ritsumei.ac.jp*)
**Takatsuki City Office, Japan
***Suita City Office, Japan
****Matsushita Electric Industrial Co., Ltd., Japan

Abstract Many strategies for water quality conservation in Lake Biwa are being carried out mainly by reducing runoff pollutant loads into the lake. But influence of the runoff load reduction on the water quality in Lake Biwa has not been clarified enough so far. This study is aimed at discussing methodology to estimate water quality distribution in Lake Biwa using runoff pollutant loads from its basin. The runoff loads from the basin are calculated by Macro Model with GIS database of the Lake Biwa basin, and the water quality distribution in the lake is estimated by the spline technique with the calculated runoff loads. As a result, it has been proved that the methodology has enough reproducibility to estimate the water quality distribution in Lake Biwa and is available to examine the water quality in the lake.
Keywords GIS; Lake Biwa; Macro Model; spline technique; water quality distribution

Introduction

Lake Biwa is the largest lake in Japan and used as a drinking water resource by 14 million people in the Kinki area of Western Japan. Water quality improvement projects in Lake Biwa are being carried out mainly on reduction of inflowing loadings, but effects of such load reduction projects on water quality in Lake Biwa have not been clarified enough so far. GIS databases around Lake Biwa have been arranged in recent years (Ichiki and Yamada, 1999a; Masuda, 2000). The authors developed a management support system of pollutant runoff into Lake Biwa by combining the GIS database with the Macro Model to calculate runoff pollutant loadings from catchment basins (Ichiki *et al.*, 1996, 2001a) and made it possible to predict the inflowing loadings into Lake Biwa which vary according to basin characteristics (Ichiki *et al.*, 2001b). However, there is no procedure to discuss a method of Lake Biwa basin managements for water quality preservation because relationships between the inflowing loadings into Lake Biwa and water quality in the lake are not clear. This study is aimed at developing a series of calculation methodologies to estimate water quality distribution in Lake Biwa based on the inflowing loadings from its catchment area calculated from the basin characteristics. So, first it estimates the inflowing loadings using the Macro Model with the basin characteristics of Lake Biwa, and then it predicts water quality distribution in the lake based on the estimated inflowing loadings. Figure 1 shows a procedure of the inflowing loadings and water quality calculations. The inflowing loadings are calculated using the Macro Model at one-day intervals, and annual water quality distribution in the lake is estimated by a spline technique model. The calculations are done for TN and TP. Nagare implemented 3 dimension surveys on water quality in Lake Biwa (Nagare *et al.*, 2002), but in most

doi: 10.2166/wst.2006.035

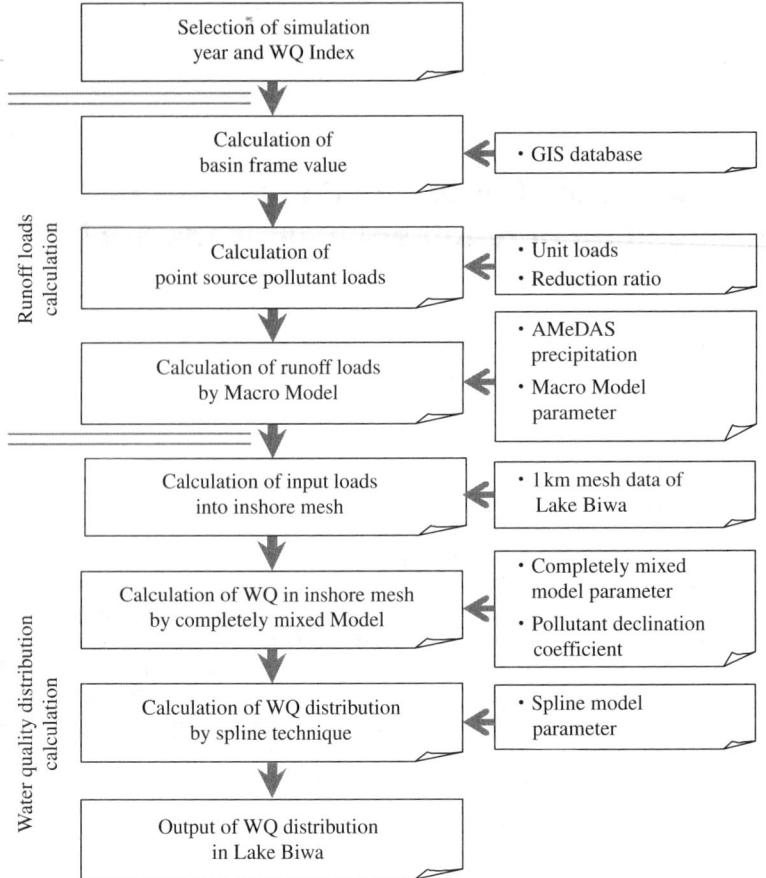

Figure 1 Estimation procedure of runoff loads and water quality distribution in Lake Biwa

cases, water quality assessments are done based on observations in the upper water layer (0.5 m depth) at environmental standard points. Therefore in this study, the calculations of water quality are implemented for the surface water in Lake Biwa.

Estimation of inflowing pollutant loadings into Lake Biwa

In the section of inflowing loadings calculation in Figure 1, the inflowing loadings from the Lake Biwa basin are calculated respectively for each tributary basin of the lake. The Macro Model is developed taking consideration of pollutant behavior which differs greatly with its source and weather. The authors showed the model constitution and parameter identification process in previous literature (Ichiki *et al.*, 1996, 1998, 2001a, 2001b; Ichiki and Yamada, 1999b). Figure 2 and Equations (1)–(11) show outlines of the Macro Model, and Table 1 shows the model parameters. Because Lake Biwa has a certain ratio of paddy field areas in its catchment basin, the model is improved to calculate runoff loadings from the paddy fields during an irrigation season (L_{ns4}).

$$L_{pv} = (L_{po1} + L_{po2} + L'_{po2})X/100 \tag{1}$$

$$L_{pu} = (L_{po1} + L_{po2} + L'_{po2})(1 - X/100) \tag{2}$$

$$L_{ps} = L_{pu}(1 - y/100) \tag{3}$$

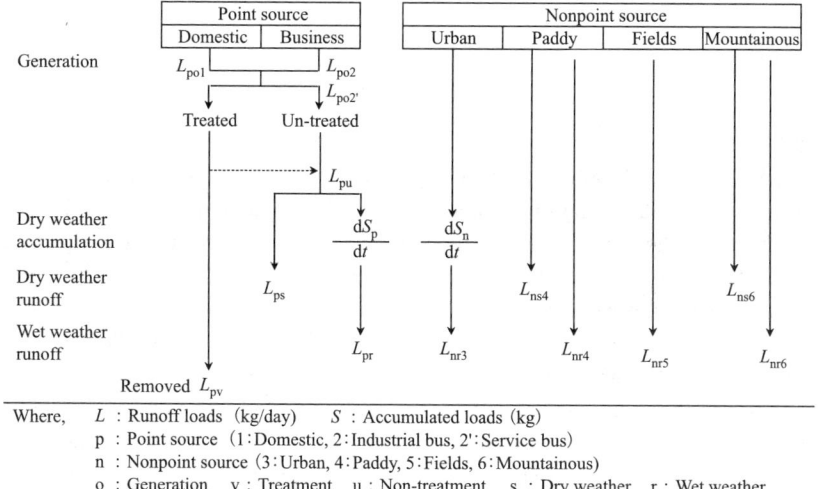

Figure 2 Outline of Macro Model

Where,
L : Runoff loads (kg/day)   S : Accumulated loads (kg)
p : Point source (1:Domestic, 2:Industrial bus, 2':Service bus)
n : Nonpoint source (3:Urban, 4:Paddy, 5:Fields, 6:Mountainous)
o : Generation   v : Treatment   u : Non-treatment   s : Dry weather   r : Wet weather

Table 1 Macro Model parameter

		Indices		TN	TP
Dry period	Accumulation	Accumulation ratio y (%)		48.60	65.10
		S_{nui}/A_{3i} (t/km^2)	Residential	0.014	0.003
			Commercial	0.027	0.001
			Industrial	0.064	0.027
			Highway	0.019	0.016
			Roof	0.007	0.001
	Runoff	K'_{n3} (1/day)		0.095	0.121
		L_{ns4}/A_4	Irrigated	2.713	0.175
		(kg/km^2 day)	Not-irrigated	0.000	0.000
		L_{ns6}/A_6	Irrigated	1.726	0.040
		(kg/km^2 day)	Not-irrigated	1.726	0.040
Wet period	Runoff	k_p ($\times 10^{-4}$/km^2)		8.969	3.360
		k_{n3} (d/mm)		0.048	0.067
		k_{n4}/A_4 ($\times 10^{-2}$/km^2)		1.938	0.272
		k_{n5}/A_5 ($\times 10^{-2}$/km^2)		6.308	6.142
		k_{n6}/A_6 ($\times 10^{-2}$/km^2)		0.915	0.131
		h_a ($\times 10^{-3}$)		2.386	3.998
		h_b		0.800	0.837

$$dS_p/dt = L_{pu}\, y/100 - L_{pr} \quad (4)$$

$$L_{pr} = k_p\, S_p^a\, Q_r^b \quad (5)$$

$$dS_n/dt = k'_{n3}\, S_{nu} \exp(-k'_{n3}\, T) \quad (6)$$

$$L_{nr3} = S_n(1 - \exp(-k_{n3}\, R)) \quad (7)$$

$$L_{nsi}(i = 4, 6) = \text{constant per unit area in each land use} \quad (8)$$

$$L_{ns4} \text{ during non-irrigation season} = 0 \quad (9)$$

$$L_{nri} = k_{ni} Q_r^b \quad (i = 4\text{-}6) \tag{10}$$

$$b = h_a X + h_b \tag{11}$$

Here, X is a removal ratio of point source pollutants (%), y is an accumulation ratio of point source pollutants during dry weather periods (%). Input data here are unit loadings, GIS database of the Lake Biwa basin and time series of precipitation observed by *Automated Meteorological Data Acquisition System* (AMeDAS).

Output loadings are calculated at one-day intervals for each tributary basin respectively. Simulations of pollutant runoff into Lake Biwa were done during the last 16 years from 1983 to 1998. Figure 3 shows the relationships between calculated loadings by the simulations and observed loadings during dry weather periods. The observed data during the dry weather periods were obtained for 24 rivers in Lake Biwa tributaries. Here, the irrigation season is from May to September, and the non-irrigation season is from October to April. Average relative errors (= |Calculated – Observed| / Observed x 100 (%)) are 68.53% for TN and 75.73% for TP during the irrigation season, while they are 76.84% for TN and 112.42% for TP during the non-irrigation season. Figure 4 shows relationships between calculated and observed loadings during wet weather periods. The observed data during the wet weather periods were obtained for five characteristic rivers during 182 storm events. While the calculated loadings are likely to be larger than the observed loadings in the Isasa R. and the Yamasina R., the differences in the other rivers

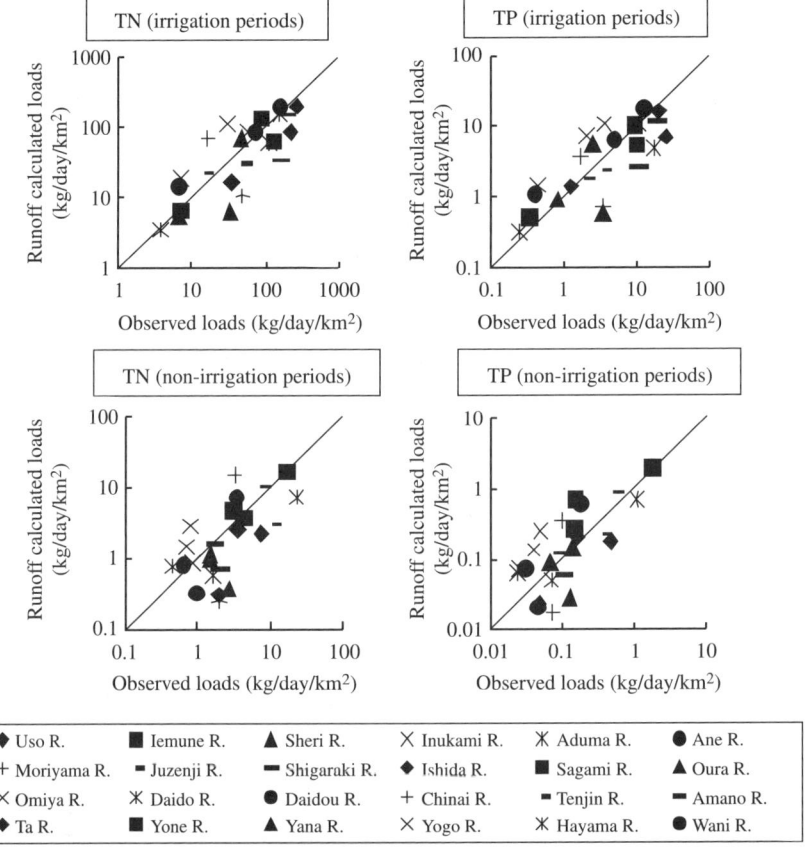

Figure 3 Relationships between calculated and observed inflowing loads during dry periods

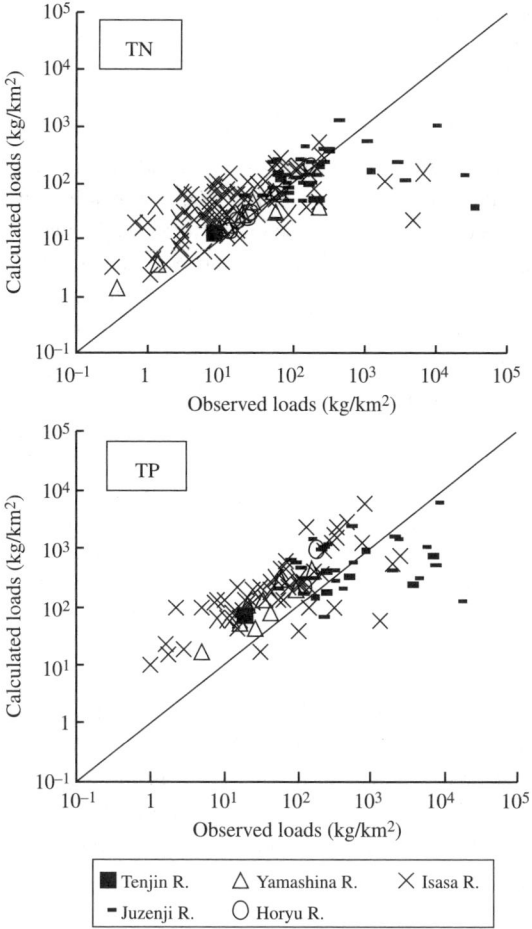

Figure 4 Relationships between calculated and observed inflowing loads during wet periods

are within single figures. Therefore, it can be regarded that the model simulation has enough replicability for rough estimations taking consideration of the dispersion of the observed data.

Estimation of water quality distribution in Lake Biwa
Estimation of water quality distribution by spline technique model

In order to understand the water quality situation in Lake Biwa, one of the most effective ways is to estimate two dimensions of water quality distribution in the lake. However, water quality data are usually monitored at some limited observation points (47 points in Lake Biwa) as shown in Shiga Environmental White Book (Shiga, 1984–1999). Therefore, it is important to select an appropriate method for interpolation of the limited observed data, in order to estimate two dimensions of water quality distribution in the lake. Fujiwara *et al.* examined an interpolation methodology for water quality distribution in Southern Lake Biwa using the spline technique, and showed an influence of the observation points on water quality distribution patterns (Fujiwara *et al.*, 1985). The spline technique makes it possible to interpolate smoothly from relatively sparse data by appropriate parameter settings, so this study also used the spline technique model for the discussion below. A basic equation of the spline technique model for water quality data

$C_{x,y}$ (mg/L) in a 2-dimensional coordinate system is shown as equation (12).

$$C_{x,y} = \{(8+nh^2)(C_{x+1,y} + C_{x-1,y} + C_{x,y+1} + C_{x,y-1}) - 2(C_{x+1,y+1} + C_{x-1,y+1} + C_{x+1,y-1} + C_{x-1,y-1}) - (C_{x+2,y} + C_{x-2,y} + C_{x,y+2} + C_{x,y-2})\}/(20+4nh^2) \quad (12)$$

Here, $C_{x,y}$ is a concentration in a mesh at a coordinate (x, y), n is a spline parameter ($n = 1.0$) and h is width of a mesh ($h = 1$(km)). The estimations were done by continual calculations which were not completed until differences between calculated concentrations before and after the calculation became less than 0.00001 mg/L for TP and 0.001 mg/L for TN in all meshes. Using the observed data during 1984–1998 at 25 observation points selected randomly from the 47 observation points in Shiga Environmental White Book, the spline technique model calculated water quality distribution. And the rest of the 22 observation points were used for verification of the spline technique model calculation. Figure 5 shows the relationships between calculated and observed annual mean concentrations in the 22 observation points. It shows a certain replicability as the correlation coefficient is 0.875 for TN and 0.864 for TP, and average relative error is 7.1% for TN and 18.8% for TP.

Estimation of water quality along lakeshore by complete mixing model

In the water quality distribution calculation section in the estimation procedure of runoff loads and water distribution (Figure 1), the water surface of Lake Biwa was divided into 1 km² meshes (Figure 6). And the runoff loads/water from the basin and discharged loads/water from wastewater treatment plants in the basin were input into 233 inshore meshes which are bounded on outlets of inflowing rivers and wastewater treatment plants. Each inshore mesh is considered to be a complete mixture box with 1 m depth, and loads/water balance in the mesh is calculated by equation (13) as shown in Figure 7.

$$CV = \{C_0 V + C_{in} Q_{in} - (C_0 V + C_{in} Q_{in})/(V + Q_{in}) Q_{out}\}(1-m) \quad (13)$$

Here, m is attenuation coefficient (1/day) of pollutants, which shows pollutant dose variation due to sedimentation/elution onto/from sediments. Matsunashi and Imamura determined the attenuation coefficient for each environmental criterion in waters by drawing a specific load curve shown by a Vollenweider type equation (14) with z for water

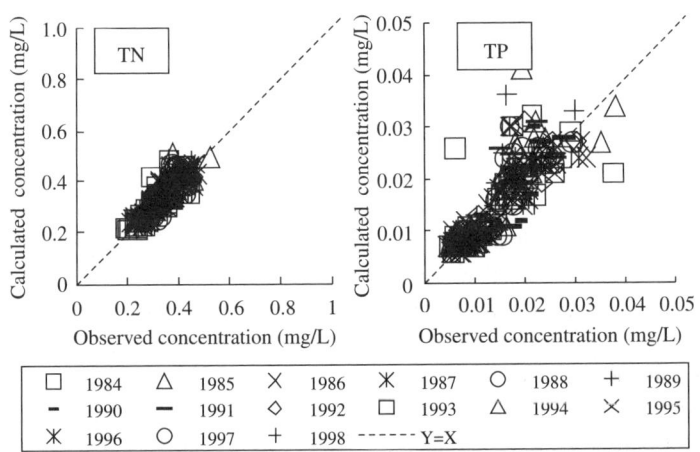

Figure 5 Relationships between calculated and observed concentrations (22 observation points) using spline technique with observed Lake Biwa data (25 observation points)

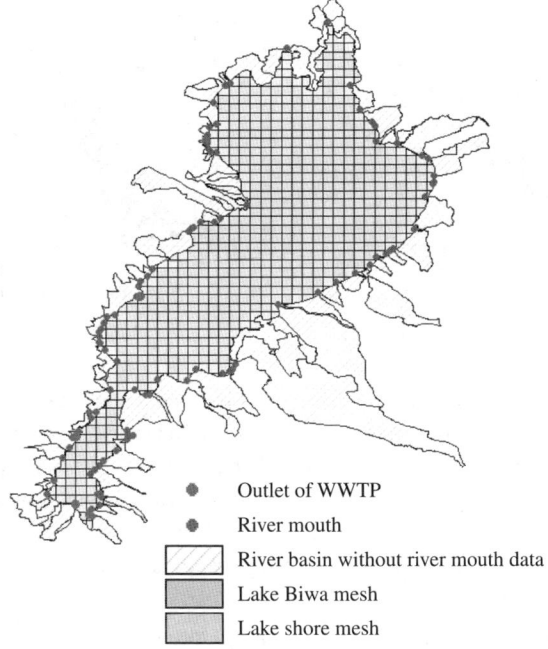

- ● Outlet of WWTP
- ● River mouth
- ▨ River basin without river mouth data
- ▦ Lake Biwa mesh
- ▤ Lake shore mesh

Figure 6 Mesh segmentation in Lake Biwa

depth (m), f for water replacement rate (1/day) and L/A for specific loads (ton/km^2/day) (Matsunashi and Imamura, 2000).

$$L/A = (C - C_0)fz + Cmz + C_0Q_{in}z/V \qquad (14)$$

Figure 8 shows relationships between the specific loads (L/A) and product of water replacement rate (f) and water depth (z) in the Lake Biwa inshore meshes. So, the attenuation coefficient (m) for each inshore mesh was determined by drawing the specific load curve to fit each inshore mesh.

Verification of estimated water quality distribution in Lake Biwa

Based on the calculated concentration in inshore meshes, the spline technique model calculates pollutant concentration in all meshes in Lake Biwa. The estimation system (Figure 1) implemented runoff loads – water quality distribution simulation for 13 years during 1986–1998, using the water quality distribution in 1985 estimated by the spline technique model with observed data as an initial condition. Figure 9 shows relationships between

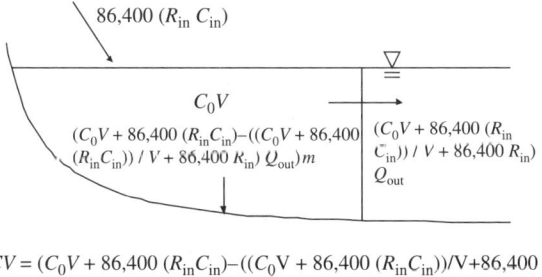

$CV = (C_0V + 86,400\ (R_{in}C_{in}) - ((C_0V + 86,400\ (R_{in}C_{in}))/V + 86,400\ R_{in})\ Q_{out})\ (1-m)$

$Q_{out} = R_{in}$

$V = \text{const.}$

Figure 7 Mass balance in a lakeshore mesh

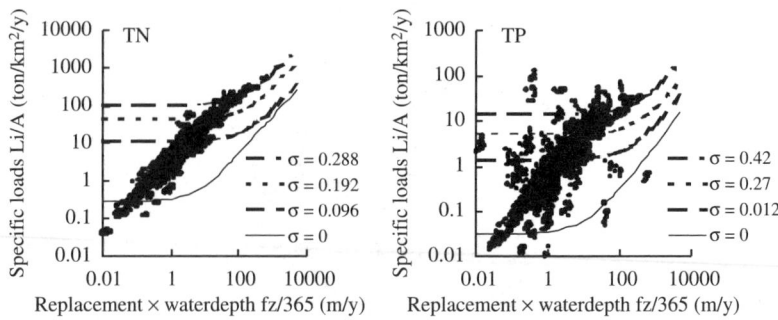

Figure 8 Inflowing load curve in lakeshore meshes

calculated concentration by the system and observed concentration at water quality observation points during the period. Correlation coefficient is 0.476 for TN and 0.638 for TP, and average relative error is 17.0% for TN and 50.6% for TP. The system replicability tends to be worse with elapsed time. However, considering that the runoff loads input into the inshore meshes already have some errors, the results of the water quality distribution estimation here can be considered to have enough reproducibility for the discussion.

Characteristics of water quality distribution in Lake Biwa

Among simulation results since 1986 to 1998, Figure 10 shows water quality distribution in Lake Biwa and inflowing loads from every tributary basin in 1994 as a drought year (1,254 mm/y of annual precipitation) and 1998 as a flood year (1,981 mm/y of annual precipitation). According to the observed data (Shiga, 1984–1999), influence of 1994 drought on annual mean concentration is remarkable for low TN concentration in Northern Lake Biwa and for high TP concentration in Southern Lake Biwa. This is caused by complex effects of decrease in concentration due to decrease of inflowing loads during dry periods and increase in concentration due to runoff loads flushed from catchment basins by storms right after the dry periods. According to the simulation for TN, both the inflowing loads and water quality in Lake Biwa have little difference between the annual precipitations, and the low concentration in Northern Lake Biwa due to the drought is not so remarkable. On the other hand, for TP, the inflowing loads decrease in the drought year, and it makes the concentration in the eastern part of Northern Lake Biwa decrease. But the concentration in Southern Lake Biwa increases in the drought year because of highly concentrated runoff from the basin and small capacity of Southern Lake Biwa. Thus, this procedure makes it possible to estimate and examine inflowing loads and water

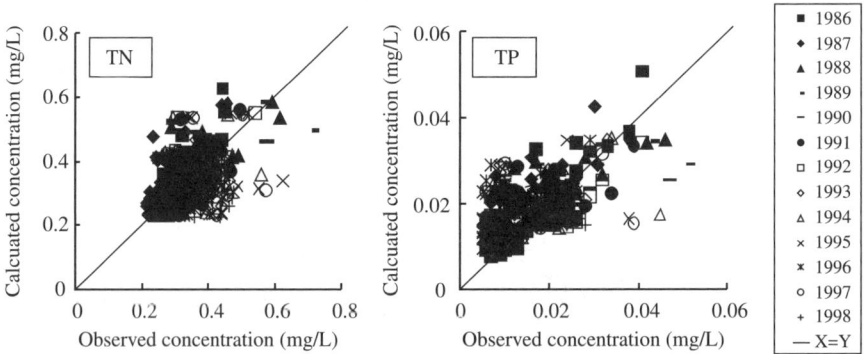

Figure 9 Relationships between calculated and observed concentrations using spline technique with calculated inflowing loads

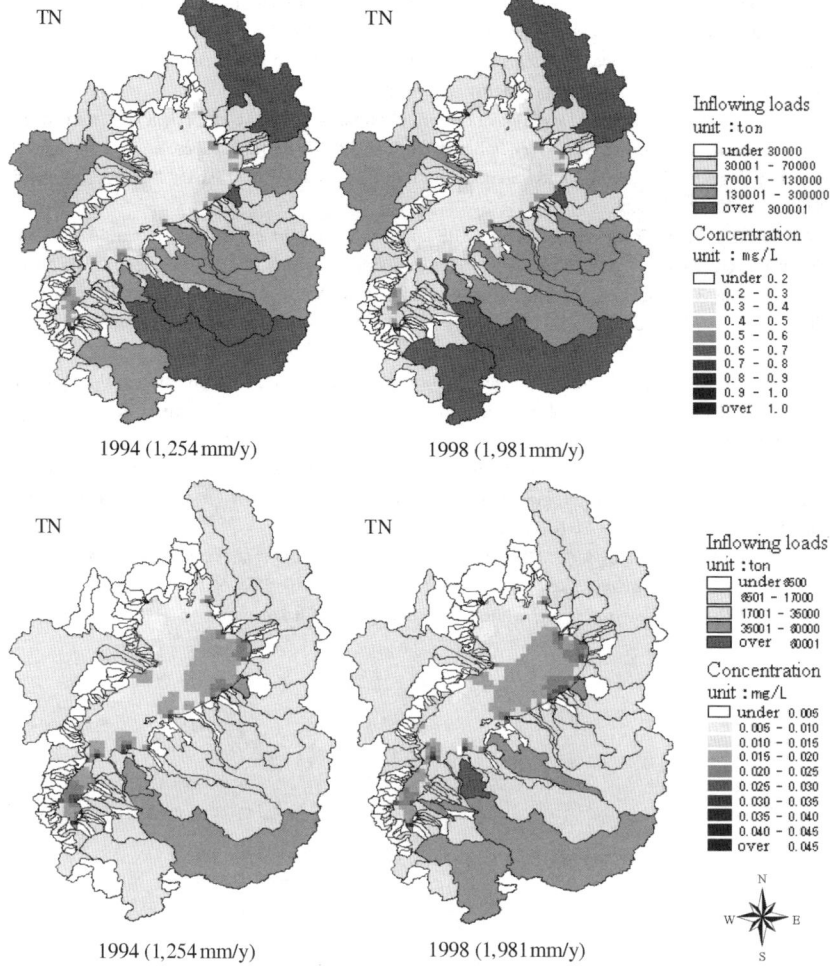

Figure 10 Results of simulation on inflowing loads and water quality distribution

quality distribution into/in Lake Biwa which vary according to change in basin characteristics and natural conditions.

Conclusions

A fundamental examination was implemented on a methodology to estimate inflowing loads into Lake Biwa using its basin characteristics and water quality distribution in the lake using the estimated inflowing loads. And it is shown that the methodology has enough reproducibility and is available to examine the water quality in the lake.

This study has been partially funded under Grants-in-Aid for Scientific Research (Prof. Somiya at Kyoto Univ. as research head) which is supported by Japan Ministry of Education, Culture, Sports, Science and Technology (MEXT). The authors would like to thank everyone whom it may concern. Especially they thank Dr Masuda at Tottori Univ. for his support on the GIS database.

References

Fujiwara, M., Somiya, I., Tuno, H. and Fujii, S. (1985). Estimate by spline technique model of water pollution concentration distribution pattern and examination of reasonable measuring point alignment. *Water Pollution Study*, **8**(2), 100–109 (in Japanese).

Ichiki, A., Yamada, K. and Onishi, T. (1996). Prediction of runoff pollutant load considering characteristics of river basin. *Water Science and Technology*, **33**(4–5), 117–126.

Ichiki, A., Onishi, T. and Yamada, K. (1998). Estimation of urban nonpoint source pollution in Lake Biwa basin. *Water Science and Technology*, **38**(10), 157–163.

Ichiki, A. and Yamada, K. (1999a). Development of pollution thing outflow control database from catch basin in Lake Biwa by using GIS. *River Information Study*, **7**, 17–23 (in Japanese).

Ichiki, A. and Yamada, K. (1999b). Study on pollutant runoff into Lake Biwa, Japan. *Water Science and Technology*, **39**(12), 17–25.

Ichiki, A., Yamada, T. and Yamada, K. (2001a). GIS application to estimate runoff pollutant loads from Lake Biwa watershed, Japan. *Urban Drainage Modeling*, American Society of Civil Engineering, pp. 701–712.

Ichiki, A., Yamada, T., Sasaki, A., Amano, K. and Yamada, K. (2001b). Availability of a GIS application for pollutant runoff from Lake Biwa watershed, Japan. In *9th International Conference on the Conservation and Management of lakes*, Session 5, pp. 345–348.

Masuda, T. (2000). Study on basin environmental information integration and application to the phenomenon analysis/plan theory by using GIS at a place of basin in Lake Biwa, Kyoto University doctoral dissertation (in Japanese).

Matunashi, S. and Imamura, M. (2000). Study on the possibility of eutrophication in closed nature sea area and permitted loading dose of nitrogen/phosphorus. *Japan Society of Civil Engineers Collected Papers*, Vol. VII-17, No. 664, pp. 11–20 (in Japanese).

Nagare, H., Somiya, I. and Fujii, S. (2002). Estimate of nutrient salts existential quantification in the water of Lake Biwa. *Journal of Japan Society of Water Environment*, **25**(10), 599–604.

Shiga (1984–1999). Shiga environmental white book - document - (in Japanese).

Estimation of annual pollutant loadings in two small catchments and examination of their differences caused by regional properties

S. Fujii*, M. Moriya**, P. Songprasert* and H. Ihara***

*Research Center for Environmental Quality Control, Kyoto University, 1-2, Yumihama, Otsu, Shiga 520-0811, Japan

** CTI Engineering Co., Ltd., 1-14-6, Uekizaki, Urawa-ku, Saitama, 3300-0071, Japan

***Department of Soils and Fertilizers, National Agricultural Research Center, 3-1-1 Kannondai, Tsukuba, 305-8666, Japan

Abstract A series of runoff surveys was conducted for more than one year in two small catchments of the Kamo River basin (75.4 km^2) and the Takano River basin (66.8 km^2) in Kyoto, Japan, which adjoin each other, and may have the same precipitation pattern. The investigation consisted of a high-frequency periodic survey, a long-term regular survey and a storm event survey. The survey results were compared with the regional properties of the basins, and the following results were obtained. (1) Pollutant loadings were successfully estimated as two portions of base discharge and storm events discharge from the survey results. (2) Estimated annual loading of the sites was 2.9–4.5, 1.3–1.8, 17–27, 1.3–2.2, 0.076–0.97 t/km^2/y, respectively for COD$_{Mn}$, DOC, SS, TN and TP. (3) 52–53% of the whole flow, which was caused by rainfall events, conveyed 81–87, 68–73, 92–95, 64–67, 76–81% of the whole loading, respectively for COD$_{Mn}$, DOC, SS, TN and TP. (4) Differences of regional properties in two basins cause different runoff patterns, but the differences in runoff patterns also depend on the rainfall patterns. In general, a more urbanized basin receives early and strong influence of precipitation on the storm event runoff.

Keywords Annual loading; regional properties; regression model; storm events; water quality

Introduction

Runoff of pollutants is a complicated phenomenon receiving various kinds of influences. Although many factors affect the runoff process, the most important factors might be regional properties of the river basins and the pattern of rainfall. The former is a spatial factor depending on conditions of the upstream basins, and includes population, land use, soil types, geographical features, and so on. The latter is a meteorological factor, and includes duration of rainfall period, maximum precipitation intensity, preceding non-rain duration, and so on. Both factors never show the same pattern if either time or place is different, so that the runoff also shows a quite different pattern in each place and each time, resulting in difficulty understanding the phenomenon. To fix one of the two factors would be a good method for evaluation of the other factor's effects. In this study, two connected river basins were selected as study fields and a series of surveys were conducted during the period for more than one year including storm events. From these results, annual runoff loadings were estimated and compared with the regional properties of two basins.

Materials and methods

Study sites

Kamo River and Takano River were selected as study fields (Figure 1). Both rivers are located in the northern part of Kyoto City, Japan and flow southward, becoming one

doi: 10.2166/wst.2006.036

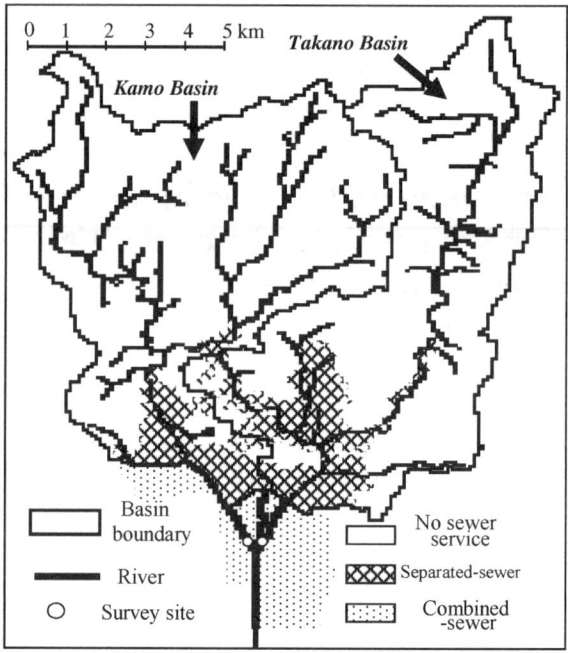

Figure 1 Research site boundaries and stream lines

stream. The observation site for each river was the point at about 20 m upstream from their juncture. Table 1 summarizes the characteristics of basins. To obtain these data, the area was divided into small rectangular cells of 3.0"SN (= 92.4 m) * 4.5"EW (= 114 m) by the latitude and longitude, and flow directions of the cells were determined with position of streams and altitude of each cell (Songprasert et al., 2003), so that the boundaries of both basins were identified. Several kinds of information were collected, and assigned to each cell, and the number of the cells corresponding to specified conditions was counted to calculate the area of the conditions.

As shown in the table, forest is major land use, and occupies 89 and 80%, respectively for the Kamo River and Takano River basins. Both basins have residential areas, and most of them receive sewer service. Combined sewer service areas were excluded from the catchment calculation because almost all water produced there is conveyed to the outside, but separated sewer areas were included. These areas are shown in Figure 1.

Table 1 Properties of Kamo and Takano Basins

Area, km² (%)		Kamo River	Takano River
Land use	Total	75.4	66.8
	Forest	67.2 (89.1)	53.2 (79.6)
	Agriculture	1.5 (2.0)	3.0 (4.5)
	Residence	5.7 (7.6)	9.5 (14.2)
	Others	1.0 (0.5)	1.1 (14.2)
	Sewer service	6.4 (8.5)	10.5 (15.7)
Population, ca (%)	Total	41,300 [548]	62,700 [939]
	Non-sewer service	3,300 (7.9)	5,300 (8.5)
	Sewer service	38,000 (92.0)	57,000 (90.9)
Mean gradient (m/km)		364	320
Mean flow distance, km		16.9	14.1

Area and population in combined sewer service are excluded from the above statistics [population density, ca/km²]

Percentages of residence are 5.7% for the Kamo River basin and 14.2% for the Takano River basin, and population densities are 548 and 939 ca/km^2, respectively. In general, the Takano is more urbanized than the Kamo.

Observation methods

The survey started from September 23, 2002. During the first 80 days, sampling and measurement were conducted every other day, and continued at intervals of 10 days until December 10, 2003. The survey is still ongoing with an extended interval of one month. The sampling time for these observations was fixed at 8:00. Table 2 summarizes these investigations. The observed items are flow rate, environmental conditions, particulate pollutants, organic matter, nutrients, anionic ions and metals. In addition to these regular observations, investigation of storm events was conducted six times as shown in Table 3. Those storm event surveys had 2–50 mm rainfall with the maximum precipitation intensity ranging from 1 to 12 mm/h. Duration of each survey ranged from 28 hours to 12 days. The number of samplings ranged from 5 to 28 times. The same items were measured in these surveys as in the regular observations.

Results and Discussion
Annual variations of flow rate and water quality

Figure 2 shows time-series variations of flow rate and some water quality indices in both sites from September 23, 2002 to December 10, 2003. Precipitation data used in the study were obtained from AmeDas (Automatic meteorological data acquisition system), and hourly data of Kyoto City observation station, which is located near the basins (Japan Meteorological Business Support Center, 2004). The figure shows summed values of hourly data in 6 hours. During the period, the basins received rainfall of 1999 mm in 281 days out of 444 days. Maximum rainfall was 28.5 mm in an hour, 53.5 mm in 6 hours, and 115 mm in a day. When one storm event is considered as a series of rainfalls with their intervals less than 12 hours, 114 storm events happened with 16.0 mm of the average precipitation in 11.6 hours of the average duration.

The flow rate fluctuated in the range of 0.16–18 m^3/s (Kamo River) and 0.076–14 m^3/s (Takano River). The pattern in each river consisted of a basic gentle change and several sudden sharp peaks. These peaks always occurred at storm events, but some storm events did not express such peaks. The durations of these peaks were usually less than one day, so that the regular sampling cannot explain all of the storm events even if it was conducted with the interval of two days. Kamo River and Takano River had very similar flow rates, but differences of about 30% were observed at the peaks, which occurred during storm events. This may be explained by the difference in land properties in both catchments.

Table 2 Outline of regular sampling

Period	Frequency
2002/09/23 ~ 2002/12/12	every 2 days
2002/12/12 ~ 2003/12/10	every 10 days
2003/12/10 ~ (continue)	every month
Observed Items	
Flow rate, Temperature, pH, DO, SS, VSS	
COD$_{Mn}$, D-COD$_{Mn}$, TN, DN, TP, DP, DOC	
IC, SiO$_2$, NH$_4$-N, NO$_2$-N, NO$_3$-N, PO$_4$-P	
SO$_4^{-2}$, Cl$^-$, D-Metals (Na, Mg, Ca, Fe, etc)	

Table 3 Outline of storm events investigation

Storm event	Sampling			Precipitation		Preceding conditions	
	Start	End	No.	Total mm	*1mm/h	*2hrs	*3mm
Se1	02/10/19 11h	10/20 15h	9	4.5	1.5	91	1.0
Se2	02/11/07 21h	11/08 12h	5	2.0	1.0	139	24.0
Se3	03/07/21 03h	07/22 12h	6	5.0	2.0	53	0.5
Se4	03/07/23 09h	07/28 08h	14	33.0	12.0	48	5.5
Se5	03/11/05 16h	11/08 08h	15	12.0	4.5	47	13.0
Se6	03/11/28 21h	12/10 08h	28	53.5	8.5	82	51.5

*1 Maximum precipitation intensity
*2 Duration from the end of previous rainfall to the storm event
*3 Cumulative rainfall during preceding 7 days

Concentrations of SS, TN and TP were shown as examples of water quality variations. These items also had a pattern consisting of a basic gentle change and several sharp sudden peaks. SS was the item that had quite high peaks. In regular sampling, 90% of SS concentrations were less than 7 mg/L (Kamo River), or 5 mg/L (Takano River), but the concentrations at storm events suddenly increased to the level of 20–100 mg/L. On the other hand, TN received less significant effects of storm events, and was less fluctuated. The coefficient of variance (standard deviation / mean) in regular samples was only 31% (Takano), or 39% (Kamo) for TN, but 135% (Kamo), or 204% (Takano) for SS.

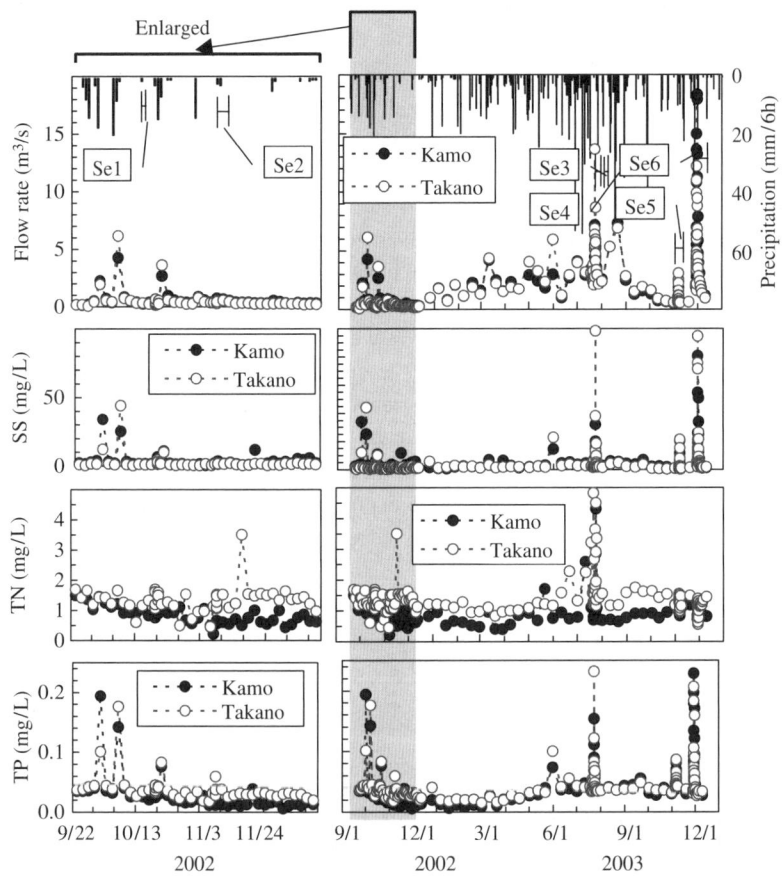

Figure 2 Fluctuations of flow rate, SS, TN and TP

Concentration differences of both rivers were, however, more obvious in TN, and Takano River showed 1.5 times the concentration of that in Kamo River. TP made an intermediate pattern between SS and TN.

Variations of flow rate and water quality in storm events

Figure 3 shows variations of the flow rate and some water quality items in three storm events. In this figure, the axis scales of flow rate and SS were changed in three events because the range of fluctuation was quite different. Flow rate was very sensitive to the rainfall, and the influences were observed even in the case of 2 mm precipitation (Se2). Flow rate increased very sharply when rainfall started, but a few hours of delay were always observed. In the case of heavy rain (Se6, 53.5 mm), obvious effects still remained a few days after the rainfall stopped. SS showed a similar pattern, but the change was more obvious than flow rate. The concentration increased to 10–100 times within a few hours, and returned back to the previous level within one day if the rainfall ceased. This pattern was also observed in COD_{Mn}, D-COD_{Mn}, DOC, VSS, TP, DP and D-Al.

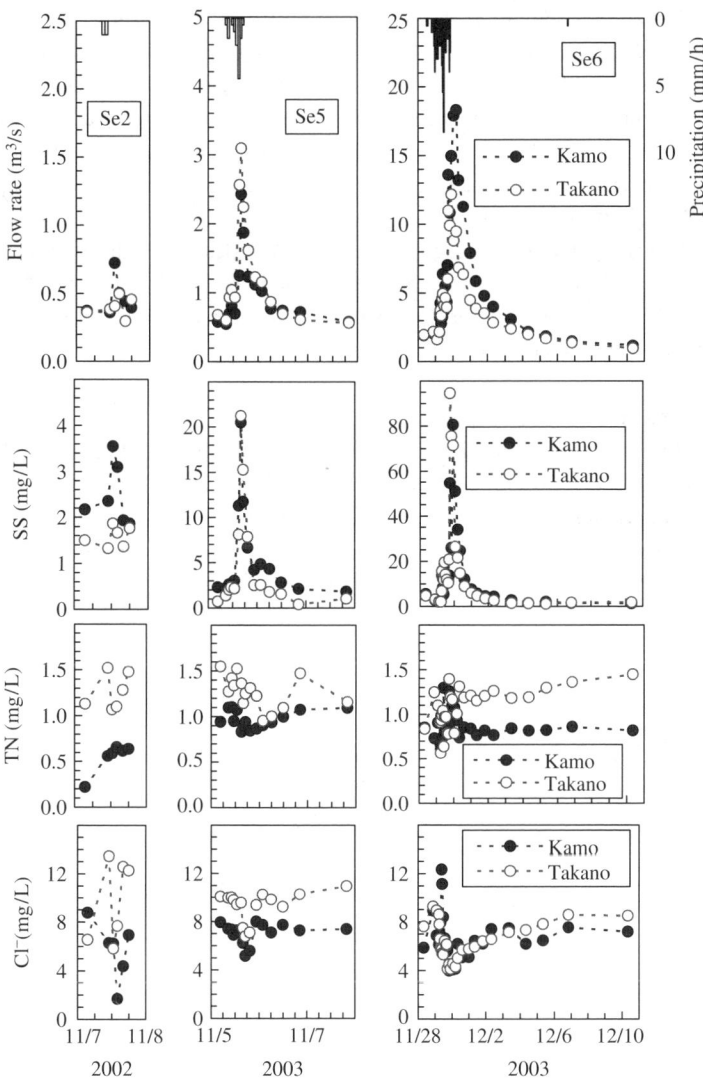

Figure 3 Fluctuations of flow rate, SS, TN and chloride during storm events

The items related to organic matter and particulate substances seem to have a similar pattern. In case of particulate substances, this pattern can be explained by washout of deposited substances, which are accumulated on riverbeds or land surface during dry weather. The increase of discharge by storm events might promote the rinse of such deposits, resulting in the increase of organic matter (D-COD$_{Mn}$, DOC) during storm events.

On the other hand, ionic species, such as chloride, showed a different pattern. They decreased with the increase of flow rate and returned back to the usual level when the flow rate was recovered. These items were IC, Cl$^-$, SO$_4^{2-}$-S, D-Na, D-Ca, D-Mg and D-Sr. These items seem to be rapidly leached from the watershed without being trapped on riverbeds or the land surface, receiving the effect of dilution of rainwater directly. TN, NO$_2^-$-N, NH$_4^+$-N, SiO$_2$-Si and K had patterns different from the above two. These items might have other discharge processes or receive both effects.

Response of flow rate and water quality to rainfall

As discussed in the previous section, flow rate increased with some time lag after rainfall started. This point was analyzed with mutual correlation analysis between precipitation and flow rate. Figure 4 shows the results of Takano River, which showed a more obvious relationship between precipitation and flow rate. In this analysis, a leaner interpolation was used to make the flow rate a continuous function from discrete measured data. As shown in the figure, obvious correlation peaks were observed in most of the storm events with the time lag of 1–4 hours. The correlation peaks in Kamo River weekly appeared with slightly longer time lag. Water quality items related to organic matter and particulate substances also showed a similar pattern.

On the other hand, the decreasing pattern after these peaks was different between flow rate and water quality items. Figure 5 shows variation of flow rate and SS in Takano River at Se6. SS decreased with an exponential pattern very rapidly. The time needed to become one tenth was about a quarter of that for flow rate.

Estimation of pollutant loading during storm events

As discussed in the previous section, data on regular sampling could hardly reflect storm events, so that an estimation method for the effects of storm events is required to evaluate annual pollutant loading from a watershed. Then, we adopted a regression equation model to estimate loading from the rainfall data (Ebise, 2004). Figure 6 shows its conceptual chart. This model has the following assumptions:

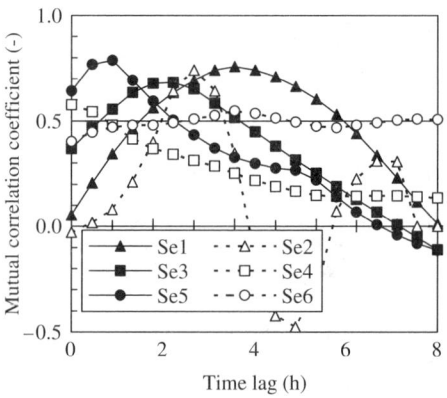

Figure 4 Mutual correlation analysis for delay of flow discharge in Takano River

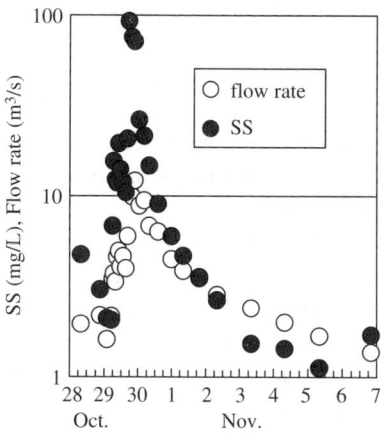

Figure 5 Change of flow rate and SS in Se6 (Takano River)

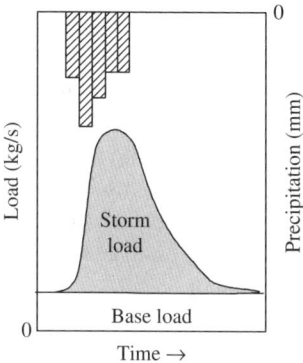

Figure 6 Concept of load structure

(1) Discharge loading consists of a base and a storm portion.
(2) The base loading part is never affected by storm events, and is assumed to be constant.
(3) The storm loading is an increment part that increases after storm events.
(4) The whole loading in one storm event is a function of the whole flow quantity, and is given in the following equation:

$$\sum_{i}^{N} Ls_i/A = a \times \left(\sum_{i}^{N} Q_i/A\right)^n \quad (1)$$

where Q_i and Ls_i are flow quantity and loading during t_{i-1} to t_i (t_0: start time, t_N: end time), A is the area of the basin and a and n are parameters.

(1) The flow quantity is considered as a function of total precipitation amount, and is given in the following equation:

$$\sum_{i}^{N} Q_i/A = b \times \left(\sum_{i}^{N} R_i/A\right)^m \quad (2)$$

where R_i is precipitation amount during t_{i-1} to t_i.

For the practical calculation of the loading in the two basins, we used the following procedures.

(1) The period from January 1, 2003 to December 31, 2003 was applied for the calculation of annual loading.
(2) Fifteen out of 34 regular samplings were used for the base loading calculation. The samples were regarded as the data that scarcely receive the effects of any storm events.
(3) A series of rainfalls with their intervals less than 12 hours was considered as one storm event. Such storm events were counted 100 times in the period.
(4) Discharge increments were calculated for six storm events (Table 3) from the special observation data and the base flow rate. These increments were accumulated during the period of each storm event. The resultant cumulative value was regressed with its cumulative precipitation amount, and the parameter values for the equation (Eq–2) were determined.
(5) Similarly, the increments of loading were calculated, and the cumulative value was regressed with the cumulative flow rate. The parameter values for the equation (1) were determined.

Figure 7 illustrates the relationship between precipitation and discharge in both rivers in a log–log chart. Cumulative discharge had a linear relationship to the whole precipitation in a logarithmic scale, and the correlation coefficient of them was 0.973 for Kamo River and 0.968 for Takano River. This means the regression equations shown in the figure can be available for the estimation of flow rate in other storm events. The slope value was obviously higher than unity, being 1.67 for Kamo, and 1.77 for Takano. This means that stronger precipitation yields higher discharge flow. For example, percentage of rainfall appearing as the storm flow is estimated to be only 3% at the precipitation of 2 mm, but to be more than 40% at that of 50 mm.

Similarly, the relationship between cumulative discharge and cumulative loading is shown in Figure 8, using COD_{Mn} as an example. In case of COD_{Mn}, a high linearity was observed in a logarithmic scale yielding a high correlation coefficient (0.987, 0.988). The slope was slightly higher than unity, indicating that a bigger storm might cause much higher effects.

The relationship between discharge and loading was different among water quality items. Table 4 summarizes the characteristics of such a relationship with parameter values in equation 1 and correlation coefficient values. The values of n, which expresses

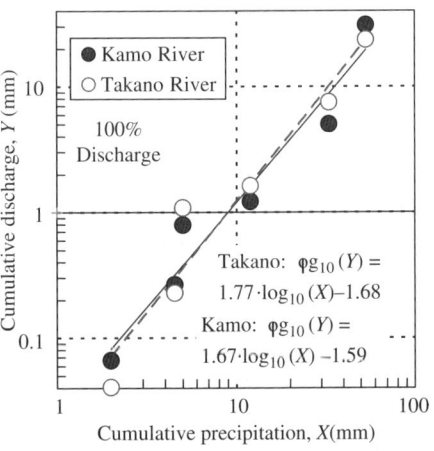

Figure 7 Relationship between precipitation and discharge

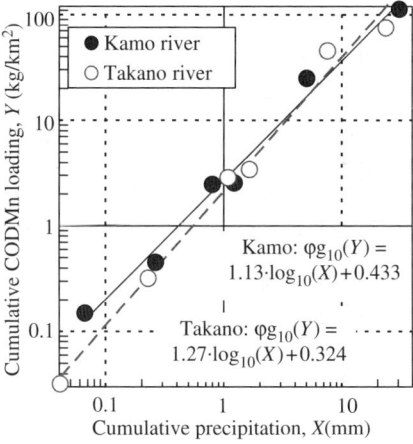

Figure 8 Relationship between precipitation and load

the magnitude of flow influence, were ranging from 0.77 to 1.4. The differences between water quality items were higher than those between the two rivers. The items having n more than 1 in both rivers were indices related to organic matter or particulate substrate such as COD_{Mn}, DOC, SS, and TP, while those having n less than 1 were indices related to ionic portions such as IC, Cl^- and Na. SS and VSS had very high values of n in the range of 1.3–1.4. This means if cumulative flow rate became 100 times, the loading

Table 4 Parameter values of storm runoff equations and estimation of loadings

	Parameters of regression Equation[1]						Loading[2]	
	Kamo River			Takano River			Kamo River	Takano River
	a	n	R^2	a	n	R^2		
Flow	0.026	1.668	0.973	0.021	1.765	0.968	1,131 (51.7)	1,323 (53.2)
COD_{Mn}	2.70	1.128	0.987	2.11	1.270	0.988	2.93 (80.7)	4.46 (86.5)
$D\text{-}COD_{Mn}$	1.93	1.099	0.980	1.46	1.045	0.970	1.96 (78.5)	1.64 (72.8)
DOC	1.24	1.069	0.968	1.17	1.129	0.992	1.32 (68.1)	1.77 (72.5)
IC	6.88	0.852	0.984	5.16	0.984	0.995	6.15 (42.5)	7.59 (45.4)
SS	8.58	1.348	0.994	7.75	1.428	0.985	17.17 (92.0)	26.96 (95.7)
VSS	2.29	1.289	0.997	2.25	1.313	0.983	3.68 (93.4)	5.11 (94.6)
Cl^-	4.94	0.962	0.984	5.49	0.933	0.983	5.95 (43.3)	8.63 (36.0)
$SO_4^{2-}\text{-}S$	7.86	0.945	0.990	3.97	1.233	0.984	8.60 (45.4)	12.54 (48.8)
$SiO_2\text{-}Si$	3.32	1.104	0.996	3.36	0.940	0.991	3.91 (45.5)	4.34 (44.8)
Ca	8.66	0.811	0.899	6.34	1.000	0.980	8.98 (32.8)	11.77 (37.9)
Mg	1.33	0.782	0.834	1.24	0.842	0.925	1.35 (30.9)	1.88 (28.1)
Na	4.51	0.825	0.843	4.25	0.845	0.958	4.55 (34.9)	6.07 (30.1)
K	1.02	0.845	0.943	0.93	0.979	0.990	0.78 (48.4)	1.36 (45.2)
TN	0.74	1.210	0.894	1.22	1.149	0.818	1,275 (66.6)	2,247 (63.8)
DN	0.89	0.918	0.987	0.57	1.149	0.948	773 (52.6)	1,456 (46.4)
$NO_3^-\text{-}N$	0.68	1.012	0.941	0.73	1.113	0.969	748 (54.8)	1,438 (51.9)
$NO_2^-\text{-}N$	0.0046	0.772	0.994	0.0046	0.903	0.979	3.29 (42.7)	5.48 (43.4)
$NH_4^+\text{-}N$	0.026	1.278	0.766	0.035	1.081	0.801	45.7 (82.3)	40.5 (79.8)
TP	0.049	1.241	0.992	0.051	1.223	0.988	76.2 (81.8)	94.7 (81.0)
DP	0.026	1.234	0.944	0.035	1.073	0.977	42.8 (75.4)	47.5 (65.6)
D-Al	0.026	1.006	0.798	0.009	1.260	0.869	21.0 (73.6)	20.6 (74.4)
D-Fe	0.045	0.912	0.980	0.031	0.945	0.993	42.6 (53.2)	41.1 (51.1)
D-Sr	0.045	0.797	0.859	0.031	0.941	0.787	48.3 (30.7)	56.5 (31.9)

[1] Parameters in Eq-1 (in flow, $a \Rightarrow b$, $n \Rightarrow m$ in Eq-2), $Z = a \cdot Y^n$; $Y = b \cdot X^m$, X: cumulative precipitation (mm), Y: Cumulative discharge (mm, $= 10^3 m^3/km^2$), X: cumulative loading (km/km²)
[2] Unit; Flow (mm) COD_{Mn} to K, (t/km²/y), TN to D-Sr (kg/km²/y), ()Storm events %

Table 5 Comparison of unit loading factors for pollutant discharge runoff in Japanese rivers

		This study		Takashima et al., 1995		Ebise, 1989		Ichiki et al., 1999 (storm events)				
		Kamo	Takano	Ohbori	Ohtsu	San-oh	Koise	Tanjin	Yama-shina	Juzenji	Isasa	Houryu
Area (km^2)		75.4	66.8	31.5	36.7	12.4	151.5	25.8	32.2	0.9	4.2	7.1
Land use (%)	Forest (%)	89	80	28	23	15	46					
	Agriculture (%)	2	5	16	29	46	44					
	Residence (%)	8	14	56	48	39	9					
Population density (ca/km^2)		548	939	5,488	5,207	1,772	247	54	34	80	72	26
Precipitation (mm/y)		1,814	1,814	1,751	1,576	1,193	1,193	6,058	4,409	2,128	2,410	992
Flow rate (mm/y)		1,131	1,323	1,428	1,271	1,157	601	1,562	1,562	1,562	1,562	1,562
Unit loading factor (kg/km^2/y)	COD$_{Mn}$	2,930	4,460	17,500	17,400	9,650	4,580	17,500	6,060	2,280	5,750	7,500
	TN	1,270	2,250	9,660	8,610	3,411	1,794		2,070	1,990	1,720	3,500
	TP	76	97	1964	998	399	125		230	480	310	410

would become 430–600 times. "a" is the parameter value indicating strength of the loading for each index, and gives discharge concentration at the total precipitation of 1 mm.

Estimation of annual loading

Based on the above parameter values, annual loadings in 2003 (from January 1 to December 31) were estimated in both rivers. The results are shown in Table 4. In that year, 100 storm events (a series of rainfalls with their intervals less than 12 hours) occurred, and 76 events out of 100 had precipitations of 2.0–53.5 mm, which was the interpolating range of the regression equation. Eighteen events happened in the extrapolating range less than 2 mm, but they must not cause any important estimation errors because the sum of their cumulative precipitation amounts is 13.5 mm, only 0.75% of the whole precipitation amount ($=1{,}810$ mm). On the other hand, storm events in the outer extrapolating range (>53.5 mm) may have important influences on the estimation of annual loadings. Such storm events happened six times (58.5, 64.5, 67, 105.5, 123.5, 127 mm), and their sum became 30% of annual precipitation in 2003. The sum of the estimated discharge in the outer range occupied 51% (Kamo River) and 54% (Takano River) of the whole flows in 2003.

Estimated annual loading of the sites was 2.9–4.5, 1.3–1.8, 17–27, 1.3–2.2, 0.076–0.97 t/km^2/y, respectively for COD$_{Mn}$, DOC, SS, TN and TP. Comparison of both rivers indicates the Takano River basin has a structure to discharge more pollutants with increased flow rate, probably due to more urbanized land use. Organic matter (COD$_{Mn}$, SS) and nitrogenous substances (TN, DN and NO$_3^-$-N) showed high differences of more than 1.5 in the ratios (Takano/Kamo). On the other hand, loadings of NH$_4^+$-N, D-Al and D-Fe, which may be related to soil components, were smaller in Takano River than in Kamo River although the specific flow rate in Takano River was 1.2 times that in Kamo River. The storm events caused 52 (Kamo) or 53 (Takano) % of the whole flow, and conveyed 81–87, 68–73, 92–95, 64–67, 76–81% of the whole loading, respectively for COD$_{Mn}$, DOC, SS, TN and TP.

These values (unit loading factors) were compared with previous studies (Table 5). Since land use pattern and population density are different, the values are also different. The values in the study are roughly half or less of the values in other studies. Our surveyed areas have less population with a majority of forest cover. Therefore, our values may be considered as rather background values for river basin runoff.

Conclusions

In this study, pollutant runoff patterns were investigated in two small catchments, the Kamo River basin and the Takano River basin, based on a series of surveys. The main results obtained are as follows.

(1) Pollutant loadings were successfully estimated as two portions of base discharge and storm events discharge from the survey results.
(2) Estimated annual loadings of the sites were 2.9–4.5, 1.3–1.8, 17–27, 1.3–2.2, 0.076–0.97 t/km^2/y, respectively for COD$_{Mn}$, DOC, SS, TN and TP.
(3) 52–53% of the whole flow which was caused by rainfall events conveyed 81–87, 68–73, 92–95, 64–67, 76–81% of the whole loading, respectively for COD$_{Mn}$, DOC, SS, TN and TP.
(4) Differences of regional properties in the two basins cause different runoff patterns, but the differences in runoff patterns also depend on the rainfall patterns. In general, a more urbanized basin receives an early and strong influence of precipitation on the storm event runoff.

References

Ebise, S. (1989). Amount and change of annual discharge loadings from different land-use areas. *Water Pollution Research*, **12**, 497–505.

Ebise, S. (2004). Nonpoint pollution source, in "Formulas, models and tables in environmental engineering" (edited by the Publication Task Group for Environmental Engineering), *Japan Society of Civil Engineers*, pp. 333–344.

Ichiki, A., Yamada, K., Nakatani, Y., Yamada, T. and Okubo, T. (1999). Comparison of pollutant runoff characteristics in several river basins. *Proc. of Environmental Engineering Study Forum*, **36**, 88–90.

Japan Meteorological Business Support Center (2004). Weather observation data in electrical data reading room, http://www.data.kishou.go.jp/.

Songprasert, P., Moriya, M., Fujii, S., Nagare, H., Ihara, H. and Shimizu, Y. (2003). Influence of basin properties on river pollution loading in fine weather. *CD Proceedings of IWA Asia-Pacific Regional Conference*, **9**, p.1Q5A04.

Takashima, E., *et al.* (1995). Characteristics of pollutant loadings in rivers to Teganuma Lake, and their estimation of annual loadings. *Jour. of Japan Society on Wat. Environ.*, **18**, 297–306.

Effects of soil erosion on water quality and water uses in the upper Phong watershed

S. Sthiannopkao*, S. Takizawa* and W. Wirojanagud**

*The University of Tokyo, Department of Urban Engineering, Graduate School of Engineering, 7-3-1, Hongo, Bunkyo-ku, Tokyo 113-8656 Japan (E-mail: *suthisuthi@hotmail.com*)
**Faculty of Engineering, Khon Kaen University, Khon Kaen, 40002, Thailand

Abstract The main objective of this paper is to simulate the effects of soil erosion on river water quality and on agricultural production as a result of the transformation of forestlands in the catchment of the upstream Phong River. Suspended solids carry down attached nutrients and agricultural chemicals causing water pollution in the downstream. There are four different types of land use in this simulation, namely forestlands, flatland and highland sugarcane plantation areas, and paddy fields. The highest mean annual amount of soil erosion is from paddy fields (585,700 tons/year), followed by highland (73,800 tons/year) and flatland (63,950 tons/year) sugarcane plantation areas and forestlands (41,800 tons/year), respectively. However, as most of paddy fields are located in a low land and are wet type cultivations, the soil erosion occurred has less impact on river water quality and its production compared to the soil erosion from the steeper slopes of highland plantation areas. Under the resource-based agriculture, the sugarcane production is mainly increased by expanding the plantation areas leading to a significant loss of topsoil and a considerable reduction of agricultural production. Soil erosion contributes to an increase in the average annual suspended solids concentration by 72 mg/l.
Keywords Land use change; soil erosion; sugarcane production; system dynamic model; water quality

Introduction

The Phong watershed is an important watershed of the upper northeast region of Thailand. The Phong River is a tributary of the Chi River system, which flows into the Mekong River. The Phong watershed covers 1.5 million ha, extending to the five main provinces of Chaiyaphum, Khon Kaen, Loei, Nong Bua Lumphu, and Petchaboon. There are four main rivers in the upstream watershed area, namely the Phrom, the Choen, the Pa Niang, and the upstream Phong. These rivers run into the Ubolratana Dam, located in the middle section of the watershed. The downstream Pong River is divided into two river sections, namely sections from the Ubonratana Dam to the Nong Wai irrigation weir and from the wier to Mahasarakam dam, respectively.

The upper Phong watershed shares 79.59% of the total watershed area and is composed of 27.33% forest, 64.05% agriculture (mainly rain-fed paddy fields and plantations producing crops such as cassava, corn and sugarcane), and 8.62% other uses. The majority of the agricultural land has been transformed from forestland in the last few decades. This land transformation, together with farming practices without soil conservation, causes soil erosion and increased amounts of suspended solids in rivers, which silt up reservoirs, raise the riverbeds and affect water quality and water uses due to elevated turbidity levels in the rainy season. In addition, suspended solids carry down attached nutrients and agricultural chemicals, which causes water pollution in the downstream.

According to the Bank of Thailand (2004), use of natural resources for economic development in the northeast Thailand is extravagant. Soil, water, and forest resources are depleting at staggering rates. More than 1.6 million ha of forest were destroyed during the past 20 years, from approximately 4 million ha in 1975 to only 2.24 million ha in 1995. 0.32 million ha was encroached and turned to paddy and 0.64 million ha was turned into crop fields. Transformation of land use from the forest areas into agricultural areas for the past 30 years within the Phong watershed has left only 10–15% of the total watershed as natural forest and caused many problems such as flooding, soil erosion and sedimentation, soil salinization, and water quality deterioration (KKU, 2003). This is in accordance with Babel *et al.* (2004) stating that in the past four decades, deforestation in Thailand has been very rapid and the forestland has been converted into agricultural land, which has increased erosion from these watersheds. Soil erosion from agricultural areas results in loss of not only productive soil, but also plant nutrients, and organic and inorganic matter causing reduction in soil fertility. Sediment, a product of soil erosion, becomes a pollutant in rivers. According to the land development department, Thailand, some 33% of the 51.3 million hectares of the total geographical area is moderately to severely eroded. Suspended sediments from all the watersheds in Thailand are estimated to be 27 million tons annually. Cropland expansion through exploitation of forested hilly regions in the North and the utilization of the marginal uplands in the East and Northeast have been major contributors.

Among the four main rivers located in the upper Phong watershed area, the upstream Phong River is facing the most severe soil erosion due to the transformation of forestlands into agricultural areas. The effects of forestlands transformation into sugarcane plantation areas in the upstream Phong River are simulated in this study. Plantation of sugarcane in the highland of the upstream Phong River is encouraged by the Thai Government because of suitable climate, altitude, soil conditions, price stability and sugar mills situated nearby. There are three sugar mills located within the Phong watershed with a total processing capacity of 480,000 tons of sugarcane/day (KKU, 2003). This is in accordance with Sneddon (2002) reporting that the Thai state had long targeted the northeast as a region amenable to agro-industrial development.

As the phenomena occurring in the Phong watershed are multifaceted, interrelated and difficult to understand, a system dynamic model is proposed here as a research tool. According to Forrester (1961), system dynamics is a theory of system structure and a set of tools for representing complex systems and analyzing their dynamic behavior. Simonovic (2002) states that perhaps the most important aim of system dynamics is to elucidate the endogenous structure of the system under study, to see how the different elements of the system actually relate to one another, and to experiment with changing relations within the system when different decisions are included.

Methods

First, the current situation of land use and its effects on soil erosion and on water quality in the upper Phong watershed was identified by studying reports on the Phong Watershed Management conducted by the Khon Kaen University (KKU). Second, the causal loop diagram was drawn according to the outcome obtained in the first step. Finally, a system dynamic model was constructed using the STELLA program to simulate the soil erosion effects on water quality and on agricultural production in the upstream Phong River as a result of the forestland transformation. The following assumptions were made for the modeling based on the data obtained from Khon Kaen University (2003): 30 cm topsoil depth, average monthly rainfall of 100 mm during the rainy season, i.e. June to September, and no forestland rehabilitation. The rate of soil erosion was calculated based on the Universal Soil Loss Equation (MSU, 2004). The simulation was run for a 30-year period. The simulation model is composed of four sub-models, namely the soil erosion, the land use transformation, the water quality, and the agricultural production.

Results and discussion
Identifying problems in the upper Phong watershed

The conversion of forestlands into agricultural areas has caused a severe problem of soil erosion. According to the Office of Agricultural Economics (2004), among five main provinces located within the Phong watershed, four provinces, i.e. Chaiyaphum, Nong Bua Lumphu, Khon Kaen and Petchaboon, clearly show the increasing trends of agricultural lands during the period of 1994–2002 at mean annual rates of 0.77, 3.28, 4.71, and 10.66%, respectively. On the contrary all of the forest areas in the provinces of Chaiyaphum, Nong Bua Lumphu, Khon Kaen, Petchaboon and Loei had decreased during the period of 1988–1999 at mean annual rates of 0.84, 1.00, 2.34, 3.21, and 2.07%, respectively. Khon Kaen University (2003) monitored suspended solids in the main rivers of upper Phong watershed and estimated amounts of transported suspended solids in the rivers of Phrom, Choen, upstream Phong, and Pa Niang as 0.06, 0.32, 1.04, 0.19 tons/ha/year, respectively. According to KKU (2003), the highest amount of suspended solids was found in the upstream Phong River (1.04 tons/ha/year) because of the lesser forest coverage compared to the higher forest coverage in other rivers.

Promotion of agriculture is often considered as the only tool to eradicate rural poverty and boost exports to get foreign currency in many developing countries, including Thailand. The national policy of exporting agricultural produce is the main driving force for the intensive use of land resources in Thailand. Thailand is a main exporter of both rice and sugar. According to Food Market Exchange (2004), Thailand was the world's largest rice exporting country between 1995–1998. Although at the end of the year 1998 Thailand's export market share has declined by 4.2% from its original share in 1995, Thailand is projected to keep the largest market share at approximately 26–28% of the world rice production.

Besides rice, Thailand is the world's sixth largest sugar producer and the twelfth largest consumer. According to Food Market Exchange (2004), there are 107,000 small farmers growing sugarcane in Thailand. Mills do not grow cane themselves, but have contract farming with growers. In recent years, growth in sugar production has come largely from area expansion in the north and northeast of the country. The Thai sugar industry has done extremely well in the past decades, thanks to high cane prices, greater stability and confidence in the industry, successful government initiatives in mill relocation and expansion and favorable weather. Moreover, Thailand is presently one of the five largest global sugar exporters, with relatively small domestic demands for sugar and low shipping costs, especially to growing regional markets. The government policy of maintaining high domestic sugar prices has supported increased production, dampened growth in use and increased exportable surpluses. Thai sugar production in 1999/2000 was 5.72 million tons, an increase of 6% from the previous year. Domestic consumption was 1.6 million tons, a decrease by 8.3% from 1.8 million tons in the previous year, leaving plenty of sugar to be exported. In fact, Thailand does not import sugar, but exports about 3.3–3.4 million tons per year, being ranked as the world's fourth largest exporter.

Based on these reports and observations a causal loop was developed as shown in Figure 1. Forestland transformation into agricultural fields increases the rate of soil erosion, which leads to a higher concentration of suspended solids in rivers and a faster rate of siltation in reservoirs. When topsoil is lost due to soil erosion, the agricultural production, especially sugarcane production in this study, is diminished. The sugar stock kept at sugar mills is determined by the difference between agricultural (sugarcane) production and consumption, which comprises domestic consumption and exports. When the sugar stock exceeds a certain sugar stock goal, then the sugar mills will try to reduce the production next year. On the contrary, if the sugar stock depletes then the sugar mills will encourage

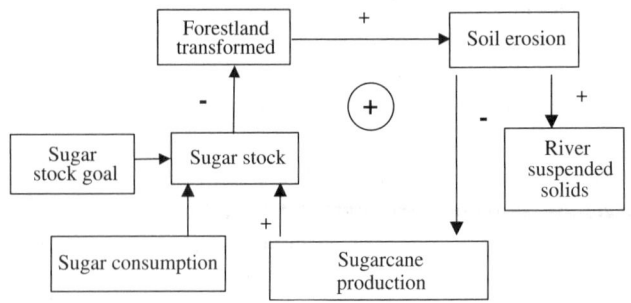

Figure 1 A causal loop diagram of soil erosion effects as a result of forestlands transformation

expansion of the sugarcane production, very often by transforming forestlands into sugarcane fields.

Simulation results

Effects of soil erosion on sugarcane production. As shown in Figure 2, the goal of sugar stock is set at 1,217,000 tons/year, which is 30% of the total mean annual sugarcane productions in the provinces of Loei, Nong Bua Lumphu and Khon Kaen. Therefore, sugarcane production is yearly adjusted to make the sugar stock reach this target as far as possible. Due to one-year delay in production adjustment, sugar stock fluctuates slightly every year. As the forestland conversion rate is determined by the pressure for increase in sugarcane production, this rate also fluctuates year by year. Although, the results shown in Figure 2 fluctuate yearly, the data are summarized in three decades of the simulation period to see the long-term changes (Table 1). The mean annual rates of change displayed in Table 1 are obtained by taking an average of the percent changes from previous years for each parameter. Due to the decreasing flatland sugarcane production at mean annual rates of 1.20%, 1.95%, and 11.01% in the 1st, 2nd, and 3rd decades of the simulation period, respectively, caused by topsoil loss, together with traditional nature-based agriculture, the sugarcane production can be increased only by converting forestlands into highland sugarcane plantations. The land conversion is

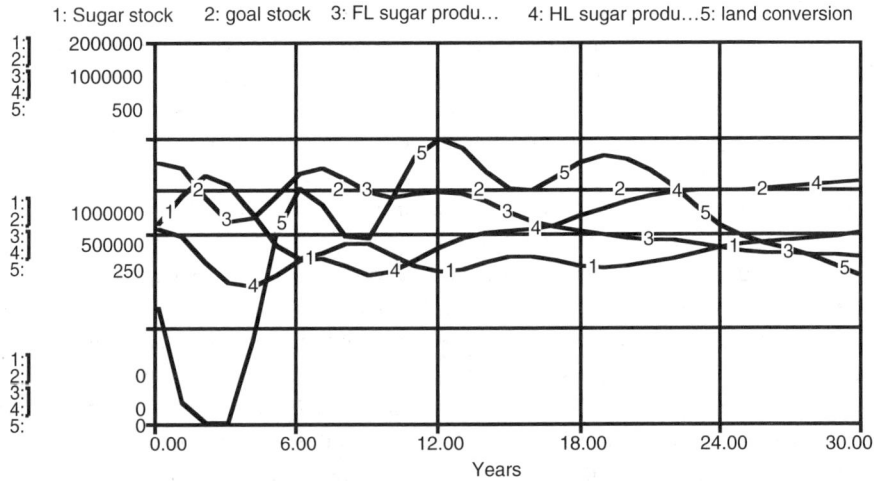

Figure 2 Sugarcane production (tons/year), sugar stock (tons) and land conversion (ha/year) (FL = flatland, HL = highland)

Table 1 Mean annual rates of changes in different simulated parameters (% per year)

Parameters	Year periods		
	1–10	11–20	21–30
Soil erosion–forest (tons/year)	−1.37	−3.61	−4.29
Soil erosion–highland sugarcane plantation (tons/year)	−7.95	−1.95	−11.69
Soil erosion–flatland sugarcane plantation (tons/year)	0	0	−24.99
Soil erosion–paddy (tons/year)	0	0	0
Total soil erosion (tons/year)	−1.55	−1.37	−0.83
Topsoil–forest (cm)	−2.80	−3.90	−6.49
Topsoil–highland sugarcane plantation (cm)	−37.87	0	0
Topsoil–flatland sugarcane plantation (cm)	−6.60	−26.72	0
Topsoil–paddy fields (cm)	−1.17	−5.61	−7.37
Forestland (ha)	−1.37	−3.61	−4.29
Highland sugarcane plantation area (ha)	1.90	3.07	1.92
Land conversion rate	9.85	2.00	−15.04
Sugar stock (tons)	−1.30	−0.67	1.96
Flatland sugarcane production (tons/year)	−1.20	−1.95	−11.01
Highland sugarcane production (tons/year)	−2.11	3.93	−9.22
Paddy production (tons/year)	−0.49	−0.45	−0.46

increasing at mean annual rates of 9.85% and 2.00% during the first and second decades leading to an increase of highland sugarcane production at a mean annual rate of 3.93% in the second decade of simulation. However, as the topsoil of the highland is easily eroded because of the steeper slope, this causes the decrease of highland sugarcane production in the third decade at a mean annual rate of 9.22%. Consequently, the sugar stock is affected by the decrease of sugarcane production in both flatlands and highlands. The mean annual sugar stock is decreased at rates of 1.30% and 0.67% during the first and second periods of simulation, respectively (Figure 2).

Effects of soil erosion on the topsoil depth and on water quality. As shown in Table 1 and Figure 3, the mean annual soil erosion from the forestland is gradually decreasing every year by 1.37%, 3.61% and 4.29% in the first, second and third decades, respectively. The main reason is that more forestlands are transformed into highland

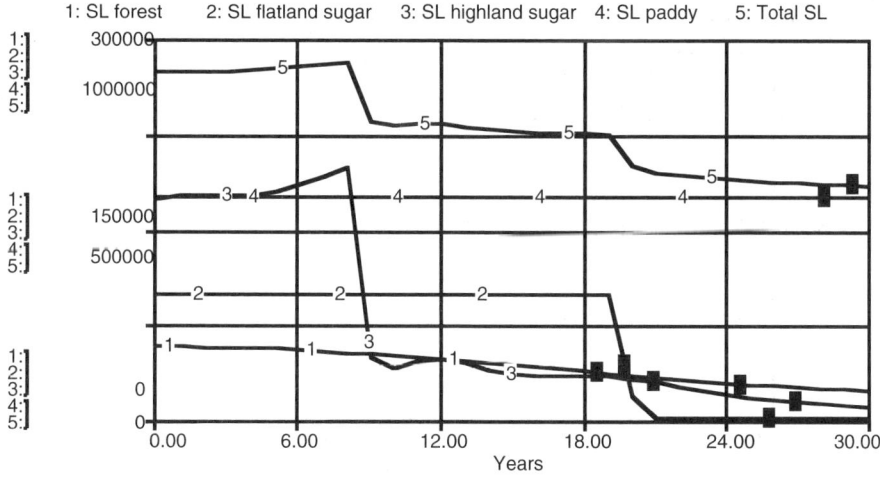

Figure 3 Amounts of soil erosion (tons/year) from different land uses (SL: soil loss)

sugarcane plantation areas at annual rates of 1.90%, 3.07%, and 1.92% in each decade of the 30-year simulation period, respectively. Consequently, the 30 cm topsoil depth from highland sugarcane plantation areas sharply decreases at a mean annual rate of 37.87% in the first decade leading to a complete topsoil loss in the second and third decades. However, the mean annual rates of soil erosion of the highland sugarcane plantations decrease at 7.95%, 1.95% and 11.69% during the first, second and third decades, respectively. This clearly indicates continuing conversions of forestlands to highland sugarcane plantation areas.

Table 1 and Figure 3 show that, in the first and second decades, the mean annual soil loss of flatland sugarcane plantation areas is stable and then it decreases during the third decades annually by 24.99%. This is because the 30 cm topsoil is continuously eroded during the first and second decades annually by 6.60% and 26.72% respectively and is totally depleted during the third decade of simulation. However, in the case of paddy fields, the annual rate of soil erosion is stable during a 30-year simulation period. This is because lowland paddy fields are wet type cultivations. Water storage in a field at the water level ranging from 5–10 cm helps to protect soil being transported far away from its original sources. In the case of deep-water rice plantations, the water level kept in the fields ranges from 50 cm to more than 300 cm.

As shown in Table 1 and Figure 3, the total soil erosion decreases annually by 1.55%, 1.37% and 0.83% during the first, second and third decades of simulation, respectively. This is because of the significant depleting of topsoil in the flatland and highland sugar plantation areas during the second and third decades. Hazarika and Honda (2001) estimated the rate of soil erosion in 1992 and 1996 using remote sensing data and GIS. According to their study the maximum rate of soil loss in the northern Thailand reached 25 mm/year; and the average rates of soil loss decreased from 1.24 mm/year in 1992 to 0.94 mm/year in 1996. The mean amounts of soil erosion from forestlands, highland and flatland plantation areas and paddy fields during a 30-year simulation period are 41,800, 73,800, 63,950, and 585,700 tons/year, respectively. The mean amount of soil erosion from paddy fields is the highest because of the largest areas of land cultivated (281,848 ha).

Figure 4 shows that, assuming 30% of eroded soil reaches the river, suspended solids in the upper Phong River increase by mean annual concentrations of 84.76, 71.26, and

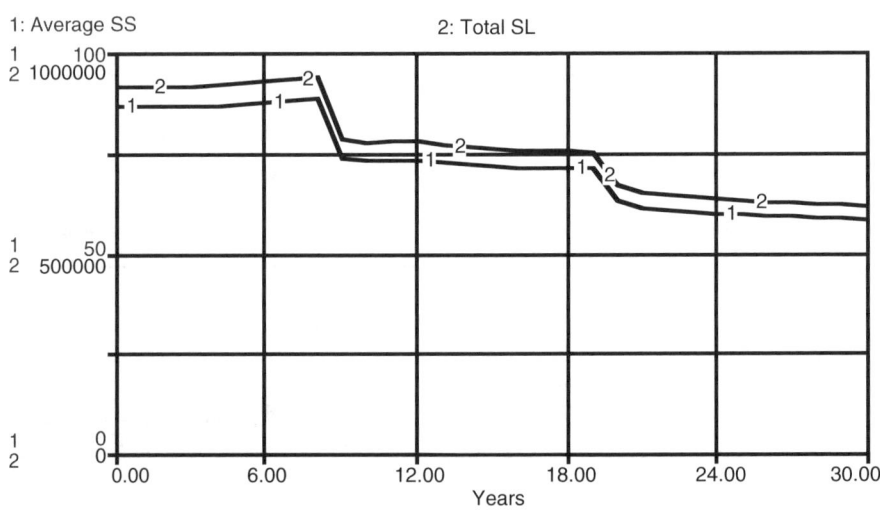

Figure 4 Suspended solid concentrations (mg/l) in the upstream Phong River

59.43 mg/l during the first, second and third decades of simulation, respectively. Soil erosion occurs in the upstream Phong River so that the suspended solids concentration in a river gradually decreases due to the depletion of topsoil in the next three decades. This result is obtained by assuming the topsoil depth as 30 cm. We may obtain different results if the topsoil depth is deeper or shallower than 30 cm. In addition, the soil erosion rate and suspended solids concentration are calculated as annual rates; depletion of topsoil may change the run-off rate and run-off pattern so that the fluctuations of suspended solids concentration in a river may be higher in the future than the present level.

Management application of the model produced. The simulation model presented in this study shows how sugar industry and soil erosion are interconnected. Export-driven economy prospers sometimes at the cost of the domestic environment. This simulation model can be used as one of the effective tools for environmental management, especially at the watershed level. It can easily be modified to examine the effectiveness of, e.g., the land conservation program on reduction of soil erosion and loss of agricultural production. The relationships and feedback built among sub-models of land use transformation, soil erosion, water quality, and agricultural production are comprehensively linked so as to see the effects of running different environmental policies on these different sub-models. With this model alternative policies can be compared and the result of simulation can be used to demonstrate the consequences of different policy options to the stakeholders.

Conclusions

The national policy of export-driven agriculture is a main driving force for the intensive land use in Thailand. In the upper Phong Watershed, the agricultural sector contributes to about 50% of the local economy, which is greater than two other sectors of manufacturing and wholesale. Together with suitable soil property, favorable climate and the establishment of the Sugarcane Association, sugarcane has become one of the preferred cash crops in the region. In addition, there are three sugar mills located within the Phong Watershed with a total processing capacity of 480,000 tons of sugarcane/day. A system dynamic model was developed and used to simulate soil erosion, deforestation and also loss of agricultural production. The increased forestlands conversion into sugarcane plantation areas accelerated the rate of soil erosion in the highland and the topsoil of the farmlands became completely eroded during the third decade of the simulation period. Soil erosion in the upstream Phong River was estimated to increase the suspended solids in a river by the annual concentration of 72 mg/l during the 30-year simulation period. In addition, soil erosion led to the decrease in sugarcane production of both flatlands and highlands mainly because of the decrease of topsoil set at 30 cm in this simulation. The system dynamic model developed in this study can be a useful tool to compare different policy options, and thus to come up with the best management practice.

References

Babel, M.S., Mujithaba, M., Najim, M. and Loof, R. (2004). Assessment of agricultural nonpoint source model for a watershed in tropical environment. *Journal of Environmental Engineering*, **3**(9), 1032–1041.

Bank of Thailand (2004). The northeast economy: Isaan region and social, economic, and environmental development. www.bot.or.th.

Foodmarketexchange (2004). Thai sugar. www.marketexchange.com.

Foodmarketexchange (2004). Thai rice. www.marketexchange.com.

Forrester, J.W. (1961). *Industrial dynamics*. MIT Press, Cambridge, Massachusetts.

Hazarika, M.K. and Honda, K. (2001). Estimation of soil erosion using remote sensing and GIS, Its valuation and Economic Implications on Agricultural Production. In *Sustaining the Global Farm*, Stott, D.E. (ed.), pp. 1090–1093.

Khon Kaen University (KKU) (2003). *Environment management of the Phong watershed.*

Michigan State University (2004). On-line soil erosion assessment tool. www.iwr.msu.edu.

Office of Agricultural Economics (2004). Agricultural Statistics of Thailand. www.oae.go.th.

Simonovic, S. (2002). World water dynamics: global modeling of water resources. *Environmental Management* **66**, 249–267.

Sneddon, C. (2002). Water conflicts and river basins: The contradictions of co-management and scale in Northeast Thailand. *Society & Natural Resources*, **15**(8), 725–741.

Long-term fluctuation and regional variation of nutrient loads from the atmosphere to lakes

T. Kunimatsu* and M. Sudo**

*Department of Ecosystem Studies, School of Environmental Science, University of Shiga Prefecture, Hassaka, Hikone, Shiga 522-8533, Japan (E-mail: *kunimatu@ses.usp.ac.jp*)

**Department of Biological Resources Management, School of Environmental Science, University of Shiga Prefecture, Hassaka, Hikone, Shiga 522-8533, Japan (E-mail: *sudo@ses.usp.ac.jp*)

Abstract The atmospheric depositions were collected by ordinary bulk-samplers mounted with a glass or polyethylene funnel of 30 cm orifice. A long-term observation was carried out at a site (35°01′30″N, 135°58′07″E) in the urbanized area for 21 years since 1974. The annual volume-weighted average concentrations of total nitrogen (TN) and phosphorus (TP) were 1.02 ± 0.30 and 0.031 ± 0.015 mg l^{-1}, respectively, and the loading rates were 14.5 ± 2.8 and 0.43 ± 0.16 kg ha^{-1} yr^{-1}. The rates neither had a relationship with the precipitation ($1,492 \pm 343$ mm yr^{-1}) nor showed any diachronic tendencies. In order to obtain the loading rate of TN and TP within 10% uncertainty under the significant level of 0.01, there is no way but to continue the observation for seven and eleven years or more, respectively. In order to clarify the regional variation of the loading amounts of the depositions, the samplers were set at 12 sites distributing throughout the Kinki District for two years. The distance from northernmost (2322 mm yr^{-1}) to southernmost (1242 mm yr^{-1}) is about 150 km. The average loading rates of the 12 sites were 16.2 ± 2.5 kg ha^{-1} yr^{-1} of TN and 0.730 ± 0.247 kg ha^{-1} yr^{-1} of TP. The depositions of NO_3-N, NH_4-N, and TN as well as TP showed tendencies with distances neither from a big city nor from Japan Sea.

Keywords Atmospheric deposition; long term monitoring; nitrogen; phosphorus

Introduction

It has been well documented that atmospheric depositions as well as rivers and the groundwaters play a significant role in supplying nutrients to lake ecosystems. In the case of Lake Biwa, having 27.5 km^3 of volume, 675 km^2 of surface area and 3,144 km^2 of catchment area, the amount of the direct deposition on the lake surface was estimated to be 33% of the total load of nitrogen (Kunimatsu, 1998), and 8% of that of phosphorus (Kunimatsu and Kitamura, 1981). In order to study the material balances of the lake ecosystem as well as to decide the priority of measures protecting the lake from eutrophication, it is necessary to evaluate the precise loading rates of nutrients from the atmosphere. Although there have been many studies of the depositions of nitrate and/or phosphate, those of the total nitrogen and/or phosphorus have been restricted. There are at least four major problems in evaluation of the accurate and comprehensive loading rates; first there have not been any standardized sampling devices, namely the material of the sampler, separation of litter and insects, on site storage, etc., secondly there is no way to separate materials circulating on site, namely, dust, pollen, volatile materials, etc. from the soils, fauna and flora from the bulk deposition (Ahn and James, 1999; Tsukuda *et al.*, 2004), thirdly the problem of how to appreciate the long-term fluctuation of the depositions mainly due to annual changes of meteorological conditions, and fourthly the lack of any criteria to select the observation site to evaluate the true amount of the deposition supplying to a large lake, namely distance from roads, smoke-emitting factories, agricultural fields applied with slurry manure, coast, etc.

doi: 10.2166/wst.2006.038

This paper focuses mainly on the latter two problems to protect such a large lake as Lake Biwa against eutrophication, and shows the long-term fluctuations of the loading rates of the total nitrogen and phosphorus as well as their chemical components from the observation for 21 years at a site and the regional variations from the data measured at 12 sites for two years.

Methods

Experimental sites

A long-term experiment to measure the bulk deposition of nutrients was carried out at Site 1 shown in Figure 1, which is located at 35° 01′ 30″ N in latitude, and 135° 58′ 07″ E in longitude. The site is in an urbanized area of Kusatsu City (population; 64,825 in 1975 and 114,009 in 2003, city area; 48.22 km^2) surrounded by paddy fields and forests in the basin of Lake Biwa, which is the largest lake and lies around the central part of the main island of Japan. In the city area, there were neither large factories vomiting thick smoke nor upland fields applied with a lot of manure, compost and/or dairy slurry, which were the anthropogenic sources of nitrogen, sulfur, and chlorine in gaseous and particulate forms. The long-term observation was carried out for 21 years from October 1974 to December 2003 with an interruption for eight years from January 1982 to June 1989. According to the record for the 21 years, the average temperature and precipitation were 11.1°C and 1,492 mm y^{-1}, respectively.

Regional variations of the bulk depositions were investigated by measuring at 12 sites distributed throughout the Kinki District, as shown in Figure 1. Sites 2 and 3 were set in northernmost rural mountain areas and Site 12 in southernmost urbanized areas, namely the Osaka megalopolis area (8.56 million people live in 1,881 km^2). The experiment was carried out for two years from January 1992 to December 1993.

Bulk deposit samplers

A bulk sampling method was used to collect the atmospheric wet and dry deposition. Bulk samplers have been widely used, because they were inexpensive and did not require a power supply (Kopacek *et al.*, 1997; Herut *et al.*, 1999; Campo *et al.*, 2001). Two types of the samplers were used in the study; Type 1 consisting of a glass cylindrical funnel with 30-cm orifice and 15-cm wall height and a light-shielded 20-litre glass reservoir,

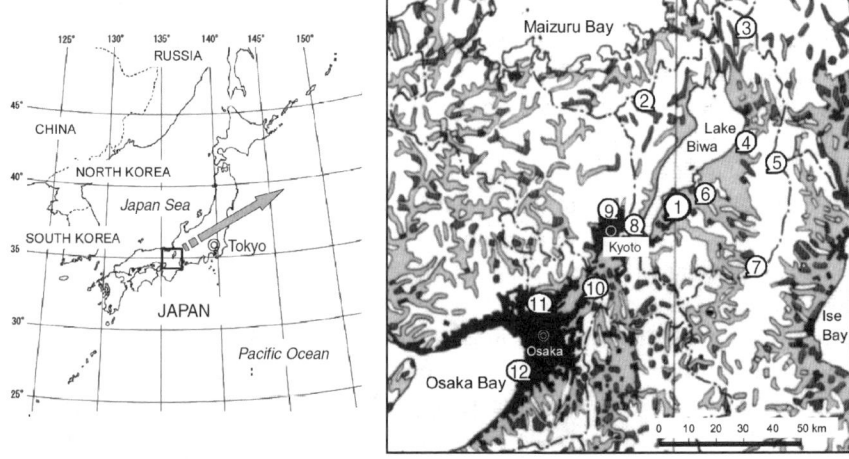

Figure 1 Experimental sites. The numbers typed inside circles show the sampling sites for the observation of regional variation. The site for long-term observation was carried out at Site 1. Areas painted white, gray and black in the right-hand map show mountain areas, agricultural areas and urban areas, respectively

and Type 2 of a conical-shaped polyethylene (PE) funnel with 20 or 30 cm orifice and a 20 L opaque grey PE-reservoir. It is generally recognized that bulk sampling is not always adequate in monitoring the amount of deposition, because substantial biochemical changes and physicochemical aggregation occur inevitably in the reservoirs during the deposition period. In this study, 2 ml of concentrated H_2SO_4 were contained in the reservoir of the sampler of 20-cm orifice to avoid these problems, and the water samples collected by the sampler of 30-cm orifice were used only to analyze pH, EC, and SO_4-S.

Sampling methods

Site 1 was on the roof of a building of four stories in the campus of Junior College of Shiga Prefecture, where the long-term observations were carried out. Before the interruption, Type 1 was used, and water samples were collected once a fortnight. After that, the sampler was changed to a set of Type 2, and water samples were collected monthly for two years. After May 1991, water samples were collected by Type 1 at the same time-interval as before. Water samples were collected monthly in 2-L PE-bottles after measuring the water volumes, from which precipitations were calculated. A rain gauge, as standardized by the Japan Weather Bureau with a 20.3-cm orifice, was placed at the experimental site. Precipitation was also measured at the Otsu station located about 15 km south of the site, which belongs to the Hikone Local Meteorological Observatory. In the regional investigation, a set of Type 2 samplers was set at each site. Sites 2 and 3 were located in an area normally covered with snow from December through early April, where discharge from a rain–snow gauge was pooled in a reservoir during the period.

Water chemistry

A certain volume of a water sample was filtered through a 1-μm pore-size glass-fiber filter with a 45-mm diameter. The total amounts of nitrogen (TN) and phosphorus (TP), and the dissolved components of those (DN and DP) were analyzed in the water sample before filtration and in the filtered sample, respectively. The concentrations of these elements in particulate forms (PN and PP) were obtained from the differences between the concentrations of each of those samples. Chemical analyses were carried out mainly with Japan Industrial Standard Methods (1986), which had been revised for years. So the methods for some items were changed during the long study-period, as follows: Kjeldahl digestion-indophenol blue method for TN and DN changed to basic potassium peroxodisulfate digestion (120°C, 1.5 atm.) – UV absorbance (260 nm) method; HNO_3–$KClO_4$ digestion-molybdenum blue method for TP and DP to acidic potassium peroxodisulfate digestion-molybdenum blue method; sodium salicylate method for nitrate nitrogen (NO_3-N), mercuric nitrate method for chloride ion (Cl^-), flame analysis for sodium ion (Na^+) to suppressed ion chromatographic method (HPLC) using an HIC-6A attached to Shimpac IC-A1 and IC-C2 (Shimadzu, Kyoto, Japan). Nitrite nitrogen (NO_2-N), ammonium nitrogen (NH_4-N) and phosphate phosphorus (PO_4-P) were assayed with diazonation method, the indophenol blue method and the molybdenum blue method, respectively.

Results and discussion

Fluctuations of loading rates

Seasonal fluctuations. The daily loading rates calculated from data analyzing each sample (every two weeks or month) were plotted with time in Figure 2, in which curves smoothed by the moving average of 5 data were also shown. The curve of TN decreased in winter and increased from summer in autumn every year. The period of the high loading rate coincided with the rainy season and that of the low rate with the dry season in the region. A similar phenomenon was found in the curve of TP. From the result, it

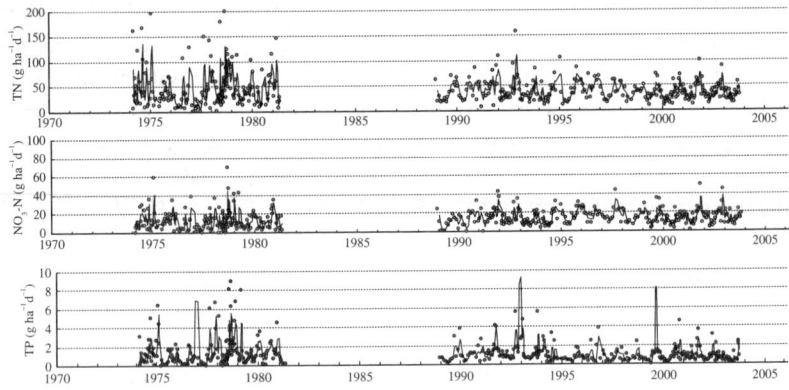

Figure 2 Seasonal changes of depositions of total nitrogen, nitrate nitrogen and total phosphorus at Kusatsu City. Lines were drawn by the moving average of 2 data

should be memorized that we cannot evaluate the annual rate of the atmospheric deposition from observation for less than one year, because the rates of nutrient depositions usually fluctuate more than two times in a year.

Long-term fluctuations. The annual volume-weighted average concentrations were calculated from the long-term data, and plotted with year in Figure 3. No definite tendency was found in the concentrations of TN. As mentioned above, there were no large factories discharging thick smoke and the surrounding agricultural fields were mainly paddy fields, so the changes of the concentrations of NO_3-N and NH_4-N, both of which might be strongly affected by anthropogenic sources, did not show any definite tendency. However, slightly decreasing tendencies were found in the changes of DP and PO_4-P.

The mean values of the annual volume-weighted average concentrations for 21 years and their coefficient of variation are shown in Table 1. The concentrations of TN and TP were 1.02 mg l^{-1} and 0.0310 mg l^{-1}, respectively, of which CV-values were 29 and 47%. The variation of the concentration of phosphorus was greater than that of nitrogen. The ratio of N/P was 33. These results indicate that the sources of nitrogen and phosphorus might be different from each other.

Annual loading rates

The mean values of the annual loading rate for 21 years and its coefficient of variation are also shown in Table 1. The loading rate of TN was 14.5 ± 2.82 kg ha^{-1} yr^{-1}. NO_3-N,

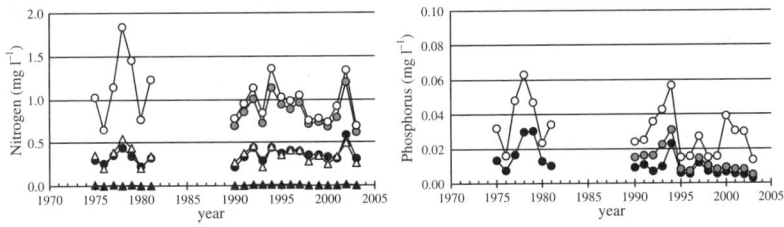

Figure 3 Long-term fluctuations of nitrogen and phosphorus concentrations of atmospheric deposition measured at Kusatsu City. The concentrations were the annual volume-weighted average concentrations calculated from the long-term data obtained once every two weeks before 1982 and once a month after 1989 with the bulk sampler. Nitrogen: ○;TN, ◉;DN, ●; NO_3-N, △; NH_4-N, ▲; NO_2-N. Phosphorus: ○; TP, ◉; DP, ●; PO_4-P

Table 1 Mean values of the annual volume-weighted average concentrations and the annual loads measured at Kusatsu City for 21 years

Items	Concentrations		Loading rates		
	mg l^{-1}	CV (%)	kg ha^{-1} yr^{-1}	CV (%)	Composition (%)
TN	1.02	29	14.5	19	100
DN	0.854	21	12.8	15	88
NH$_4$-N	0.348	28	4.97	22	34
NO$_3$-N	0.351	24	5.05	21	35
NO$_2$-N	0.006	56	0.08	59	1
TP	0.0310	47	0.429	37	100
DP	0.0127	56	0.190	53	44
PO$_4$-P	0.0112	70	0.155	57	36
Precipitation (mm yr^{-1})	–	–	1,492	23	–

NH$_4$-N and Org-N occupied about a third of TN each. PN contained only 12%. The variations of each component except for NO$_2$-N were around 20%, which were almost the same as that of the annual precipitation. The loading rate of TP was 0.429 ± 0.158 kg ha^{-1} yr^{-1}, of which PO$_4$-P occupied only about one third. The characteristic of phosphorus was that PP contained more than a half, and that the variations were much larger than those of nitrogen. These compositions were not largely different from those measured previously at three different sites in mountain forests in the Lake Biwa basin (Kunimatsu *et al.*, 2001), except that the ratios of NO$_3$-N (23–31%) were slightly lower and those of PO$_4$-P (46–57%) were higher than those measured at Kusatsu City. It should be noted from these results that the loading rate measured only inorganic substances largely underestimating the loads of nitrogen and phosphorus from the atmosphere to the lakes.

It was clearly shown from the above investigation that the loading rates of nutrients from the atmosphere changed significantly from season to season as well as from year to year. Changes of important sources for the period may be one of the reasons. The rates, however, did not show any tendencies related with the progress of years in Figure 2. It is also necessary to examine how the yearly changes of precipitation may have affected the rates. The annual loading rates versus the annual precipitations were plotted in Figure 4, in which the average values for 21 years with error spans were also shown. It became clear that any relationships between these parameters were not found on the scatter diagrams of TN or TP.

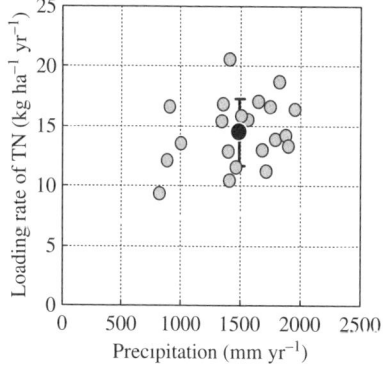 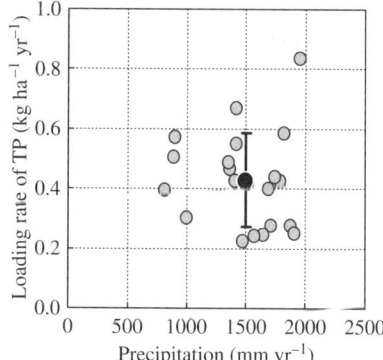

Figure 4 Relationships between the annual loading rates of the atmospheric deposition and the annual precipitations measured at Kusatsu City. The data were the same as those used in Figure 2 and Table 1.
● shows the average value for 21 years with error span

Accuracy of observed annual loading rates

It was left behind as a subject to identify each source, so any suitable methods to obtain reliable loading rates could not be recommended, besides carrying out continuous measurement as long as possible. Then, how long should we continue measuring atmospheric depositions in order to obtain reliable loading rates? Then we examined it using the data measured for 21 years. The data were assumed to be continuous data, although they included an interruption for 8 years from 1981. Suppose that observation was continued for n years and the annual average value x_n was obtained. Every continuous n data were picked up from the data as shifting by a year, and a set of $(20 - n + 1)$ averages x_n were obtained. On the basis of central limit theorem, x_n should distribute the normal distribution around their mean value X_n with the standard deviation s_n in spite of the type of the distribution of the original population. So the confidence interval is calculated as following;

$$\mu = X_n \pm t(s_n/n^{0.5}) \quad (1)$$

where μ and t are the true value and t-value, respectively. t is read in the t-distribution table under a given confidence limit and the degree of freedom ($n - 1$ in this case). The ratios α of $t(s_n/n^{0.5})$ corresponding to 90%- and 95%-confidence limit to X_n were calculated in the long-term data and plotted versus n in Figure 5.

$$\alpha = t(s_n/n^{0.5})/X_n \quad (2)$$

The curves of TP in the figure were obviously distorted in mid course. This is because an unusually large value obtained in 1985 was contained in the 21 years data, as shown in Figure 4. However any reasons for rejecting such a datum have not been found as yet.

If n gives α_{99} smaller than 0.1, x_n within the confidence interval of X_n (1 ± 0.1) will be obtained under 0.01 significance level. In other words, 99% of the average values obtained by measuring for n years may be included within $\pm 10\%$ of X_n, which is the estimated value of true value of the annual loading rate. The curves of α_{95} and α_{99} in the figure indicate that the mean value of the loading rate of TN satisfying the above condition is obtained by continuous measuring over five and seven years, respectively. The rate of precipitation having the same accuracy may be obtained by measurement over five and six years. It is necessary to continue measuring for more than ten and eleven

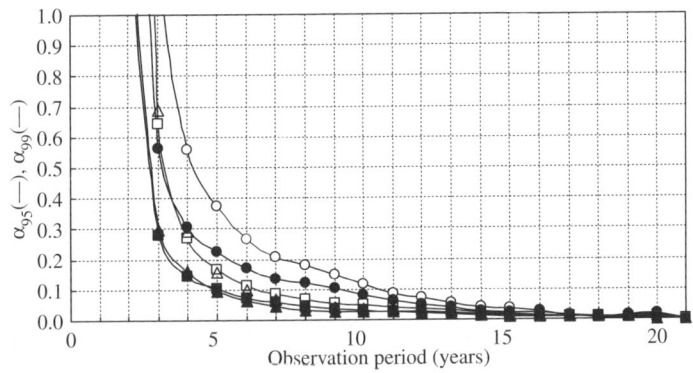

Figure 5 Relationships between the continuous observation periods (n years) and the coefficient of 90% and 95% confidence limits (α_{90}, α_{95}) of the average values. Every n data were picked up from the data obtained for 21 years, and $(20 - n + 1)$ average values x_n were calculated. α is $t(s_n/n^{0.5})/X_n$, where s_n and X_n were standard deviation and mean values of x_n. □; TN, ○; TP, △; precipitation

years, respectively, in order to obtain the loading rate of TP satisfying the same condition.

Regional variations

It is well known that the loading rates of the atmospheric depositions are strongly affected by winds blowing across the sea or the different land-use areas. Then, the site effects on the rates were investigated by setting the bulk samplers at 12 sites distributed throughout the Kinki District. As shown in Figure 1, the northern and southern coasts of Kinki District are facing the Japan Sea and the Pacific Ocean, respectively. From the Pacific Ocean, two or three typhoons accompanied by intense winds as well as intense rains usually pass through the district during summer and autumn. During winter, strong monsoon winds often blow from a reverse direction across the Japan Sea accompanied by snow.

First, in order to clarify geographical influence, the average annual depositions were plotted against shortest distances from the Maizuru Bay and from the center of Osaka City in Figure 6. The depositions of Na^+ decreased as the distance from the bay increased, as expected. The seasonal changes of loading rates of Na^+ measured at Site 2 set in a northern rural mountain area and Site 12 at the southern urban area near Osaka City are shown in Figure 7. The deposition of Na^+ obviously increased from November to April at Site 2 every year. However, the seasonal change was found just slightly on the graph of Site 12, although it is located near the sea. The results indicate that the influences of the sea salts are not estimated simply by a function of the distances from the sea. Secondly, it seems to be common knowledge that urbanization strongly affects the deposition of nitrate nitrogen. However, as shown in Figure 6, the effects of the distances from the City Hall of the Osaka megalopolis on the depositions of $nss-NO_3-N$, which were connected to the supply from the sea salt, were not clear. Moreover, essential differences were not found between Site 2 and Site 12 in Figure 7. Phosphorus also did not show any tendencies.

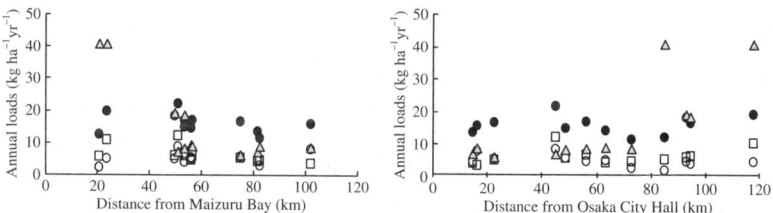

Figure 6 Relationship between the annual loading rates of deposition of nitrogen, phosphorus and Na^+ and shortest distances from Maizuru Bay and from the center of Osaka City. ●; TN, ○; NO_3-N, □; TP (x10), ∆; Na^+

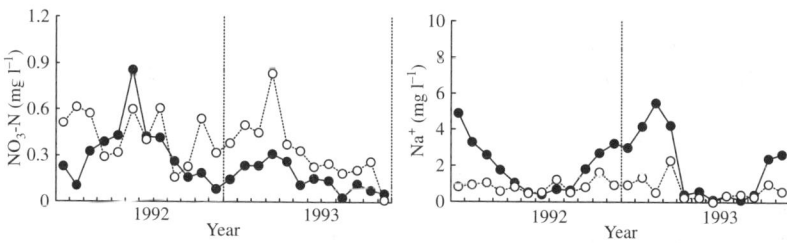

Figure 7 Monthly depositions of NO_3-N and Na^+. Site 2 (●) is located at the northern mountain area, and Site 12 (○) at the Osaka megalopolis area in the southern area of Kinki District

Table 2 Regional variation of the annual loading rates of atmospheric depositions[1] (kg ha^{-1} yr^{-1})

No.	Sites	TN	DN	NH$_4$-N	NO$_2$-N	NO$_3$-N	TP	DP	PO$_4$-P	Precipitation[3]
1	Kusatsu	18.2	16.9	7.0	0.01	5.4	0.75	0.56	0.42	1678
2	Kutsuki[2]	13.4	11.0	3.8	0.03	3.1	0.85	0.68	0.46	2133
3	Yogo	19.4	17.3	5.5	0.01	4.9	1.10	0.83	0.49	2322
4	Hikone	16.1	15.3	4.8	0.01	4.1	0.52	0.40	0.25	1240
5	Taga[2]	17.0	16.5	4.8	0.00	4.3	0.68	0.60	0.36	2452
6	Yasu	14.0	12.3	4.8	0.02	5.5	0.81	0.63	0.35	1716
7	Koga	12.1	10.3	3.4	0.00	3.0	0.62	0.49	0.35	1737
8	Otsu	18.8	17.7	6.8	0.01	6.0	0.99	0.85	0.62	1660
9	Kyoto	19.4	17.8	7.0	0.02	7.2	1.09	0.97	0.74	1696
10	Hirakata	15.7	13.8	4.9	0.03	5.4	0.57	0.41	0.31	1462
11	Suita	13.5	11.6	3.7	0.02	4.8	0.42	0.28	0.18	1515
12	Sakai	16.7	15.6	6.3	0.04	7.0	0.36	0.25	0.16	1242
Average		16.2	14.7	5.23	0.02	5.06	0.730	0.579	0.391	1738
CV (%)		15.5	18.8	24.7	73.9	26.2	33.9	39.3	43.5	22.2

[1] The bulk depositions were continuously measured once a month from January 1992 to December 1993
[2] Those were measured from January 1993 to December 1993
[3] mm yr^{-1}

The average annual loading rates are shown in Table 2. Precipitations ranged from 1240 mm yr^{-1} at Site 4 to 2452 mm yr^{-1} at Site 5. The average loading rate of TN of all sites was 16.2 kg ha^{-1} yr^{-1} varying from 12.1 kg ha^{-1} yr^{-1} to 19.4 kg ha^{-1} yr^{-1}. The average composition of chemical species was NO$_2$-N 0.1, NO$_3$-N 31, NH$_4$-N 32, Org-N 36 and PN 9 in percent. There were tendencies that the deposition of TN increased in urban areas as well as in high precipitation areas. DIN tended to increase in the urbanized areas. The average loading rate of TP was 0.730 kg ha^{-1} yr^{-1}, varying from 0.36 kg ha^{-1} yr^{-1} to 1.10 kg ha^{-1} yr^{-1}. The average chemical compositions were 54, 46 and 21% for PO$_4$-P, Org-P and PP, respectively. Phosphorus showed a tendency to decrease in the urbanized areas. These results also indicated that the source of phosphorus was different from that of nitrogen.

Conclusion

From the results of the long-term observation for 21 years at the same site, the loading rates of nutrients from the atmosphere varied significantly from year to year, namely in the cases of TN and TP from 9.37–20.6 kg ha^{-1} yr^{-1} and 0.226–0.838 kg ha^{-1} yr^{-1}, respectively, and had no correlations with annual precipitations. It was shown from the data that in order to obtain the loading rates within 10% of uncertainty under the significant level of 0.01 it was necessary to continue observation for seven and eleven years or more for TN and TP, respectively. It was indicated that the sources of these elements may be different from each other. However, there are so many factors giving more or less annual change to the loading rates, that it was difficult to identify any particular or major factors.

It was also shown that the rates varied from site to site, distributed to not so large an area (150 km from north to south, 60 km from east to west), from 12.1–19.4 kg ha^{-1} yr^{-1} for TN, and from 0.36–1.10 kg ha^{-1} yr^{-1} for TP. However, it was ambiguous whether these variations resulted from anthropogenic pollutions, meteorological conditions, and/or biotic conditions. Thus it seemed difficult to develop any methods or models for predicting the loading rate by using a number of these parameters. In conclusion, there would not be any other way but to make observations at as many sites as possible and to continue for seven years or more in order to obtain accurate and comprehensive loading rates of nutrients from the atmosphere to a lake.

References

Ahn, H. and James, R.T. (1999). Outlier detection in phosphorus dry deposition rates measured in South Florida. *Atmospheric Environment*, **33**, 5123–5131.

Japan Industrial Standard Methods (1986). *Testing Methods for Industrial Wastewater*, JIS K-0102.

Kunimatsu, T. (1988). Loading Mechanism of Nutrients from Rivers and Estimation of Nutrient Budget of Lake Biwa: a review. In *Research on Lake Biwa -from the watershed through the lake water-*, Lake Biwa Research Institute (ed.), pp. 49–63 (in Japanese).

Kunimatsu, T., Hamabata, E., Sudo, M. and Hida, Y. (2001). Comparison of nutrient budgets between three forested mountain watersheds on granite bedrock. *Wat. Sci. Tech.*, **44**(7), 129–140.

Kunimatsu, T. and Kitamura, G. (1981). Phosphorus Balance of Lake Biwa. *Verh. Internat. Verein. Limnol.*, **21**, 539–544.

Swank, W.T. and Waide, J.B. (1987). Characterization of baseline precipitation and stream chemistry and nutrient budgets for control watersheds. In *Forest Hydrology and Ecology at Coweeta*, Swank, W.T. and Crossley, D.A. (eds), Springer-Verlag, New York, pp. 57–79.

Tsukuda, S., Sugiyama, M., Harita, Y. and Nishimura, K. (2004). A methodological re-examination of atmospheric phosphorus input estimates based on special microheterogeneity. *Water Air Soil Pollution*, **152**, 333–347.

Water chemistry gradient in a degraded bog area

R. Iqbal*, S. Akimoto*, K. Tokutake*, T. Inoue** and H. Tachibana*

*Graduate School of Engineering, Hokkaido University, Kita 13 Nishi 8, Kita-ku, Sapporo, Japan
(E-mail: *iqbal@rofiq.net; saorin@eng.hokudai.ac.jp; takeyan710@hotmail.com; harukuni@eng.hokudai.ac.jp*)
**Graduate School of Agriculture, Hokkaido University, Kita 9 Nishi 9, Kita-ku, Sapporo, Japan
(E-mail: *tino@env.agr.hokudai.ac.jp*)

Abstract Surface and ground water was sampled in a degraded bog area 36 times during 1993–2003 at Five representative points: point E (natural area with *Sphagnum* as the main vegetal cover), point W (boundary between the natural and degraded areas), point W' (area installed with vinyl sheeting), point WW (area where *Sasa* thrives), and point NC (area with naturally formed ditches). Analysis of variance (ANOVA) was conducted for parameters measured in surface water and ground water at 0.5, 1.0, 1.5, and 2.0 m depths. "Sampling point" (i.e. locations along the degradation gradient) accounted for most of the variation in surface and ground water chemistry. It accounted for 30–80% of the total variation in pH, electrical conductivity, ammonia, dissolved nitrogen, major cations (Na^+, K^+, Ca^{2+}, Mg^{2+}), alkalinity and dissolved organic carbon. "Year" accounted for more variation in nitrate, nitrite, chloride, and sulfate than the sampling point did, but the variation in dissolved reactive phosphorus and dissolved phosphorus concentrations was not based on any of the calculated variables.
Keywords Bog; land-use change; plant diversity; water chemistry; water level

Introduction

Water chemistry is one of the most important aspects of wetland ecosystems due to its strong correlation with vegetation types (Vitt *et al.*, 1995; Hotes *et al.*, 2001) and its important role in maintaining the original diversity of wetland plant species. This role is more pronounced in bogs, which are rain-fed peaty wetlands. Local species thrive under a bog's acidic, nutrient-poor conditions and are highly sensitive to any changes in the chemical conditions at the mire surface (Bragg and Tallis, 2001). However, adjoining land-use changes such as drainage development for agriculture have great impact on its hydrology and hydrochemistry. Sarobetsu Mire is a case in point. Such changes have been occurring on the western side of the mire, where its ecosystem is gradually losing its natural state due to the development of agricultural land. This has led to invasion by the non-native species of dwarf bamboo (*Sasa palmata*) that prefer dry conditions (Inoue *et al.*, 1992) in place of the native *Sphagnum* spp. This paper addresses the changes in water chemistry in this degraded bog that have resulted from human disturbances in the surrounding areas and discusses the mechanisms behind the observed degradation.

Material and methods

Site description

Sarobetsu Mire, a coastal bog in northern Hokkaido, measures 23,000 ha. The peat thickness ranges from 5 to 7 m. Most of the surface is less than 10 m a.s.l., the figure being 6 m at the central mire area. The mire is topographically very flat, with a gentle slope of about 1/3,000 toward the Sarobetsu River. The vegetation of the bog is characterized by moss cover of *Sphagnum papillosum*, *Sphagnum riparium*, and *Drepanocladus exannulatus*. The vegetation in the shrub layer is predominantly *Rubus chamaemorus*,

doi: 10.2166/wst.2006.039

Scirpus wichurae, Drosera anglica, Hemerocallis middendorffii and *Myrica gale*. Sedges of *Phragmites communis* and *Carex lasiocarpa* var. *occultans* are also present. Agricultural development in Sarobetsu Mire started in anti-hunger projects after World War II. At first, immigrant farmers gave up their attempts to cultivate the land, due to frequent flooding. During 1961–1969, the Hokkaido Development Bureau constructed a main shortcut channel to prevent flooding by lowering the water level. This construction succeeded in changing the hydrological conditions of the area and increased the productivity of the farmland. However, the emergence of invasive *Sasa* spp. endangered the unique habitat preserved in the raised bog. This threat brought attention to the importance of conserving the remaining wetland. In 1974, 7,000 ha of this area were designated as Rishiri Rebun Sarobetsu National Park, including the study area.

Sampling points and field methods

To examine the broadest possible range of mire degradation, five sampling points (Figure 1: E, W, W', WW and NC) were chosen. Point E is on the eastern side of the mire, where the water level is still high and the natural bog vegetation is preserved. The ground surface is covered by *Sphagnum nemoreum* Scop. and *S. papillosum* Lindb. Point W, approximately 180 m west of point E, represents the *Sasa* front, i.e. the boundary between the natural and degraded areas. The ground cover is predominantly *Sphagnum*, as well as *Oxycoccus quadripetalus,* and *Carex middendorffii*. Point W' is 150 m west of point W. This is where 0.3-mm-thick and 1.3-m-width vinyl sheeting was installed in 1991 to raise the water level and re-establish the bog vegetation. The leaf area index (LAI) of *Sasa* in this area is 0.6, slightly higher than at point W, where the LAI is 0.5. Point WW is the area with a high distribution of *Sasa*: the LAI value is 1.8, the highest among all sampling points. The only natural plant observed on the ground surface is *Ilex crenata* var. *paludosa*. The water level is also very low, and surface water is not present. Point NC is the area where a naturally formed gully system (natural channel) drains from the mire into the Sarobetsu River. The *Sasa* distribution is also high here, and the LAI value is 1.3, with no *Sphagnum* moss or other bog vegetation present.

Water samples were collected during 1993–2003. Using a polyvinyl jug that was rinsed before each sampling, surface water samples were collected directly from 30-cm-diameter PVC pipes that had been installed for the experiment. Groundwater was collected from 7.5-cm-diameter PVC pipes by suction after initially purging the water at the start of each sampling, to obtain fresh water samples. Each point had four pipes, and

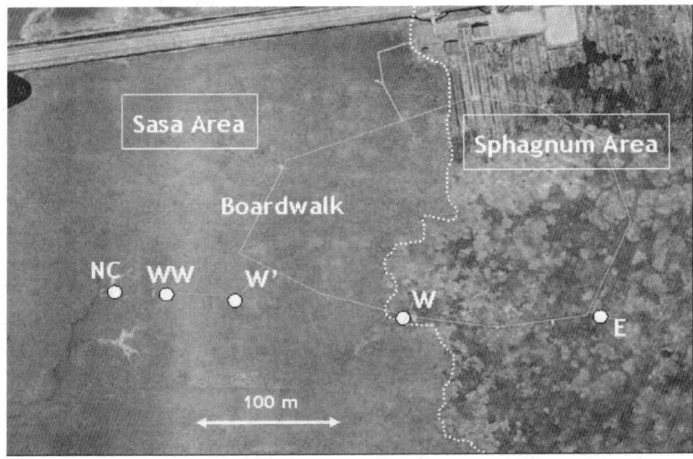

Figure 1 Sampling points (aerial view)

each pipe had 5.0-mm-diameter holes punched around the circumference at various depths relative to the mire surface: 0.5, 1.0, 1.5 or 2.0 m. Between sampling periods the PVC pipes were capped to avoid contamination. The water levels at points E, W, W' and WW were monitored bi-hourly by water level logger during 1993–1997 and during November 2002 –November 2003.

Analytical and statistical methods

Water samples were chilled immediately to 5 °C after sampling to minimize microbial reactions. Within 2 days, samples were filtrated through a 0.45-μm membrane filter. Chemical analyses were conducted according to analytical methods recommended in Water Analysis (2000). The following were analyzed: pH, electrical conductivity (EC), alkalinity 4.3 Bx (50 ml of each sample was titrated with 0.02 mol l^{-1} sulphuric acid, with pH = 4.3 as the endpoint), dissolved organic carbon (DOC) measured by combustion-infraspectrometric method (Shimadzu TOC-5000), and major inorganic ions (Na$^+$, K$^+$, Mg^{2+}, Ca^{2+}, Cl$^-$, SO$_4^{2-}$) measured by ion chromatography with ECD (Hewlett-Packard IC-7000RP). Phosphorus content was measured by spectrophotometry (measured in the form of molybdenum blue) after persulfate digestion, where ammonia (NH$_4^+$-N) was determined by spectrophotometry (phenate method), nitrite (NO$_2^-$-N) was determined by spectrophotometry (Griss reagent), nitrate (NO$_3^-$-N) was measured by ion chromatography with ECD, and dissolved nitrogen (DN) was measured by spectrophotometry (phenol method) after persulfate digestion. For data analysis, precipitation data are the average daily precipitation. For better comparison of meteorological data, data for July through November were calculated together with data from other months as well as separately, because they were available for each monitored year. Three statistics of water level were used for the analysis; mean water level (WL$_{mean}$), water level range (WL$_{mm}$) and water level standard deviation (WL$_{sd}$). WL$_{sd}$ and WL$_{mm}$ were used to indicate the fluctuation of water levels. Surface water and groundwater chemical data were compared in terms of sampling point, year, month and depth. Surface water samples were compared in terms of sampling point, year and month. Analyses of variance (ANOVA) were performed using the Data Analysis add-in of Microsoft Excel. Results are reported in percent of the sum of variance components for all factors.

Results and discussion

Variation in water level

The average annual precipitation in this area is 1,049 mm (Takahashi, 1988), which corresponds to 2.87 mm average daily precipitation. Generally, the precipitation was higher during spring to autumn, and in 1993 and 2003 it was slightly less than in other years (Table 1). However, the water levels during the period from May through November were slightly lower than during the entire year, as can be seen from the 2003 data, especially at point E. This was because the winter snowfall did not flow directly out from the mire, but remained in the mire and kept the water levels high. According to Takahashi (1988), the average depth of snow cover during winter (January and February) is about 1 m. Water levels generally responded quickly to precipitation events over 10 mm and when periods of little or no precipitation exceeded two weeks, water levels

Table 1 Average daily precipitation (mm)

	1993	1994	1995	1996	1997	2003
July–November	2.16	3.09	3.63	3.55	3.90	2.67
Entire year	n.a.	2.66	3.89	2.66	3.20	n.a.

Table 2 Summary of water level variations (in cm)

	E	W	W'	WW
Ground surface level	606.6	600.9	586.7	519.0
WL_{mean} (May–Nov 2003)	−4.4	−8.4	−18.5	−21.5
WL_{mm} (May–Nov 2003)	18.3	16.5	41.8	43.9
WL_{sd} (May–Nov 2003)	4.4	3.4	11.0	12.5
WL_{mean} (1993–1997)	−1.3	−9.0	−18.9	−21.5
WL_{mm} (1993–1997)	28.1	27.7	44.0	65.0
WL_{sd} (1993–1997)	5.1	4.8	8.8	14.3
WL_{mean} (1993–1997 & 2003)	−1.4	−8.9	−17.7	−21.2
WL_{mm} (1993–1997 & 2003)	28.1	27.7	47.0	65.0
WL_{sd} (1993–1997 & 2003)	5.5	4.6	9.2	13.7

fell noticeably (Figure 2). The responses of water levels at the four monitored points to each precipitation event showed greater instability in the westward direction. WL_{sd} and WL_{mm} were smaller at points E and W than at points W' and WW. The maximum difference between the highest and the lowest water level was 65 cm at point WW, and only 28 cm at points E and W. The small WL_{sd} and WL_{mm} indicate that the water level in the eastern part of the mire is more stable and has less fluctuation than that of the western part. The variable WL_{mean} shows a similar trend: it decreased gradually at point W', and the lowest value was at point WW, at 21.5 cm in depth. The patterns of WL_{sd}, WL_{mm} and WL_{mean} for these four sampling points show a degradation gradient toward the western part of the mire. This result shows that the soil in the degraded area has poor water retentivity. These observations are consistent with the fact that *Sphagnum*, which usually acts as a huge sponge in a natural bog, is absent. The low water retentivity of the soil is considered to be the result of the low water level, which exposes the bare peat after the loss of *Sphagnum* cover.

Water chemistry gradient

Water chemical analysis is summarized in Figure 3. In general, there is an ombrotrophic-minerotrophic gradient as one moves from east to west. Moreover, water chemistry characteristics of point NC and, to a lesser extent, point WW are noticeably different from those of the other sampling points. In terms of water temperature, there is no evidence of different patterns from surface to 2 m depth between all sampling points. The insulating ability of a natural bog ecosystem (Vitt et al., 1995) is not evident. At all sampling points the temperature at the surface is higher than that of the groundwater, gradually decreasing in the depth direction. The surface water pH of all five sampling

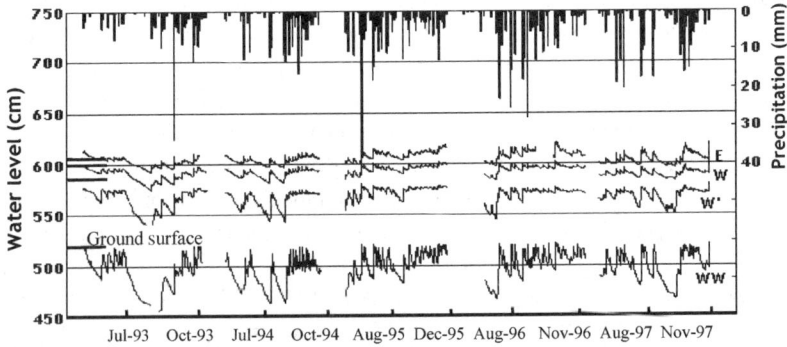

Figure 2 Water level fluctuations (cm) and precipitation data (mm) for the 1993–1997 sampling period. Due to space limitations, water level data for 2003 are not shown in the graph. Precipitation and water level data for 1993–2003 is summarized in Tables 1 and 2

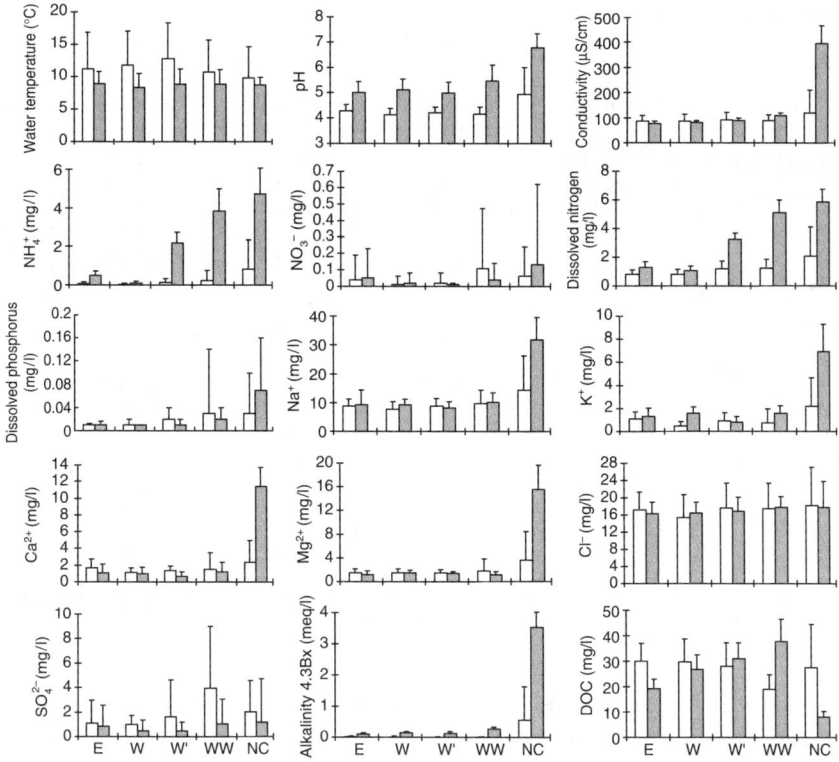

Figure 3 Chemical properties of water during the 1993–2003 sampling period. Means are given, with their standard deviations. Only data for surface water (□) and groundwater at 2-m depth (■) are presented and summarized, in light of the relatively consistent pattern of the chemical properties with respect to depth (Tables 3 and 4)

Table 3 Analysis of variance for the parameters measured in surface water and groundwater at 0.5, 1.0, 1.5, and 2.0 m depth. Significance level $*P < 0.05$; $**P < 0.01$; $***P < 0.001$; ns, not significant $P > 0.05$. "Point" indicates sampling point, and it also applies to point-depth and point-year

Variable	% Total variation						
	Point	Year	Point-depth	Depth	Month	Point-year	Error
Water temperature	0.4ns	11.2**	10.4ns	7.7ns	71.6***	2.3ns	4.4
pH	52.0***	7.6***	18.1***	18.1***	2.6*	2.5***	9.1
EC	68.8***	9.2***	11.7***	2.0*	4.1**	1.3***	2.8
NH_4^+-N	46.9***	5.3***	16.3***	18.3***	1.7ns	−0.9***	12.4
NO_2^--N	2.9**	30.4***	1.1ns	0.0ns	5.5***	18.6***	41.4
NO_3^--N	2.2*	69.0***	1.2ns	0.2ns	11.4***	7.0***	9.2
DN	51.3***	3.3*	14.2***	17.7***	1.9ns	−0.2***	11.8
DRP	16.1***	5.6***	6.0***	1.1ns	1.9ns	13.1***	56.3
DP	13.0***	3.2*	2.9***	0.3ns	1.1ns	11.4***	68.1
Na^+	60.3***	15.2***	7.3***	1.3ns	3.3*	−5.3***	17.9
K^+	53.5***	9.5***	8.1***	2.8**	8.2***	0.0***	17.9
Ca^{2+}	64.7***	11.7***	11.9***	1.0ns	3.7**	−1.1***	8.1
Mg^{2+}	66.8***	8.6***	11.1***	1.3ns	2.9*	−0.9***	10.1
Cl^-	2.8**	54.3***	1.5ns	1.1ns	15.4***	11.7***	13.0
SO_4^{2-}	3.5***	44.1***	4.2***	5.1***	8.1***	3.0***	32.0
4.3 Bx	76.1***	6.8***	9.4***	2.9**	2.7*	−0.5***	2.6
DOC	31.7***	7.4***	32.2***	2.2*	4.5**	2.4***	19.6

Table 4 Analysis of variance for the parameters measured in surface water

Variable	% Total variation				
	Point	Year	Point-year	Month	Error
Water temperature	3.9ns	–	–	92.8***	3.3
pH	26.3***	11.0ns	26.4***	4.8ns	31.5
EC	6.5ns	23.0***	50.6***	14.7*	5.2
NH_4^+-N	15.6***	8.7ns	43.1***	4.3ns	28.2
NO_2^--N	2.8ns	34.3***	37.3***	6.2ns	19.5
NO_3^--N	3.5ns	65.1***	–	21.0**	10.5
DN	20.4***	11.3ns	28.3***	5.0ns	35.0
DRP	5.2ns	4.9ns	77.0***	3.2ns	9.8
DP	3.9ns	9.1ns	31.6ns	3.5ns	52.0
Na^+	12.9**	38.1***	36.2ns	12.8***	0.0
K^+	17.6***	12.1*	41.9***	10.9ns	17.4
Ca^{2+}	5.8ns	32.5***	50.2***	–	11.4
Mg^{2+}	11.4*	26.4***	58.6***	–	3.6
Cl^-	2.5ns	61.1***	–	32.4***	3.9
SO_4^{2-}	12.8**	36.1***	9.2**	10.3ns	31.5
4.3 Bx	18.0***	14.0ns	65.1***	3.7ns	0.0
DOC	17.1**	14.88ns	29.5**	20.9**	17.7

Note: For some parameters (water temperature, nitrate, calcium, magnesium, and chloride) ANOVA were not performed for all variables, due to their close correlations.

points studied was still low (pH 4.13 to 4.28), but it was slightly higher at point NC (pH 4.94). The surface water pH at point NC showed greater variation (SD = 1.05) than those at the other points, perhaps due to precipitation events, since during the periods of little or no precipitation, pH rose to 7 (data not shown). This shows that the bog is no longer a natural bog. The pH at 2-m depth was higher than at the surface, and it gradually increased from point E toward point NC. Sampling point accounts for 52% of the variation in the surface water and groundwater but only 26.3% of the variation in surface water alone. Alkalinity and conductivity showed trends similar to those of pH, with point NC having higher values for these at the surface than any other point, and alkalinity and EC gradually increased from point E toward point NC for samples at the 2-m depth.

The high alkalinity at point NC may be a result of the high pH value, because if the pH increases beyond 6, then carbonic acid dissociates into hydrogen ions (H^+) and bicarbonate ions (HCO_3^-). Therefore the carbonate–bicarbonate buffering system typical of surface waters replaces the humic substances' contribution to the buffering system at pH values less than 5 (Gorham et al., 1984).

Nitrite concentrations were usually below the detection limit, while the highest values for nitrate concentration were at point WW at the surface (0.11 mg/l) and at point NC at the 2-m depth (0.13 mg/l). Sampling point alone accounts for less than 3% of the total variation in nitrite and nitrate in surface water and at all sampling depths. For nitrate, the variation for yearly difference is more evident than that for other variations, accounting for 65% ($P < 0.001$) of the variation in surface water and 69% ($P < 0.001$) of the variation for all sampling depths, showing that there was instability in nitrate concentration during the sampling period (1993–2003). Monthly variation was slightly pronounced at the surface, accounting for 21% of total variation. DRP concentrations were usually below the detection limit, and the concentration of other nutrient forms (NH_4^+-N, DN, and DP) both at the surface and at the 2-m depth showed consistent increases according to degradation gradient, from point E toward point NC. NH_4^+-N and DN concentrations started to increase at point W' (2.18 mg/l for NH_4^+-N and 3.24 mg/l for DN at 2-m depth), and DP concentration showed noticeable increase at point WW, especially at the surface with the value of 0.03 mg/l, similar to that at point NC. Sampling point accounts for 47% and 51% of the variation in NH_4^+-N and DN concentrations, respectively, at all sampling depths. Moving water at the natural channel and unstable water levels of point WW seemed to be responsible for the high nutrient content, since moving water is able to supply much more nutrient by desorption from peat soil, even if the concentrations at any one time are fairly low. In terms of phosphorus content, however, the degradation gradient alone accounts for 16% of DRP concentration and 13% of DP concentration. The errors for these two parameters are 56% and 68%, perhaps due to their low concentrations, which may make the parameters prone to quick response to each environmental event and to measurement error caused by disturbed samples.

In terms of major cation contents, Na^+ was the dominant ion and K^+ had the lowest content of any ion, which is similar to the case of other coastal mires in Hokkaido (Hotes et al., 2001). The cation concentrations are consistent with the EC values, with only point NC showing great differences from the other sampling points. Both surface water and groundwater at 2-m depth showed considerably higher concentrations at point NC than at other points, and the concentration at 2 m depth was higher than that at the surface. There were slight differences in major cation contents among the other four sampling points, either in surface water or in groundwater, but these differences did not show a degradation gradient toward the western side. Ca^{2+} and Mg^{2+} showed a steady concentration decrease with depth, which was also seen at 0.5, 1.0, and 1.5 m depths, although there were slight exceptions at point W (data not presented). In addition to being caused by evaporative effects, the higher concentration at the surface was probably caused by cation exchanges with the peat, since high Na^+ input from rain led to high concentrations of other cations, including Ca^{2+} and Mg^{2+}. The four points (E to WW) showed a tendency for Ca^{2+} and Mg^{2+} concentrations to decrease with depth and for Na^+ and K^+ concentrations to increase with depth. This shows that the cation exchange capacity of the peat at the surface is higher than that below the surface, which is also consistent with the fact that the surface water has lower pH than the ground, since less humification of organic content at lower pH enables the soil to have a higher capacity to absorb exchangeable ions. In contrast, point NC, which had higher pH than other points, showed less cation exchangeability, and the ions contained in the water there were not much reduced. The source of variations in surface and ground water for all major cations was mostly sampling point, which accounted for between 54% (for K^+) and 67% (for Mg^{2+}) of variation. However, point-year interaction was the most important source for almost all major cation variations in surface water, except for Na^+, and the difference between years was more easily seen than for other variations.

DOC concentration showed a degradation gradient for both surface water and groundwater. Surface DOC concentration decreased from point E toward point WW, whereas at 2-m depth the DOC concentration increased in the westward direction. Point NC showed irregularity in DOC concentration at both depths, with highest DOC concentration (27.52 mg/l) being at the surface and the lowest (7.94 mg/l) being at the 2-m depth. That irregularity at point NC may be attributable to the presence of gullies that provide and transport particulate organic matter from eroded peat soil. This also explains the high variation in DOC concentration at the surface (SD = 16.91), since the leaching of eroded peat soils depends on each precipitation event. The low DOC concentration at the 2-m depth suggests that there was high microbial activity, which converted the organic matter into inorganic form; this conversion also occurred to nutrients. Point-depth interaction accounted for 32.2% of the variation in surface and ground water DOC concentrations, and sampling point accounted for 31.7% of such variation. As for Cl^- content, its overall range within the mire is small, ranging from 15.48 mg/l to 18.28 mg/l. There is no obvious difference in Cl⁻ concentration between sampling points and depth, showing that the main source of water is from precipitation and that the mire has not experienced seawater inundation. Sulfate concentration showed significant difference among the sampling points. The concentrations at point WW and NC were higher than at other points, all of the concentrations having a pattern in which the concentration decreased with depth. However, difference between years was the most significant source for variation in Cl⁻ and in SO_4^{2-}, accounting for 61% and 36% of the variation for surface water and 54% and 44% of the variation for all sampling depths.

Mire degradation and water chemistry changes

In the degraded area (points WW and NC) the soil has poor water retentivity. Much of this may be due to the low levels of organic matter present. Organic matter from dead plants creates protective mulch that reduces soil erosion and water evaporation and acts as a sponge to absorb water in the soil. Many authors have shown that hydrologic parameters control the chemical and biotic processes in wetland ecosystems, and that those parameters may be the most important factors in wetland ecosystem development, regulating natural processes. A stable and high water level will keep the soil from drying (as seen at points E and W), and when soil and peat layer becomes dry, it is prone to erosion and decomposition. Decomposition of peat soil affects the surface water and groundwater chemistry, particularly after leaching when the decomposed soil is washed by rainwater, which yields a high concentration of DOC (such as that observed at point NC, at the surface). The decline of water level in the mire will make inflow from surrounding areas possible. This inflow can transport mineral-rich soil and sediment, which can cause changes in water chemistry. According to Bragg and Tallis (2001), in response to adjoining land-use changes, many species of typical blanket mire vegetation (including *Sphagnum*) had declined substantially in abundance or had become extinct during a 28-year period. Once bare peat is exposed, it is prone to the adverse affects of natural weather events, which is followed by erosion of the peat. After bare peat is exposed, the next degradation stage will occur, ranging from the development of gully systems to the large-scale removal of the peat blanket. Considering the fact that development in the land surrounding Sarobetsu Mire started in 1961, about 40 years ago, it is considered that the gully systems found in the vicinity of point NC have resulted from degradation. Without concerted efforts to conserve the mire, the gullies will spread. Gully systems result from fundamental changes in the hydrology of the peat mass, where the lateral transport of excess water towards the peat margins through the permeable acrotelm is replaced, in part or in whole, by transport over the bog surface. This change in flow direction may lead to the changes in water chemistry, since the surface flow at the eroded peat surface will transport nutrients and minerals that are incorporated into the suspended sediments.

Conclusions

- Land-use changes in surrounding areas are affecting the hydrological regime of Sarobetsu Mire. The lowered water level is causing hydrochemistry changes and is allowing nutrient and mineral loading from adjacent agricultural areas. Groundwater chemistry at all sampling points shows the influence of mineral-rich groundwater inflow through the mire basin, which yields an ombrotrophic-minerotrophic gradient seen in degradation on the western side of the mire.
- ANOVA analysis showed that the water chemistry parameters which most closely correlate to the degradation gradient are pH, EC, NH_4^+-N, DN, Na^+, K^+, Ca^{2+}, Mg^{2+}, 4.3 Bx, and DOC concentrations, and sampling points accounted for 30–80% of total

variation. NO_3^--N, NO_2^--N, Cl^-, and SO_4^{2-} were more closely related to year than to other variation, and the variation in DRP and DP concentrations was not based on any of the calculated variables.
- The restoration of water chemistry for the conservation of the mire needs to address all aspects that govern the hydrological regime, such as topographical and morphological mire surface.

References

Bragg, O.M. and Tallis, J.H. (2001). The sensitivity of peat-covered upland landscapes. *Catena*, **42**, 345–360.

Gorham, E., Bailey, S. and Schindler, D. (1984). Ecological effects of Acid Deposition upon Peatlands: A Neglected Field in Acid Rain Research. *Can. J. Fish. Aquat. Sci.*, **41**, 1256–1268.

Hotes, S., Poschold, P., Sakai, H. and Inoue, T. (2001). Vegetation, hydrology, and development of a coastal mire in Hokkaido, Japan, affected by flooding and tephra deposition. *Can. J. Bot.*, **79**, 341–361.

Inoue, T., Umeda, Y. and Nagasawa, T. (1992). Some experiments on restoring the hydrological conditions of peatland in Hokkaido, Japan. In *Land Reclamation: Advances in Research & Technology, Proceedings of the International Symposium*, Younus, T., Mostaghimi, S. and Diplas, P. (eds), American Society of Agricultural Engineers, Tennessee, pp. 196–203.

Takahashi, H. (1988). *Vegetation of Japan*, Vol. 9, *Hokkaido*, Miyawaki, A. (ed.), Shibundo, Tokyo, pp. 65–71.

Vitt, D.H., Bayley, S.E. and Jin, T.L. (1995). Seasonal variation in water chemistry over a bog-rich fen gradient in Continental Western Canada. *Can. J. Fish. Aquat. Sci.*, **52**, 587–606.

Water Analysis (2000). 4th edn., Hokkaido Branch of Japan Society of Analytical Chemistry, Kagaku Dojin Publishing, Tokyo, Japan.

Characteristics of sediments in a newly constructed reservoir in Japan

Y. Sakurai and S. Haruta

Faculty of Agriculture, Ehime University, 3-5-7 Tarumi, Matsuyama, Japan
(E-mail: *ysakurai@agr.ehime-u.ac.jp*; *haruta@agr.ehime-u.ac.jp*)

Abstract The sediment formation mechanisms of a newly constructed reservoir in Ehime, Japan were evaluated by characterizing the soil particles (SP) and particulate phosphorus (PP) in the runoff and reservoir sediments. The SP and PP loads from the runoffs of the main river in the watershed considerably increased, when the specific discharge rates were over 300 l/s/km^2 (high flow conditions). When the specific discharge rates exceeded over 300 l/s/km^2, 19% of the watershed generated over 80% of the SP and PP loads. When the specific discharge rates were under 300 l/s/km^2 (low flow conditions), the contributions of the previously mentioned 19% area to the SP and PP loads were smaller. Significant amounts of smectite were found in the sediments in the reservoir and in the soil samples obtained at the forest exposed area in this 19% area while it was negligible in citrus orchards and paddy fields that constituted the remaining land surfaces. The forest area exposed by recent landslides was significant for the SP and PP in the reservoir. Judging from the outcomes, land use information alone may not be sufficient to detect critical sources of SP and PP in the runoffs and reservoirs. To identify and confirm crucial areas for the SP and PP in the runoffs, the investigations should be conducted under high flow conditions and the composition of clay minerals in the sediments should be checked against the clay mineral distributions of soils in the watershed.
Keywords Newly constructed reservoir; phosphorus; prevention of eutrophication; runoff characteristics; sediment; soil particle

Introduction

Protecting reservoirs from eutrophication is essential to maintain the suitability of water for irrigation and recreational uses (Ryding and Walter, 1989). It is widely recognized that nutrient releases from the sediments are the primary causes accelerating the progression of the eutrophication processes in reservoirs (Correll, 1998). In existing reservoirs, sediments gradually accumulated overtime and the sediment layer had already formed (Bennett *et al.*, 2005). The origin of the existing sediments, the relative contribution of sediment and nutrient loads from sub units of the catchment basin, and the sediment deposition patterns in the reservoirs are not fully known. Therefore, it is imperative to understand the sediment formation and deposition processes when the reservoirs are first constructed. In this manner, protective measures and management strategies may be formulated to prevent the subsequent nutrient enrichment in the water body.

When a new reservoir for irrigation in Ehime, Japan was completed and began filling in 2001, it was an opportunity to investigate the water quality changes, chemical and physical condition of the reservoir sediments and relative contributions of sediment and nutrient loads of sub units of the catchment basin. It was established through initial investigations that the phosphorus release from the sediments would be the most important factor in eutrophication in the new reservoir. To devise suitable measures to minimize the generation of sediments and release of phosphorus, it is essential to understand the mechanisms of sediment formation in the catchment basin.

doi: 10.2166/wst.2006.040

This study examined the sediment formation mechanisms and the runoff characteristics of soil particles (SP) and particulate phosphorus (PP) and identified the critical sources of SP and PP in the watershed of a newly constructed reservoir in Japan.

Methods

Study site

The study site, a newly completed reservoir and its catchment basin, is located in the Ehime Prefecture of southern Japan (Figure 1).

The reservoir (designated as S-reservoir hereafter) has 1,100,000 m^3 of storage volume, and had been filled since January of 2001. The high water level of the reservoir is at 25.1 m, and the low water level is at 9.2 m. The surface area of the watershed surrounding S-reservoir is 4.05 km^2, consisting of claystone and shale formations with partly tuff. The catchment basin is 93% forest, 2.0% paddy field, and 1.6% upland field and the catchment has only eight households. The vegetation in the forest consists primarily of Japanese redwood and Japanese cypress. Almost all of upland fields are citrus orchards. The major river and its tributaries (designated as S-river hereafter), which has a catchment basin of 2.70 km^2, supplied the majority of the inflows to the reservoir. The watershed outside of the S-river catchment basin is covered entirely by forest and is uninhabited.

Sampling and analytical methods

Investigations were conducted in three parts.
1. To define the runoff characteristics of SP and PP of the S-river watershed, the water was sampled and the corresponding flows were measured simultaneously at the point where the S-river is emptied into the S-reservoir (Station A in Figure 1). The water sampling and the flow rate measurement were made for 20 rainfall events during the 2001 to 2003 period. The water samples were filtered by Whatman GF/B glass fiber filters with 1 μm pore size, and the spent filters were heated at 600°C for 30 min to obtain the non-volatile solids content. The non-volatile solids were regarded as the soil particles, SP in the water. Total phosphorus (TP) and dissolved total phosphorus (DTP) of the water were analyzed by the molybdenum-blue method after the aliquots of filtered and non-filtered water were digested with potassium peroxodisulfate according to procedures outlined by the Japanese Industrial Standards Committee (1993), and the particulate phosphorus, PP, was obtained from the difference between TP and

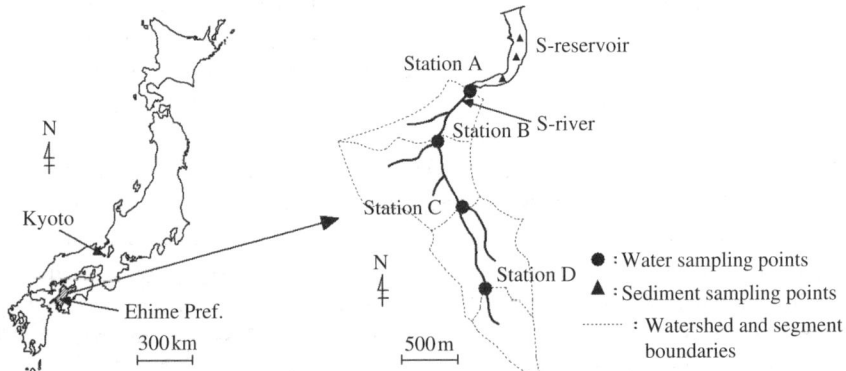

Figure 1 Maps showing the location of the S-reservoir and sites of water and sediment sampling

DTP. Flow rates were calculated from measurements of water depth, stream cross sectional area and flow velocity.

2. To delineate the origins of SP and PP in the catchment basin, water sampling and flow rate measurement were simultaneously made at the mid point of three rainfall events in which 19, 23, and 55 mm of rain had fallan, respectively at the four sites along the S-river, namely Stations A through D in the map in Figure 1. Table 1 shows the land use patterns of the entire watershed of S-river and the sub-catchments covered by each gauging station. Water sampling was collected at 0.5 to 1 hour before the peak of rains.

3. To identify the critical sources of SP and PP in the watershed, the clay minerals in the reservoir sediments and in the soils representing land uses of the watershed were analyzed by the X-ray diffraction method. In the reservoir, the sediment samples (surface 1 cm of the reservoir bottom) were taken at the three sites as shown in Figure 1 in the autumn of 2001. In the watershed, samples of surface soils (the 0 to 1 cm layer) were taken at a paddy field, an orchard, and an exposed area of the forest that had experienced recent land slides adjacent to the S-river.

Results and discussion

Sediment and phosphorus loads and runoff characteristics

Figure 2 shows the effects of specific discharge rates on the specific loads of SP and PP at Station A, runoff entry point to the S-reservoir.

The SP and PP loads of the reservoir inflows during the rainfall events (measured as mass of particulates and mass of phosphorus per second per km^2) changed dynamically with the changes of the specific discharge rates. Both of the SP and PP loads from the S-river increased considerably, when the specific discharge rates exceeded $300\,l/s/km^2$. For example, the SP and PP specific loads reached $310\,g/s/km^2$ and $190\,mg/s/km^2$, respectively at the specific discharge rate of $1,100\,l/s/km^2$. In contrast, when the specific discharge rates were under $300\,l/s/km^2$, the SP and PP specific loads were less than $10\,g/s/km^2$ and $15\,mg/s/km^2$, respectively.

Table 1 Land use of the S-river watershed

Catchment	Catchment area (km²)	Orchard (%)	Paddy field (%)	Upland field* (%)	Forest (%)	Others (%)
S-river	2.70	2.1	3.0	0.2	83.2	11.5
> Station D	0.39	0.0	1.4	0.0	87.4	11.2
Station C-D	0.44	0.4	7.2	0.9	73.4	18.0
Station B-C	1.37	2.3	2.9	0.0	91.7	3.1
Station A-B	0.50	4.7	0.8	0.2	65.2	29.1

*Except for orchard

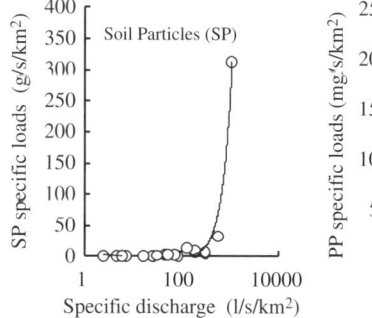
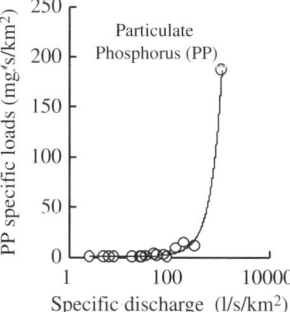

Figure 2 Relationship of specific discharge rates and specific loads of SP and PP

The sediment loads of the S-river were associated primarily with rains that generated discharges exceeding 300 l/s/km². Therefore, the sources contributing to the SP and PP loads must be identified with specific discharge rates that are greater than 300 l/s/km².

Contribution of sub-catchments

Figure 3 shows the SP and PP loads of the sub-catchments defined by the four gauging stations along the S-river. Under 55 mm of rainfall, the specific discharge rate was 1,100 l/s/km². Following 23 mm and 19 mm of rainfall, the specific discharge rates were 55 l/s/km² and 33 l/s/km², respectively.

Data plotted in Figure 3 demonstrated that the contribution of each sub-catchment of the watershed to the SP and PP loads of the reservoir varied with the antecedent rain conditions and the specific discharge rates. Had the scale not been expanded by 25 times, the SP and PP loads of the 19 and 23 mm rainfalls would not be distinguishable from the x-axis. Under the rainfall event of 55 mm, the SP loads at Station B were only 15% of the SP loads at Station A, and the PP loads at Station B were only 16% of the PP loads at Station A.

In contrast, the SP loads at Station B were over 40% of the SP loads at Station A, and the PP loads at Station B were over 35% of the PP loads at Station A when the rain was 23 mm. When the rain was 19 mm, the SP loads at Station B were over 50% of the SP loads at Station A, and PP loads at Station B were over 45% of the PP loads at Station A.

The sub-catchments of gauging Station A covered 19% of the watershed, yet the contributions to the SP and PP loads to S-reservoir varied considerably with the increase of discharge rates. When the specific discharge rates were over 300 l/s/km², the contribution of the sub-catchment from Station B to Station A were several orders of magnitude greater than those from all of the upstream sub-catchments added together. When the specific discharges were less than 300 l/s/km², the relative contribution of the sub-catchment defined by Station A was considerably lower. These results indicate that the critical sources of SP and PP were present in the sub-catchment from Station B to Station A.

Nishimura *et al.* (2002) showed that the citrus production areas were significant generators of the non-point source pollutant loads of the downstream water bodies. The sub-catchment from Station B to Station A contained slightly larger citrus production area than other sub-catchments (Table 1). Therefore, the results shown in Figure 3 appeared to imply that the citrus orchard could be a significant source of SP and PP.

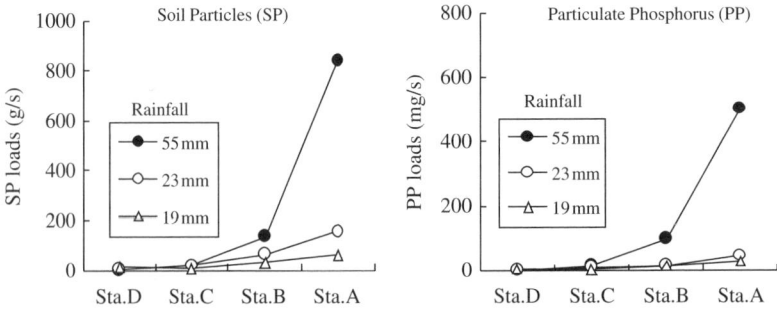

Figure 3 SP and PP loads in S-River following 19, 23 and 55 mm of rainfall (the points indicating 19 and 23 mm rainfall are plotted 25x of their actual values)

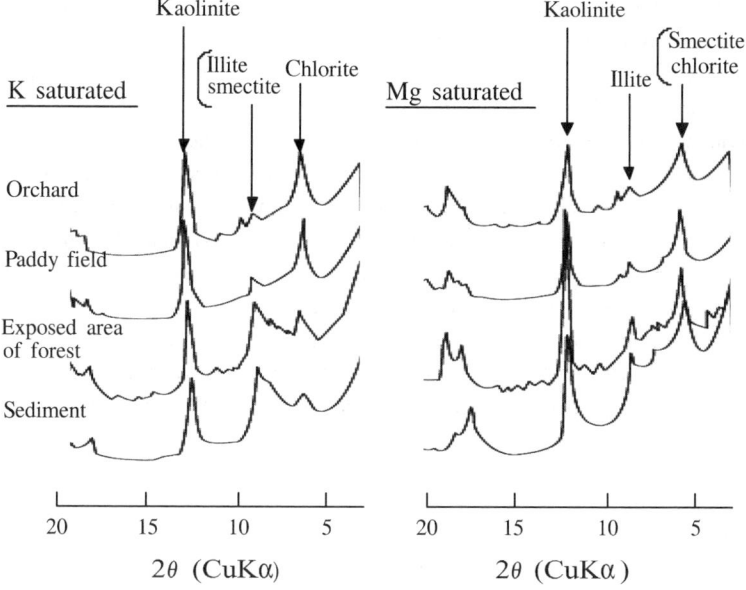

Figure 4 X-ray diffraction patterns for soils and sediments

Clay minerals in soils and sediment

The X-ray diffraction patterns showed that mineral compositions of three sediment samples were similar and only one tracing for the sediment was illustrated (Figure 4). It is difficult to separate the illite and smectite peak in the K saturated samples, but in the Mg saturated samples, illite and smectite peaks can be separated.

The shape of the X-ray diffraction patterns showed that mineral compositions of the sediment were very similar to that of soils obtained at the exposed area of the forest. Smectite in this case was detected in a large quantity relative to other clay minerals in the sediments. At the exposed area of the forest, smectite was also detected in a relatively large quantity. Conversely, in the paddy field and the orchard, the presence of smectite was considerably less prominent. In our survey, the sub-catchment from Station B to Station A contains a much more exposed area of forest. Kaolinite was detected in a large quantity in all samples. It is indicative that instead of the citrus orchards the exposed forest area in sub-catchment of Station A was the most significant source of sediments in S-reservoir. The mineralogical analysis of the clays accurately identified the source sediments in the S-reservoir.

Summary and conclusion

1. The soil particles (SP) and particulate phosphorus (PP) loads of the main river flowing into the newly constructed reservoir increased considerably, when the specific discharge rates exceeded 300 l/s/km^2. For example, the SP and PP specific loads reached 310 g/s/km^2 and 190 mg/s/km^2 respectively at the specific discharge rate of 1,100 l/s/km^2. These results indicate that the SP and PP loads, when the specific discharge rates were over 300 l/s/km^2, had significant effects on the generation of sediments and the sediment phosphorus in the reservoir.

2. When the specific discharge rates were over 300 l/s/km^2, 19% of the catchment basin generated over 80% of the SP and PP loads to the reservoir. The remainder of the watershed had great contributions to the SP and PP loads of the reservoir only under

the low rainfall conditions. This indicates that the major source of the SP and PP loads was associated with releases from the 19% area.
3. The X-ray diffraction patterns of minerals in the sediments were near identical to that of the surface soil sample obtained at the exposed area of the forest. Significant amounts of smectite were detected in the sediments and the soil samples representing the exposed forest area. Conversely, smectite was less prominent in the soil samples obtained from paddy fields and the citrus orchards that constituted the remainder. The results indicate that the exposed area in the forest is the significant SP and PP source.

To assess the SP and PP loads to reservoirs, land use information alone may not be sufficient. The critical source areas of SP and PP in a watershed are best investigated under the high flow rate conditions during and following significant rainfall events. To identify and confirm the critical source origins, it may be essential to check the clay mineral distribution of the sediment in the reservoir against the clay mineral distributions of soils in the watersheds.

Acknowledgement

We would like to thank the Chugoku-Shikoku Regional Agricultural Administration Office, Ministry of Agriculture, Forestry and Fisheries, Japan for the support to this study.

References

Bennett, S.J., Rhoton, F.E. and Dunbar, J.A. (2005). Texture, spatial distribution, and rate of reservoir sedimentation within a highly erosive, cultivated watershed: Grenada Lake, Mississippi. *Water Resources Research*, **41**(1), Art. No. W01005, doi:10.1029/2004WR003645.

Correll, D.L. (1998). The role of phosphorus in the eutrophication of receiving waters. *Journal of Environmental Quality*, **27**(2), 261–266.

Nishimura, F., Watanabe, M., Takahashi, R. and Akase, T. (2002). Runoff characteristics of nutrients from citrus fruit grove and its effect on water area. *Wat. Sci. Tech.*, **45**(12), 37–44.

Ryding, S.O. and Walter, R. (1989). *The Control of eutrophication of lakes and reservoirs*, UNESCO, Paris.

The Japanese Industrial Standards Committee (1993). Testing Methods for Industrial wastewater. JIS K-0102-1993.

Evaluation of nutrient loads from a mountain forest including storm runoff loads

T. Kunimatsu*, T. Otomori*, K. Osaka**, E. Hamabata*** and Y. Komai****

*Department of Ecosystem Studies, School of Environmental Sciences, University of Shiga Prefecture, Hassaka, Hikone, Shiga 522-8533, Japan (E-mail: *kunimatu@ses.usp.ac.jp, otomori@ses.usp.ac.jp*)

**Division of Environmental Science and Technology, Graduate School of Agriculture, Kyoto University, Sakyo, Kyoto, 606-8502, Japan (E-mail: *osaka@kais.kyoto-u.ac.jp*)

***Lake Biwa Environmental Research Institute, 5-34 Yanagasaki, Otsu, Shiga 520-0806, Japan (E-mail: *hamabata-e@lberi.jp*)

****Hyogo Prefectural Institute of Public Health and Environmental Sciences, 2-1-29 Arata, Hyogo, Kobe, Hyogo 652-0032, Japan (E-mail: *yukio komai@pref. hyogo.jp*)

Abstract Water quality and flow rates at a weir installed on the end of Aburahi-S Experimental Watershed (3.34 ha) were measured once a week from 2001 to 2003 and in appropriate intervals from 30 min to 6 h during five storm runoff events caused by each rainfall from 8 mm to 417 mm. The average annual loads of total nitrogen (TN) and total phosphorus (TP) were calculated to be 19.0 and 0.339 kg ha^{-1} y^{-1} from the periodical data by using the integration interval-loads method (ILM), which did not properly account for storm runoff loads. Three types of $L(Q)$ equations ($L = aQ^b$) were derived from correlations between loading rates L and flow rates Q obtained from the periodic observation and from storm runoff observation. $L(Q)$ equation method (LQM), which was derived from the storm runoff observation and allowed for the hysteresis of discharge of materials, gave 9.68 and 0.159 kg ha^{-1} y^{-1}, respectively, by substitution of the sequential hourly data of flow rates. $L(R)$ equation ($L = c(R - r)^d$) was derived from the correlations between the loads and the effective rainfall depth ($R - r$) measured during the storm runoff events, and $L(R)$ equation method (LRM) calculated 9.83 ± 1.68 and 0.175 ± 0.0761 kg ha^{-1} y^{-1}, respectively, by using the rainfall data for the past 16 years. The atmospheric input-fluxes of TN and TP were 16.5 and 0.791 kg ha^{-1} y^{-1}.

Keywords Atmospheric deposition; forest; nitrogen; nutrient loads; phosphorus; storm runoff

Introduction

Natural forests as well as planted forests have been believed to reduce pollutant loads deposited from the atmosphere, and to discharge a clean and steady stream water percolated through the biogeochemical filter of the forest ecosystem. This is true during dry days when the discharge mainly consists of groundwater. When intensive rainfalls cause storm runoffs, however, the streams discharge a large volume of muddy water accompanying large amounts of nutrients. A lot of studies of the nutrient loads from mountain forests have been reported in North America and Europe as well as in Japan (Martin *et al.*, 2000; Kunimatsu *et al.*, 2001). The small watershed method has been used in the objection, and flow rates were continuously recorded at the weir on the end of the watershed, where the concentrations of nutrient were measured periodically, for example weekly or once a month. These previous studies, however, have some substantial defects; first, they measured only inorganic nitrogen and phosphorus with few exceptions (Kunimatsu *et al.*, 2001), secondly, storm runoff loads were not properly accounted for (Smith and Stewart, 1977), thirdly, they did not involve any information on the fluctuations of the annual loads caused by the changes of hydrological conditions in every year, and fourthly, any reasonable methods for calculating the material loads fluctuating from day to day as well as from year to year have not been confirmed. It is necessary to evaluate the

doi: 10.2166/wst.2006.041

comprehensive amounts of the total loads of nitrogen and phosphorus accounting for the storm runoff loads in order to assess the state of the eutrophication of a certain lake or inland sea lying in a mountainous and rainy region such as the Asian monsoon area.

The integration interval-loads method has been widely, in other words unconsciously, used for evaluating the material loads of a river. However, the method does not contain procedures for evaluating storm runoff loads. That is the fatal defect of the method in the rainy area. The $L(Q)$ equation method had proposed for more than a quarter-century the evaluation of loading rates (Johnson, 1979; Yamaguchi et al., 1980). However, it has been scarcely used for the purpose, and besides could not calculate the loads for a period when flow rates were not measured. In order to simulate the water quality of a large lake with some kind of mathematical model, it is necessary to evaluate the changes of loading rates for long periods before and after the eutrophication of the lake had began. We had proposed the $L(R)$ equation method, which makes that possible by using the past rainfall data, which are readily supplied from some suitable public or private establishments who measure precipitation in the region.

At the Aburahi-S experimental watershed, we obtained the weekly concentrations and flow-rate data, the data of five storm runoffs and the sequential flow rate data of every 10 minutes in three years after 2001, and the rainfall data for 16 years since 1988. In this study, these data were used to assess that the $L(R)$ equation method is most appropriate for the evaluation of the comprehensive annual loads of the nutrients.

Methods

Experimental conditions

The experimental forest observed in this study is Aburahi-S, which lies in the basin of Lake Biwa located around the central part of the main island of Japan, as shown in Figure 1. Aburahi-S (N 34°51′41″ and E 136°16′08″) is the small watershed of 3.34 ha on the granite bedrock and at an altitude from 312 m to 479 m. Japanese cypress (*Chamaecyparis obtusa* Sieb. et Zucc) were planted on 58% of the watershed in 1965, and in September 1999 the second thinning was carried out. Deciduous broadleaf trees including Japanese red pines (*Pinus desiflor* Sieb. et Zucc) grew on the remaining upper part of the watershed. A small number of the pines were also left in the stand of the planted Japanese cypress.

The stream draining the Aburahi-S Experimental Watershed is perennial and not polluted by any anthropogenic sources except for forest working. A ferroconcrete waterway (width 1.5 m, length 6 m, height 0.9 m) was installed on the stream at the end of the watershed, and fitted up with a stainless steel plate for a full-width weir on the out-flow side of the waterway. The stream water was collected once a week (as a rule, at around 15.00 every Thursday) from May 1995 to December 2003. Storm-runoff observations were carried out during the five events caused by rainfalls from 8 mm–417 mm in depth,

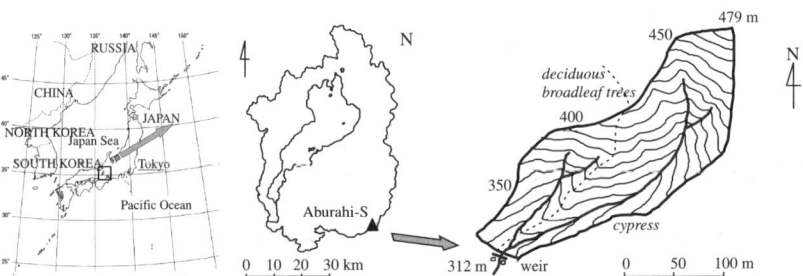

Figure 1 Aburahi-S Experimental Watershed

and water samples were taken up at appropriate intervals (0.5–6 hours) before it began to rain until the water level of the stream returned after it stopped raining. The methods of measuring atmospheric deposition and precipitation, the evaluation of flow rates, and the methods of chemical analysis were the same as those written in the previous paper (Kunimatsu et al., 2001).

Methods of evaluating the material loads

The loads L_i, namely the amount of materials flowing out downstream through a fixed site of a river, are calculated from the data of concentration C_i and flow rate Q_i measured at time t_i.

$$L_i = C_i Q_i \tag{1}$$

There are some methods of evaluating the loads during a certain interval; however, we could not find any definite methods or a way of measuring, which appreciate duly the loads caused by storm runoffs. We had proposed the $L(R)$ equation method (Kunimatsu and Sudo, 1997), which makes it possible by using base-flow data, storm runoff data and daily precipitation data.

Integration interval-loads method (ILM). The method has been used widely in the studies on nutrient balance and dynamics in the forest ecosystem. The loading amounts of materials are calculated by integrating continuously each interval-load, which is obtained from the moment loading rate L_i and the interval for one half of the period from the previous measurement t_{i-1} to the next measurement t_{i+1}.

$$L = \Sigma L_i (t_{i+1} - t_{i-1})/2 \tag{2}$$

The data of C_i and Q_i are usually obtained by periodic observations.

L(Q) equation method (LQM). It is well known that L_i are related with Q_i by the following equation.

$$L_i = a Q_i^b \tag{3}$$

where a and b are coefficients obtained from the logarithmic linear regression of the relationship between L_i and Q_i by the least-squares method. Q_i are calculated from the water-levels, which are read out at an appropriate interval t from the continuous recording media obtained at a weir. Then, L is calculated with ILM.

$$L = \Sigma a Q_i^b t \tag{4}$$

There are some ways of obtaining the data of C_i and Q_i for the leading $L(Q)$ equation. In this study, we used three kinds of data set as follows:

LQM$_1$: based on $L_1(Q)$ equation regressed on the data obtained by periodic observation.
LQM$_2$: based on $L_2(Q)$ equation regressed on the data measured during storm runoffs.
LQM$_3$: it is well known that the relationship between flow rates and loading rates during storm runoff shows a hysteresis characteristic of the material. So, the storm–runoff data were divided into two sets of the data, namely the data obtained during the increasing period of flow rate and those during the decreasing period. Then, a set of two $L_3(Q)$

equations were obtained with regression of the two data sets.

$$L_{3I}(Q) = \Sigma a_I Q_i^{bI} t \quad (Q_i - Q_{i-1} > 0) \tag{5}$$

$$L_{3D}(Q) = \Sigma a_D Q_i^{bD} t \quad (Q_i - Q_{i-1} < 0) \tag{6}$$

where subscripts of I and D represent the increasing period and the decreasing period, respectively.

L(R) equation method (LRM). The method was developed in order to evaluate the annual load including every storm runoff load for a year, and to calculate the loads of past years from the past rainfall data (Kunimatsu and Sudo, 1997). It is based on an assumption that it consists of the annual base-flow load L_B and the annual net storm-runoff load L_S. "Net" does not include base-flow load.

$$L = L_B + L_S \tag{7}$$

Observations are carried out on a day not affected by rains. In other case, the data not affected by rains are selected from the data obtained by periodic observations. After dividing a year into the dry season and the rainy season, L_B is given by the summation of the base-flow loads for the dry season L_{Bd} and the rainy season L_{Bw}, which are calculated from the lengths of each season and the average values of the concentration and the flow rate.

$$L_B = L_{Bd} + L_{Bw} \tag{8}$$

Detailed observations during storm runoffs are carried out for more than four rainfall events. Net storm runoff load L_{Si} by the event i is calculated by subtracting the base flow load from the total runoff load which is calculated by using ILM. It was shown that the net storm runoff loads are related with the depths of the rainfalls R_i by the following equation.

$$L_{Si} = c(R_i - r)^d \tag{9}$$

This equation is called the *L(R)* equation, where c and d are the coefficients obtained from the logarithmic linear regression by the least-squared method, and r is the non-effective rainfall, which is estimated from R-intercept of the equation obtained by linear regression of the relationship between R_i and net discharges Q_{Si}.

$$Q_{Si} = pR_i + r \tag{10}$$

where p and r are the coefficients obtained from the regression. The annual net storm runoff load is calculated by the summation of each storm runoff load obtained by using the *L(R)* equation and R_i larger than r for a year.

$$L_S = \Sigma c(R_i - r)^d \tag{11}$$

Results and discussion

Long-term fluctuations of loading rates

The loading rates of nutrients, which were the momentary amounts at each sampling time, were calculated from the weekly data obtained at the Aburahi-S experimental watershed, and plotted in Figure 2. The loading rates during the summer season were generally larger than those during the other seasons. This is because it is rainy in the summer in the area. Some needle-like peaks were occasionally found from early summer to autumn, when heavy rains were caused by passage of a rain front at the end of the rainy season and of

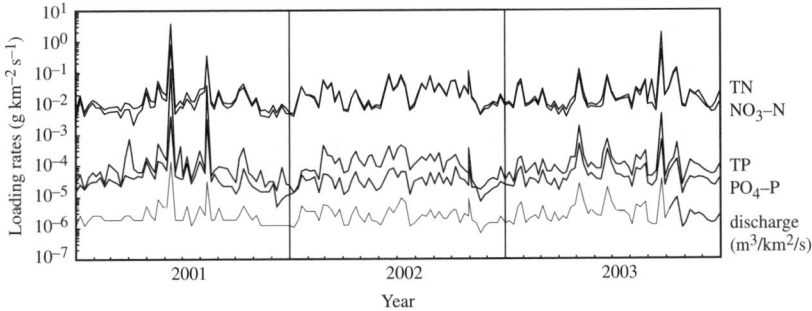

Figure 2 Fluctuations of loading rates of nutrients and discharge observed at Aburahi-S experimental watershed. The data were obtained from the weekly observations

typhoons. Namely, the loads of total nitrogen (TN) and phosphorus (TP), which include particulate constituents, increased from hundreds to a thousand times those before the heavy rain. This was because the concentration of TN changed several times and that of TP more than ten times, in addition the flow rate increasing momentarily more than a hundred times. The changing of dissolved materials appears to be within almost the same range as that of the flow rate, because the concentration of these materials was not affected so largely by the flow rate. These results show that in order to obtain an accurate estimate of annual loads, it is inevitable to evaluate the runoff loads, particularly those caused by intensive rainfalls over about hundred mm in depth, as accurately as possible.

Inaccuracy of loading rates evaluated by ILM

In order to evaluate the total load for a year by ILM, periodic observations in frequency from once a week to once a month have been conducted in many studies. However, the loading rate of a river usually changes daily, hourly or even momentarily in the rainy areas, mainly because of unpredictable rainfalls, as mentioned above. It seems to be clear from Figure 2 that the ILM, in this case based on the weekly data, could not exclude the uncertainty of underestimation or overestimation resulting from a momentary condition from which the sample was taken up only once a week. The average annual-loads calculated in three years' weekly data are shown in Table 1. The annual loads of TN and TP, and the discharge ratio were 19.0 and 0.339 kg ha^{-1} y^{-1}, and 0.75 respectively, which averaged the values of 29.3, 7.14 and 20.6 kg-N ha^{-1} y^{-1}, and 0.900, 0.039 and 0.077 kg-P ha^{-1} y^{-1}, and 1.01, 0.53 and 0.69 obtained is 2001, 2002 and 2003, respectively. The loads of TN and TP were much larger than those evaluated using another method mentioned below, especially that of TN was larger than the atmospheric deposition. When the data obtained on 14 June 2001 and 25 September 2003 were taken off as abnormal or extreme values with ignoring the inverse cases, it is a familiar scenario, with 6.25 and 7.31 kg-N ha^{-1} y^{-1}, 0.057 and 0.045 kg-P ha^{-1} y^{-1}, and 0.56 and 0.57 for discharge ratio were obtained.

The inevitable arbitrariness of ILM was shown by using the everyday data which we had measured everyday around 10:00 am for a year at Ieta Bridge crossing over Mano River (watershed area of 16.4 km^2) in Lake Biwa basin in the following. It was assumed that an observation in frequency of once per n days was carried out for a year. A virtual data set obtained by the observation could be picked up sequentially from the everyday data. The procedure was repeated n times, moving it for each one day. Then, n annual loads L_{ni} were calculated with ILM from the n data sets. In order to investigate the variation of the annual loads, the coefficients of variation cv were calculated, and plotted against the frequencies of the virtual periodical observations in Figure 3. The cv values calculated from the data of 7-days interval observations L_{7i}, for instance, were 0.54 for

Table 1 Annual average loads of Aburahi-S Experimental Watershed calculated with the methods of ILM, LQM and LRM, and of the atmospheric deposition.

Items	Concentration[2] 3 years (mg l^{-1})	Discharge from Abrahi-S experimental watershed								Atmospheric deposition 3 years (kg ha^{-1} y^{-1})
		Average annual loads[1] (kg ha^{-1} y^{-1})						LRM		
		ILM	LQM$_1$	LQM$_2$	LQM$_3$	3 years	3 years (%)[3]	16 years	(cv %)	
	3 years	3 years	3 years	3 years	3 years					
TN	0.772	19.0	7.25	9.45	9.68	9.95	(72)	9.82	(17)	15.3
DN	0.719	13.3	7.05	8.67	8.81	8.89	(71)	8.75	(15)	13.3
NH$_4^+$-N	0.013	0.208	0.116	0.078	0.073	0.182	(39)	0.181	(9.4)	6.02
NO$_2$-N	0.003	0.027	0.014	0.020	0.017	0.018	(57)	0.017	(11)	0.009
NO$_3$-N	0.597	8.47	5.27	7.24	7.42	7.24	(73)	7.12	(16)	5.24
TP	0.0057	0.339	0.0560	0.107	0.159	0.168	(82)	0.160	(39)	0.824
DP	0.0032	0.0440	0.0310	0.0300	0.0294	0.0380	(63)	0.0373	(14)	0.633
PO$_4$-P	0.0016	0.0245	0.0153	0.0155	0.0153	0.0204	(51)	0.0201	(10)	0.521
Discharge ratio[4]	–	0.75	0.57	0.57	0.57	0.48	(75)	0.50	(16)	1,702 mm y^{-1}

[1] Average values for 3 years were calculated by using the data obtained from 2001 to 2003, and those for 16 years from 1988 to 2003.
[2] Concentrations were arithmetic means of the weekly data for the 3 years.
[3] Percentage of storm runoff loads and discharge to each total values. The original data of the total discharge and the storm runoff were 852 and 647 in mm y^{-1}, respectively.
[4] Discharge ratio was discharge/precipitation, and the average precipitation for 16 years and cv were 1667 and 17%.

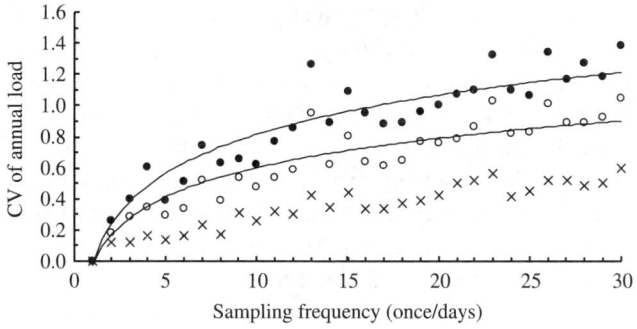

Figure 3 Relationship between sampling frequency and coefficient of variation of the annual loads calculated with ILM and data measured once a day at Ieta Bridge of Mano River for a year. ○; TN, ●; TP, ×; discharge

TN and 0.74 for TP, and that of discharge was 0.23. It appeared that the annual loads calculated with ILM vary widely depending on only the sampling frequency. Consequently in order to obtain accurate data with ILM, samples must be taken up at a higher frequency, perhaps once a day or an hour. However, it is usually impossible to collect and to analyze water samples in such high frequency for a long time. Therefore it is necessary to develop another method for making it possible to evaluate more reliably total load based on the more realistic observation.

Derivation of L(Q) and L(R)

$L(Q)$ *equations.* The relationship between the loading rates and the flow rates measured with the periodic observation at Aburahi-S for the three years were plotted in Figure 4. $L_1(Q)$ equations logarithmically regressed with the least squared method were written in the figure. The data obtained from the storm run-off observations of the five rainfalls from 8 mm to 417 mm in depth were plotted in the same way on the same figure, in which the equations of $L_2(Q)$, $L_{3I}(Q)$ and $L_{3D}(Q)$ were derived in the above procedures. About 95% of the flow rate data obtained by the periodic observation distributed in the narrower range from 0.01 to $1\,m^3\,km^{-2}\,s^{-1}$, on the other hand those obtained from the storm runoff events distributed in the wider range from 0.01 to $7\,m^3\,km^{-2}\,s^{-1}$. The coefficients of determination for $L_1(Q)$, $L_2(Q)$, $L_{3I}(Q)$ and $L_{3D}(Q)$ for TN were 0.78, 0.96, 0.98 and 0.98, respectively, and those for TP were 0.77, 0.86, 0.92 and 0.93. The coefficients for the dissolved components were larger than those for TP.

$L(R)$ *equation.* The net storm runoff loads calculated from the five rainfall events were plotted against the depths of the corresponding rainfalls in Figure 5. Nitrogen as well as phosphorus showed good linearity, but the five events were not sufficient in number of samples to derive statistically reliable $L(R)$ equations. However, the observations of storm runoff events are not so easy that it is difficult to repeat so many times. Then, the data were supplemented with values calculated with LQM_3 using hourly flow-rates of the selected storm runoff events from the data recording water level at the weir. It was shown that there were not significant differences between the five observed data and the 17 calculated data plotted by black and white circles, respectively, in the figure. Then, $L(R)$ equations regressed from all the data were written on the figure. The coefficients of determination of nitrogen and phosphorus exceeded 0.9, and these correlations were significant at 0.1% level.

Reliability of LQM and LRM

At the Aburahi-S Experimental watershed, we could use the weekly periodical data and the sequential flow rate data of every 10 minutes obtained for three years after

2001, the storm runoff data of the five rainfall events, and the rainfall data from 1988 to 2003. In order to assess the reliability of LQMs and LRM, the net storm runoff loads calculated with L(Q) and L(R) were compared with the observed values of the five rainfall events, as shown in Figure 6. LQM_1 and LQM_2 calculated smaller values than LQM_3, especially for TP. The reasons were thought from Figure 4 to be: $L_1(Q)$ were derived from the data involving few data around the high flow rate, the number of the data during decreasing period tended to be larger than that during increasing period, which affected the slope of $L_2(Q)$. The values calculated with LQM_3 were well coincident with the observed values, except TN of an unusually intensive rainfall such as 417 mm. Consequently LQM_3 takes account of the effects of the hysteresis in its procedure and $L_3(Q)$ has the high determination coefficients, so it is considered that LQM_3 is the most reliable method in LQMs. The values calculated with LRM were well coincident with the observed values, though slight deviations from the 1:1 line were found in the data of 151 mm as shown in Figure 6.

Comparison of the annual loads calculated with ILM, LQM and LRM

The annual loads calculated with ILM, LQMs and LRM by using the data of three years were compared in Table 1. As mentioned above, ILM using the discontinuous data usually includes the inevitable and unscientific arbitrariness; therefore we don't argue details here any more. Both LQM_1 and LQM_2 became clear to have the tendency of underestimating the storm-runoff loads in Figure 6, and calculated lower values than LQM_3. The effect of hysteresis on TP was larger than that on TN as shown in Figure 4,

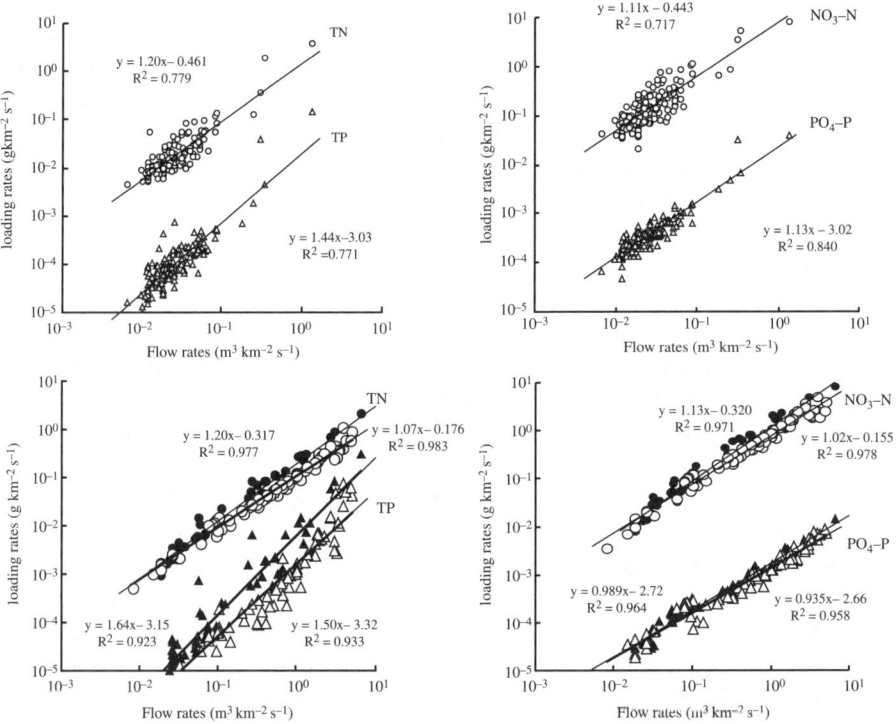

Figure 4 Relationship between the loading rates and the flow rates obtained at Aburahi-S Experimental Watershed (2001–2003). The upper two figures show the data obtained with the weekly periodical observations. The lower two ones were drawn with the data obtained from the storm-runoff observations. ●, ▲; increasing period of flow rates, ○, △; decreasing period of flow rates

so it was thought that LQM$_3$ calculates more reasonable loads than LQM$_2$. LRM gave slightly larger values than LQM$_3$, except for NO$_3$-N.

Material balance evaluated with LRM

It was shown in Table 1 that there was a distinct difference in the content of particulate form between nitrogen and phosphorus. Namely, 89% of the nitrogen flux of 9.95 kg ha^{-1} y^{-1} was discharged in the dissolved forms from the data of LRM, and 80% of the phosphorus flux of 0.187 kg ha^{-1} y^{-1} was in the particulate form. The input fluxes of nitrogen into the forest measured with the bulk deposition method for the three years was 16.5 kg ha^{-1} y^{-1} on average, so that 40% of nitrogen remained in the forest, in other words purified by the forest. The input of phosphorus was 0.791 kg ha^{-1} y^{-1}, 76% of which was eliminated through the forest ecosystem. The average precipitation was 1,702 mm y^{-1} calculated from 1770 mm y^{-1} in 2001, 1393 mm y^{-1} in 2002 and 1944 mm y^{-1} in 2003. There was rather a large difference between the discharging ratio

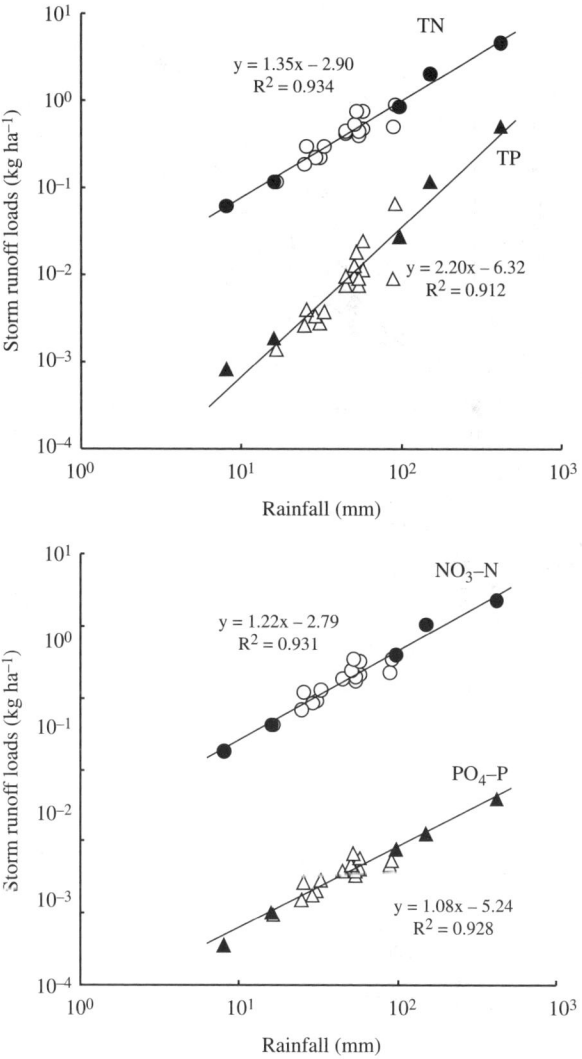

Figure 5 Relationship between the net storm runoff loads and the depths of rainfalls. ●, ▲; observed data from the five storm runoff events, ○, △; data calculated with $L_3(Q)$ equations and each set of hourly flow rate data during selected 17 storm runoffs

calculated with LRM (0.48) and those with LQMs (0.57). Judging from the annual precipitation and the evapotranspiration of the forest in these regions of 880–840 mm y^{-1} (Suzuki and Fukushima, 1985), it was thought that the ratios obtained with LQM and LRM were reasonable values and that the ratio of 0.75 with ILM was too large.

Importance of storm-runoff loads

LRM is an indirect method. However, it offers two important advantages. First LRM makes it possible to isolate storm-runoff load from the total annual load. The percentages of the storm-runoff load to the total load were calculated from the data of the observed three years and shown in Table 1. The ratio of TP (82%) was much larger than that of DP (63%) and distinctly larger than that of TN (72%). DN showed similar values to TN. Storm runoff occupied 75% of the total discharge.

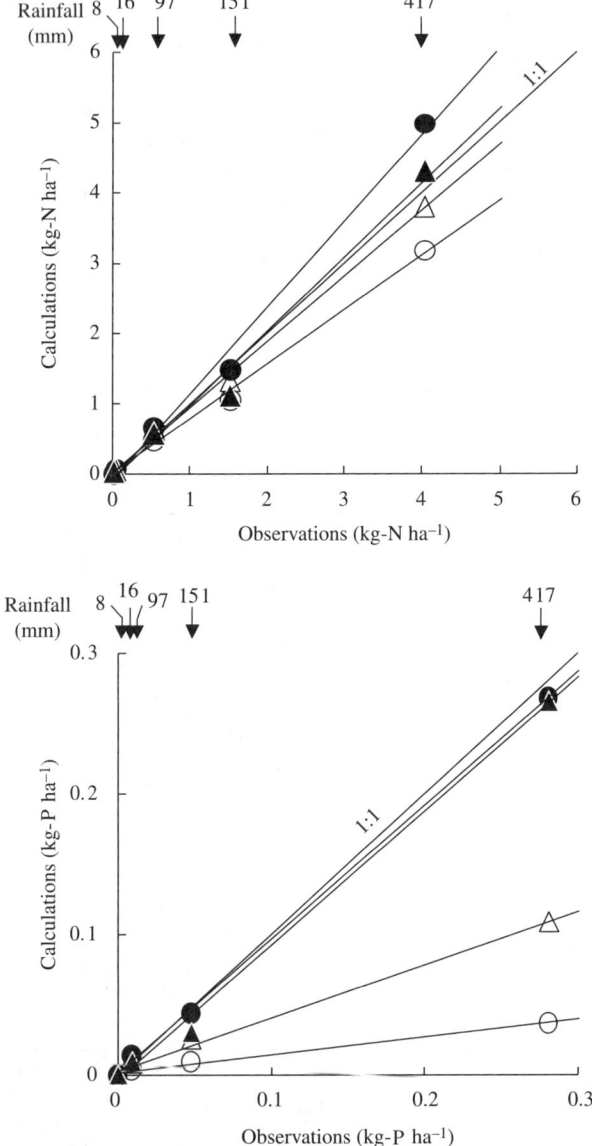

Figure 6 Identification of the storm runoff loads calculated with LQMs and LRM by the observed values. ○; LQ1, △; LQ2, ●; LQ3, ▲; LRM

Long-term fluctuation of the loading rates evaluated with LRM

Secondly LRM has a distinguished function to evaluate past annual loads from daily rainfall data measured in the past with another aim or those obtained by nearby offices such as a dam control office, a school, a firehouse and so on, if the forest has not been disturbed so intensively as to change the material flux. While LQMs do not calculate any values without the data of sequential flow rate recorded at a weir. The daily rainfalls have been measured at the Aburahi-S Experimental Watershed since 1996 and at the Aburahi-N Experimental Watershed at northwestern 2.5 km since 1988. The results calculated from these data of the past 16 years, which were not revised, are shown in Table 1 and Figure 7. The average value and CV of the precipitations were 1667 mm and 17.1%, respectively. The load of TN was 9.83 kg ha^{-1} y^{-1} and agreed well with the value calculated from the data for the three years. TP was calculated as a slightly smaller value, which seemed to be affected with the smaller value of precipitation. It seemed to result from the difference in the precipitation, which affected TP more strongly than TN. The yearly variations were not so large and on almost the same levels as that of the precipitation, except for TP. Judging from the large CV value and the violent fluctuation of TP in Figure 7, it was suggested that it might not give any reasonable results to assess the eutrophication of a water body by using the loading rate measured from one to a few years.

Figure 7 appears to show that the annual loads might have a certain relationship with the annual precipitation. Then the scatter diagrams of the annual precipitations vs. the annual loads are shown in Figure 8. We found two important facts from the Figure: first the loads of the dissolved substances such as NO$_3$-N, sodium ion and so on apparently increased proportionally to the depth of the annual precipitations. TN, which included 89% DN, showed also nearly the proportional relationship. However, such a relationship was not found on TP, which contained 79% PP. Secondly intensive rainfall such as

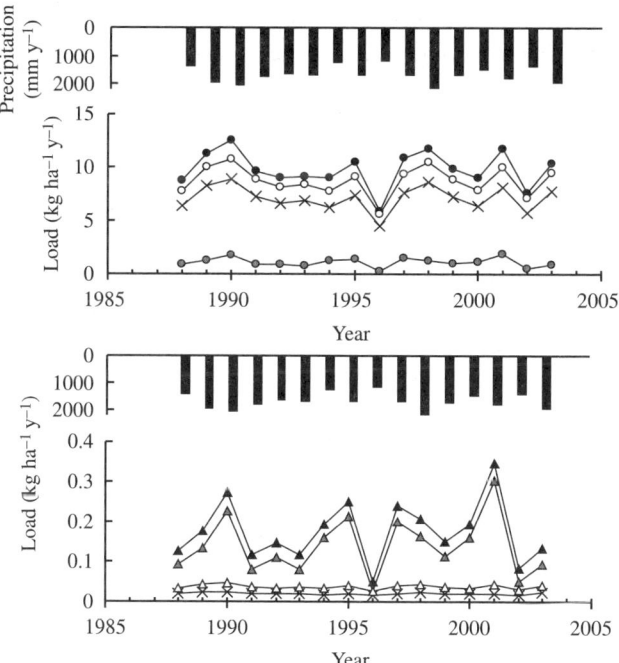

Figure 7 Long-term fluctuations of the annual loads evaluated with LRM from the daily rainfall data in the past (Aburahi-S Experimental Watershed). ●; TN, ○; DT, ●; PN, ×; NO$_3$-N, and ▲; TP, △; DP, ▲; PP, ×; PO$_4$-P

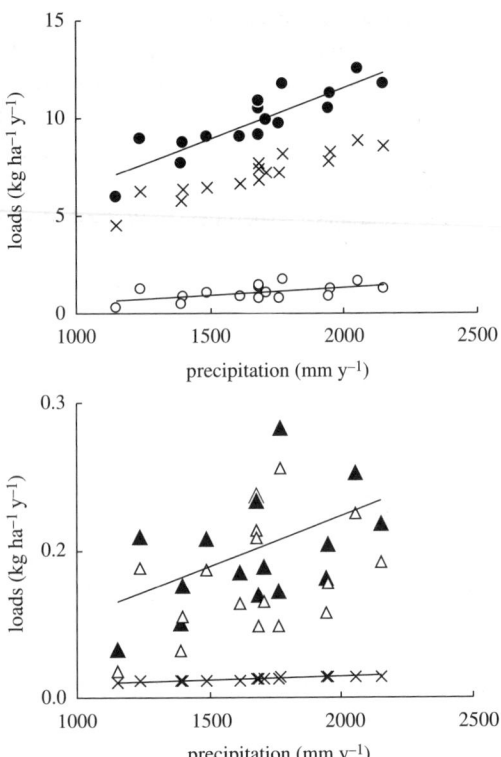

Figure 8 Relationship between the annual precipitations and the loading rates calculated with LRM and rainfall data. ●; TN, ○; PN, ×; NO_3-N, ▲; TP, △; PP, and ×; PO_4-P

exceeding 100-200 mm increased the annual loads of TP and TN much more strongly than the total amounts of the annual precipitations. Those substances contain particulate components. These important characteristics have been elucidated for the first time by developing LRM.

Conclusion

ILM has been widely used for evaluating the material loads from the forests. However, it has no scientific rationality because the flow rate always changes in a stream owing to unpredictable rainfalls. The loading rates of TN and TP of Aburahi-S Experimental Watershed calculated with LRM using the data measured for three years were 9.83, 0.175 kg ha^{-1} y^{-1}, respectively, which did not show any increasing or decreasing tendencies during 16 years since 1987. The load of TN changed every year in rather narrow range around the average value (cv ± 17%), increased in proportion to the annual precipitations, and was discharged 72% of the total during storm runoffs. That of TP did not show such a tendency, changed in wider range (cv ± 43%), and was discharged 82% of the total during storm runoffs. The eliminating rates of the atmospheric input-fluxes of TN of 16.5 kg ha^{-1} y^{-1} and TP of 0.791 kg ha^{-1} y^{-1} were estimated as 40% and 76%, respectively. Consequently, LRM, which makes it possible to calculate the annual loads and the storm-runoff loads by using the past precipitation data, was shown to be the effective method for the environmental as well as the biogeochemical studies of the forests.

References

Johnson, A.H. (1979). Estimating solute transport in streams from grab samples. *Water Resources Research*, **15**(5), 1224–1228.

Kunimatsu, T. and Sudo, M. (1997). Measurement and evaluation of nutrient runoff loads from mountain forests. *J. Japan Society Water Environment*, **20**, 810–815 (in Japanese).

Kunimatsu, T., Hamabata, E., Sudo, M. and Hida, Y. (2001). Comparison of nutrient budgets between three forested mountain watersheds on granite bedrock. *Wat. Sci. Tech.*, **44**(7), 129–140.

Martin, C.W., Hornbeck, J.W., Likens, G.E. and Buso, D.C. (2000). Impacts of intensive harvesting on hydrology and nutrient dynamics of northern hardwood forests. *Can. J. Fish. Aquat. Sci.*, **57**(2), 19–29.

Smith, R.V. and Stewart, D.A. (1977). Statistical models of river loadings of nitrogen and phosphorus in the Lough Neagh System. *Water Research*, **11**, 631–636.

Suzuki M. and Fukushima Y. (1985). Estimates of Evapotranspiration from the Land Surface of Shiga Prefecture using a Digitalized Square-Grid Map Database. *Lake Biwa Study Monographs* **2**, Lake Biwa Research Institute, Otsu, Shiga, Japan, pp 46–47.

Yamaguchi, T., Yoshikawa, K. and Koshiishi, H. (1980). A hydrologic approach to characteristics of water quality and pollutant load of the river. *Japan Society of Civil Engineers*, **293**, 49–63.

Evaluation of hydrological processes in a mountainous small basin using a quinone biomarker

M. Fujita*, H. Haga*, K. Nishida* and Y. Sakamoto*

*Interdisciplinary Graduate School of Medicine and Engineering, University of Yamanashi, 4-3-11 Takeda, Kofu, Yamanashi 400-8511, Japan (E-mail: mfujita@yamanashi.ac.jp; haga@ccn.yamanashi.ac.jp; nishida@yamanashi.ac.jp; sakamoto@ccn.yamanashi.ac.jp)

Abstract An applicability of quinone biomarker to the analysis of hillslope runoff was investigated. At first, quinone profiles of three streams as well as a hillslope runoff in a forested headwater catchment were compared. The quinone composition of hillslope runoff differed from others. Moreover, there were remarkable differences in quinone profile of hillslope runoff under different rainfall conditions. Then, the behavior of quinone biomarker during the increase and decrease of hillslope runoff after a rainfall event was examined. The fractional changes in Q-9 (H2), Q-10 (H2), Q-11, MK-6 and MK-10 suggested the effect of interflow.

Keywords Biomarker; hillslope; interflow; quinone; rainfall-runoff

Introduction

It is recognized that rainfall events cause pollutants such as organic matters and nitrogen compounds to run off from the mountainous basin. In order to understand the runoff mechanism of pollutants, it is fundamentally necessary to elucidate the hydrological processes, especially the identification of the runoff pathway in the hillslope.

Hillslope runoff is one of the classical topics in hydrological studies. Water qualities such as nitrate (Takeuchi *et al.*, 1984) and environmental isotopes (Sklash, 1990) have helped as tracers to analyze hydrological processes in the hillslope. In these analyses, mass balance equations of water quantities and tracers were made between runoff at a spring and its components. The number of unknown parameters used in both equations is larger than that of equations. Therefore, various hypotheses are introduced to reduce the number of the unknown parameters and then to solve the equations. However, sometimes these hypotheses do not agree with the actual phenomena (McDonnel, 1990). Hence, it is likely to say that the methodology for the analysis of hillslope runoff has not been established yet. In order to develop the advanced methodology, it would be necessary to focus on the new tracer that reflects the underground environment in the hillslope and has much information that is able to make the hypothesis reduced.

In this study, we focused on quinone biomarker (Hiraishi *et al.*, 1989) as a new tracer to analyze hillslope runoff. The following were examined to investigate its applicability.

(1) Quinone profiles of three streams as well as a hillslope runoff in a forested headwater catchment were compared. Especially, the differences in quinone profile of the hillslope runoff under different rainfall conditions were evaluated quantitatively.
(2) The behavior of each quinone species was examined during the increase and decrease of hillslope runoff after a rainfall event.

Materials and methods

Experimental catchment

Field observation was performed at a forested headwater catchment of approximately 0.65 ha located in the northern parts of Yamanashi prefecture, Japan (Figure 1).

Observation and sampling

Precipitation was monitored continuously by a hyetometer. Water level was measured every 5 min automatically by a water gage installed into a weir downstream (L1) from a spring. Then, the amount of runoff was estimated from the monitored water level.

In order to achieve the above objective 1), water samples of around 20 L were taken from four sampling points on June 11 and 25, July 2 and August 6, 2003. They were the weir (L1), down stream from a wetland (St.2), that from L1 (St.3) and a confluence of both the streams (St.1). To achieve the objective 2), the run off caused by a rainfall of 73.4 mm from August 14 to 16 was focused on. Then, water samples were taken from near the spring on August 15, 17 and 23, 2003. The runoff on August 15 consisted in the increase part of the hydrograph. On the other hand, that on August 23 was in the decrease part. August 17 was just after the top peak.

Quinone profile method

Quinone biomarker. Quinone is a coenzyme employed as proton carrier in electric transport chain of bacteria (Hiraishi *et al.*, 1989). Quinone structure is divided into four components: ubiquinone (Q-n (Hx)) which is used in aerobic and anoxic respiration, menaquinone (MK-n (Hx)) in anaerobic respiration, plastoquinone (PQ-n) and vitamin K1 (VK1) in photosynthesis, where n and Hx represent the length of the isoprene unit of the side chain and the number of hydrogens saturating the double bonds of the isoprene unit, respectively. Basically, a bacterium has a predominant quinone species, which is stable even though environmental conditions change. Moreover, quinone content corresponds to that of biomass. Quinone can be analyzed quantitatively by using only chemical methods without knowledge of microbiology. Therefore, it has been applied as a biomarker to complex microbial communities such as activated sludge (Furumai *et al.*, 2001) and soil (Fujie *et al.*, 1998).

Quinone analysis. The weight of the water sample was measured and then the water sample was filtrated using a glass fibre filter with 0.3 μm of pore size (GF-75,

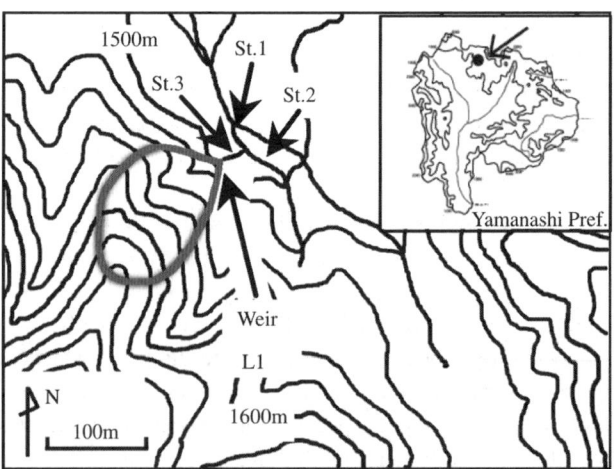

Figure 1 Experimental catchment

ADVANTEC). In order to extract lipid including quinone from the filtration residue, a chloroform–methanol mixture (2:1, v/v) and n-hexane were used in turn. Thereafter, the crude quinone extract in n-hexane was concentrated by Sep-Pak Plus Silica Cartridge (Waters) and separated to MK and Q with 2% and 10% diethylether–hexane, respectively. Quinone species were analyzed by high performance liquid chromatography and then identified by the spectrum and the equivalent number of isoprene units (Hiraishi et al., 1989) calculated from their retention time. The molar concentration of quinone species was estimated from the water sample volume converted from its weight. Furthermore, the quinone profile defined as the molar fraction of each quinone species was also determined.

Dissimilarity index. In order to investigate the difference in quinone profile of two water samples quantitatively, dissimilarity index value (D-value) was calculated according to Equation 1.

$$D(i,j) = 0.5 \sum_{k=1}^{m} |x_{i,k} - x_{j,k}| \quad (1)$$

where m is the number of quinone species and $x_{i,k}$ and $x_{j,k}$ are the molar fractions of the k quinone species for the i and j samples, respectively. D-value is in the range of 0 to 1. A value less than 0.1 indicates that microbial communities of two samples are similar. On the contrary, more than 0.2 means that both are significantly different.

Results and discussion

Precipitation and hillslope runoff

The precipitation observed from June 6 to August 26 is shown in Figure 2. In addition, the runoff at L1 is also shown in the same figure. The hillslope runoff on June 11 was larger than those on June 25 and July 2. In spite of this, it is likely that the runoff on June 11 was dominated by base flow because 90 hrs had passed from the last rainfall event. Conversely, direct flow would be predominant in those on June 25, July 2 and August 6 since only 3, 12 and 10 hrs had passed, respectively. However, the precipitation of the last rainfall event before July 2 was only 10.8 mm and the hillslope runoff was relatively low. For this reason, it is possible that the runoff on July 2 was regarded as base flow.

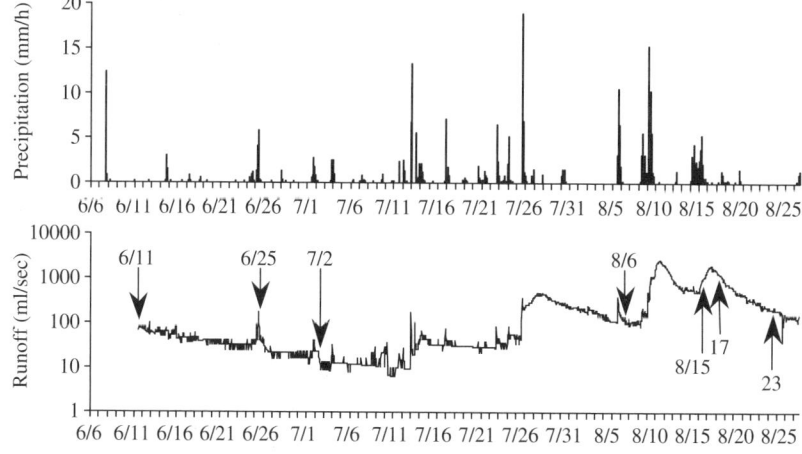

Figure 2 Precipitation and hillslope runoff

On the other hand, that on August 6 received the rainfall event under the condition that the level of original runoff was relatively high. That is why the hydrological processes in hillslope on June 25, July 2 and August 6 were probably different.

The runoff after August 14 occurred following the rainfall event of 96.2 mm on August 8 and 9. The runoff on August 15 and 17 was around 10 times as large as that on June 11. Although 80 h had passed from the last rainfall event, the runoff on August 23 was around twice as large as that on June 11.

Quinone profile of the runoff and streams around the headwater catchment

Figure 3 shows quinone concentration of the runoff and streams on June 11, 25, July 2 and August 6. Q, MK, PQ and VK1 were detected on every sampling date. The quinone concentration on June 25, July 2 and August 6 was higher than that on June 11. That is to say, the runoff dominated by direct flow contained more bacteria than that by base flow. It is suggested that bacteria ran off from the wetland that existed in the upstream at St.2 and St.3. On the other hand, it seems that the infiltrating rainwater would run off through other pathways in the hillslope where the base flow had not passed at L1. Consequently, it is inferred that bacteria that existed in the pathway ran off.

The composition of Q, MK and PQ + VK1 is shown as a triangular diagram in Figure 4. Four marks representing quinone composition of different sampling points on

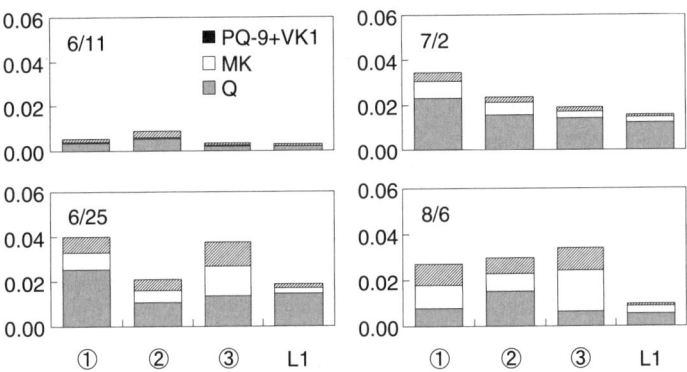

Figure 3 Quinone concentration of the runoff and streams

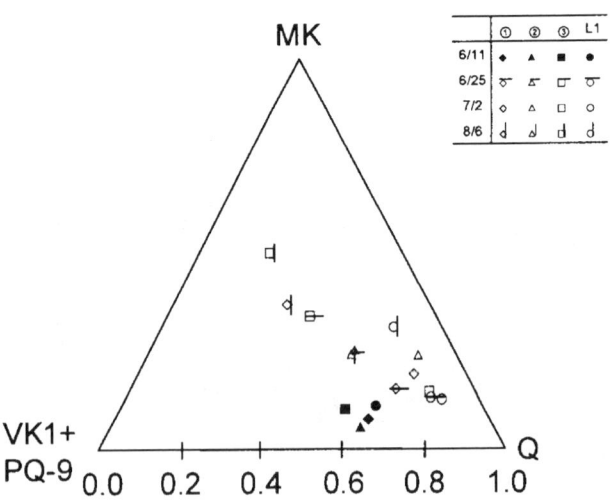

Figure 4 Triangle diagram based on Q, MK and PQ-9 + VK1

June 11 were almost converged. July 2 also showed the same tendency. On the other hand, the marks of samples on June 25 and August 6 were scattered due to their high MK composition at St.2 and St.3.

The runoff discharged from the wetland probably included MK-containing bacteria. St.1 did not indicate the middle composition between St.2 and St.3. Other runoff besides these might have flowed into St.1. Different rainfall conditions brought various quinone compositions.

Then, the molar fractions of Q, MK, PQ and VK1 species at L1 were examined (Figure 5). In every sampling date, Q was found as the major fraction of quinone species. On June 11, five quinone species were detected, which were three Q, two MK and two PQ + VK1. The quinone profile indicated that Q-8 was present as the most predominant, Q-10 was the second and Q-9 was the third, and that the most predominant MK was MK-7; MK-8 was second. On other sampling dates, in addition to the above three Q species, Q-10 (H2) and Q-11 were also detected, although the order of the major three Qs was not changed. On the other hand, in addition to the above two MK species, MK-6 and MK-9 were also detected on June 25. Furthermore, MK-8 (H4), MK-10, MK-10 (H2) and MK-10 (H4) were also detected on July 2 and August 6. The order of the major two MKs on July 2 was the same as on June 11. However, June 25 and August 6 showed MK-8 > MK-7 and MK-8 (H4) > MK-7, respectively. Different rainfall conditions caused different bacteria to run off.

In order to evaluate these differences in quinone profile, the D-value was calculated (Table 1). All D-values were more than 10%. It is interpreted that the microbial community of every sample was significantly different (Hiraishi *et al.*, 1989). In particular, June 25 and July 2 had 32.9% of the highest value. Namely, their runoff pathway in the

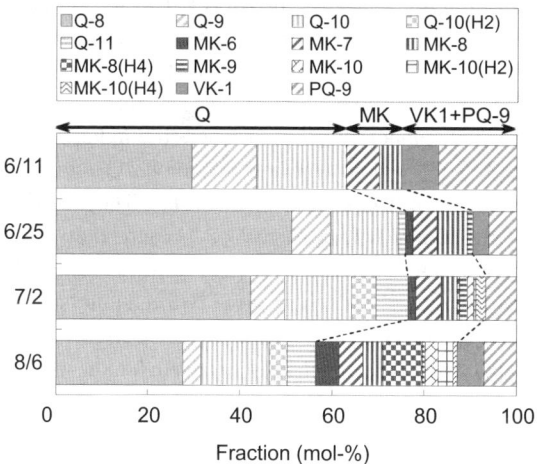

Figure 5 Quinone profile of the hillslope runoff

Table 1 Dissimilarity index

	6/11	6/25	7/2	8/6
6/11				
6/25	27.9			
7/2	32.9	16.4		
8/6	31.4	30.6	23.3	

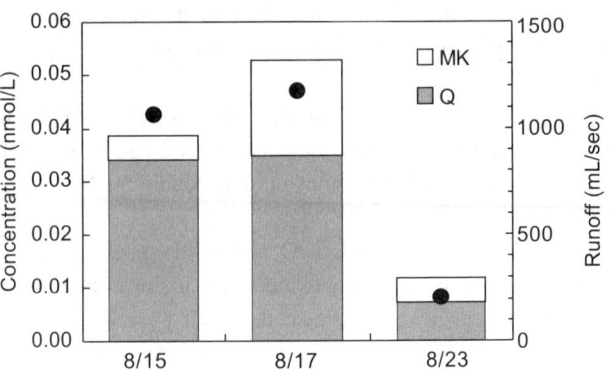

Figure 6 Change in quinone concentration during the increase and decrease of hillslope runoff

hillslope would be significantly different although both the runoff characteristics were similar.

Behavior of quinone biomarker during the increase/decrease of hillslope runoff
Here, the temporal change in quinone species was examined during the increase and decrease of hillslope runoff after a rainfall event. Quinone concentration of August 15, 17 and 23 is shown in Figure 6. Since the quinone profile of runoff was affected by surface conditions as mentioned above, water samples were taken near the spring. As a result, only Q and MK were detected. The change in quinone concentration corresponded to that in the hillslope runoff.

Figure 7 shows the fractional change in twelve detected quinone species. In order to discuss the relationship between quinone species and interflow, Q-9 (H2), Q-10 (H2), Q-11, MK-6, MK-9, MK-10 and MK-10 (H4) were focused on among the twelve quinone species, because the other five quinone species, that were detected on June 11 when base flow dominated, cannot obviously become the indices for interflow. The fraction of Q-10 (H2) and Q-11, which were the major two Q on August 15 as shown in Figure 2, decreased as the hillslope runoff increased. They could increase from the initial rising stage of the runoff before August 15 because they have never been major species at the above four sampling dates. In other words, they could be corresponding to early interflow. On the other hand, there were two quinone species that increased as the hillslope runoff increased. One is Q-9 (H2), which corresponded to decreasing part of the runoff. The other is MK-6, which was detected for the first time on August 17. Both species could reflect interflow. Furthermore, MK-10 that had a higher fraction compared with on August 15 and 17 could correspond to late interflow. While the above quinone species showed good correlation with the change in hillslope runoff, the following quinone species also existed. MK-10 (H4) showed the opposite trends with the change in runoff. MK-9 made its fraction increase during the observation period. Therefore, it was difficult to explain the relationship between them and the runoff components.

As mentioned above, quinone species detected from runoff differed in accordance with rainfall conditions. It seems that different rainfall conditions brought a different runoff pathway. Hence, the above quinone species that has good correspondence with the runoff components would not be always unique. It will be important to investigate the quinone profile of hillslope runoff and soil simultaneously in future.

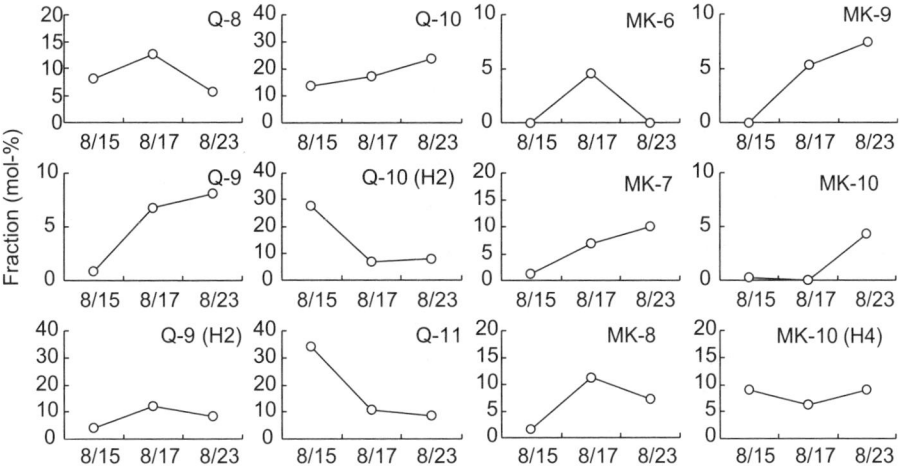

Figure 7 The behavior of quinone species

Conclusions

The quinone profile of hillslope runoff was different from that of streams in the headwater catchment. It showed significant differences under different rainfall conditions. Therefore, the behavior of quinone species was examined during the increase and decrease of hillslope runoff after a rainfall event. As a result, the fractional changes in Q-9 (H2), Q-10 (H2), Q-11, MK-6 and MK-10 suggested the effect of interflow.

Acknowledgement

We would like to thank Mr. T. Itokazu for his constant support and assistance. This research was partially funded by Research and Education on Integrated River Basin Management in Asian Monsoon Region from The 21st Century COE Program, the Ministry of Education, Culture, Sports, Science and Technology, Japan.

References

Fujie, K., Hu, H.Y., Tanaka, H., Urano, K., Saitou, K. and Katayama, A. (1998). Analysis of Respiratory Quinones in Soil for Characterization of Micro Biota. *Soil Science and Plant Nutrition*, **44**(3), 393–404.

Furumai, H., Fujita, M. and Nakajima, F. (2001). Quinone Profile Analysis of Activated Sludge in Enhanced Biological P Removal SBR Treating Actual Sewage. In *Advances in Water and Wastewater Treatment Technology*, Matsuo, T., Hanaki, K., Takizawa, S. and Satoh, H. (eds), Elsevier Science, pp. 165–174.

Hiraishi, A., Masamune, K. and Kitamura, H. (1989). Characterization of the Bacterial Population Structure in an Anaerobic-Aerobic Activated Sludge System on the Basis of Respiratory Quinone Profiles. *Applied and Environmental Microbiology*, **55**(4), 897–901.

McDonnel, J.J. (1990). A Rationale for Old Water Discharge through Macropores in a Steep, Humid Catchment. *Water Resources Research*, **26**, 2821–2832.

Sklash, M.G. (1990). Environmental Isotope Studies of Storm and Snowmelt Runoff Generation. In *Process Studies in Hillslope Hydrology*, Anderson, M.G. and Burt, T.P. (eds), John Wiley & Sons, Chichester, pp. 401–435.

Takeuchi, K., Sakamoto, Y. and Hongo, Y. (1984). Discharge characteristics of NO_3^- for the analysis of basin wide circulation of water and environmental pollutants in a small river basin. *Journal of Hydroscience and Hydraulic Engineering*, **2**(1), 73–85.

Nitrogen removal function of recycling irrigation system

T. Hitomi*, I. Yoshinaga*, Y.W. Feng** and E. Shiratani***

*National Institute for Rural Engineering, 2-1-6, Kan'nondai, Tsukuba city, Ibaraki 305-8609, Japan
(E-mail: *thitomi@nkk.affrc.go.jp*; *yoshi190@nkk.affrc.go.jp*)
**Central Research Institute of Electric Power Industry, 1646, Abiko Abiko city, 270-1194, Japan
(E-mail: *ab-feng@criepi.denken.or.jp*)
***The Ministry of Agriculture, Forestry, and Fisheries of Japan, 1-2-1, Kasumigaseki Tokyo, 100-8950, Japan
(E-mail: *eisaku_shiratani@nm.maff.go.jp*)

Abstract The purpose of this study was to clarify the nitrogen (N) purification capacity of a paddy field in a recycling irrigation system. Irrigation water was sampled at 12-h intervals during the irrigation period from April to September 2003. In addition, ponded water in a paddy field was collected at three points (inlet, centre and outlet). Total amounts of N were 30.7 kg ha^{-1} in inflow and 27.8 kg ha^{-1} in outflow. Thus, the net outflow load was −2.9 kg ha^{-1}. The N removal rate constant when N removal is expressed as a 1st-order kinetic was 0.017–0.024 m d^{-1}. This value is close to values of wetlands and paddy fields in the literature. We found a good correlation between recycling ratio and N removal effect. These results indicate that the recycling irrigation system accumulates N in the irrigation/drainage system, and thus the paddy field does a good job of water purification by removing N.
Keywords N removal rate constant; paddy field; recycling irrigation system; recycling ratio

Introduction

Recycling irrigation systems are used where there is a shortage of water. In catchments with poor water supply, especially in paddy fields, which require large amounts of irrigation water (1,000–1,500 mm), drainage water is reused in irrigation. In Japan, recycling irrigation systems are used in several agricultural areas.

Recycling irrigation systems reduce the effluent load from agricultural areas. They also reduce nutrient loss relative to non-recycling irrigation systems (Kaneki, 1991). In a catchment with a high recycling irrigation rate, the accumulation and consumption of nutrients were high (Kudo *et al.*, 1995). Shiratani *et al.* (2004) analysed scenarios in a recycling irrigation model and found that the effluent N load decreases as the recycling ratio increases.

The purification function of paddy fields varies with water management and soil type (Tabuchi and Takamura, 1985). The retention time and the N concentration of irrigation water affect the amount of N removal in paddy fields (Kunimatsu, 1993; Takeda *et al.*, 1997). A recycling irrigation system has a strong influence on the amount of N removed, because it increases the retention time and keeps the N concentration of the water high.

In Japan, paddy fields cover 55% of the agricultural area, and require large amounts of irrigation water. Thus, it is important to understand the purification function of paddy fields equipped with recycling irrigation systems for nutrient load management in catchments. To our knowledge few studies examined the N removal function of recycling irrigation systems.

We calculated the balance of water and N load in a paddy field with a recycling irrigation system, and reveal the characteristics of N removal in the paddy field. Then we

doi: 10.2166/wst.2006.043

analysed the relationship between N removal and recycling ratio, and elucidated the effect of the recycling irrigation system on N removal.

Methods

Study catchment

The Yoshinuma area lies to the north of Tsukuba city, Japan. Most of the land there is used for paddy fields; the rest is forest, upland fields and homes (Figure 1, Table 1). The Kokai River supplies irrigation water to the fields.

Yoshinuma has a recycling irrigation system (Figure 2). Water from the river is used to irrigate paddy fields and upland fields, and then the drainage flows into the main drainage canal. At the pump station, the water level of the main drainage canal is elevated by a movable barrage during the irrigation period. Some of the water from the canal is pumped up to other paddy fields (8.9 ha), from where the drainage flows back into the canal again. The remaining water flows back into the Kokai River. The recycling irrigation system thus reduces the amount of drainage flowing back into the river.

Study paddy field

We investigated a paddy field in the recycling irrigation area from April to September 2003. It had an area of 4,800 m^2 (76 m × 64 m), and had two inlets and one outlet. Irrigation water was supplied via a pipeline. Surface drainage flowed out through a drain, the level of which was controlled by the farmer. The farming schedule of the study field is shown in Table 2. The irrigation period went from the middle of April to the end of August. Transplanting occurred at the end of April; harvesting occurred in the middle of September.

Hydrological measurement

The flow rates at the two inlet points were measured with a cumulative flow meter. Changes of ponding water depth in the paddy field were recorded with an automatic water level register. Precipitation was estimated from the automated meteorological data acquisition system of the Japan Meteorological Agency (AMeDAS) collected at Shimotsuma observatory located 5 km northwest of the study field. Evapotranspiration was calculated by the Makkink method (Nagai, 1993).

The outflow rate was calculated from the water levels recorded by the automatic register. The flow rate of reused water through the recycling irrigation system was estimated from the working time of the pump station.

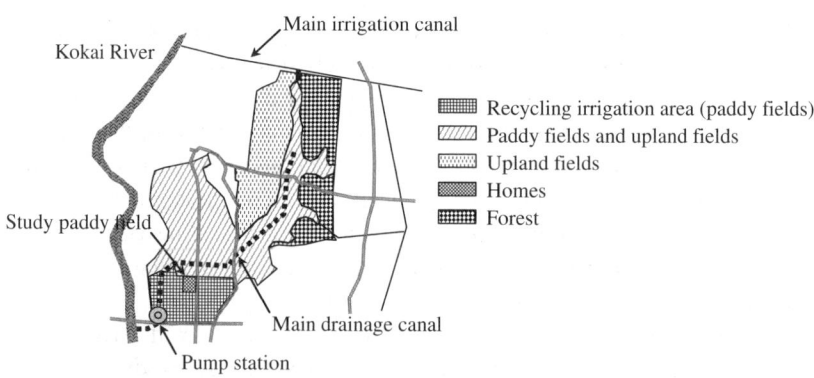

Figure 1 Outline of the study catchment

Table 1 Land uses

Land use	Area (ha)	Percentage (%)
Paddy fields	36.5	57.1
Recycling irrigation area	8.9	13.9
Upland fields	11.6	18.1
Homes	2.7	4.2
Forest	13.1	20.5
Total	63.9	

Sampling and water quality measurements

River water, ponding water in the study paddy field at three points (inlet, centre and outlet), and the outflow water at the pump station were sampled once a week during the irrigation period. The outflow water in the main drainage canal at the pump station, which was the irrigation water applied to the recycling irrigation area, was sampled at 12-h intervals by an automatic sampler (ISCO 6700, ISCO, USA). Water temperature, pH, electrical conductivity (EC), dissolved oxygen (DO) and turbidity were measured at each sampling point with a water quality meter (U-21XD, HORIBA, Japan). The concentration of total nitrogen (T-N) was measured with a T-N analyser (TN-301P, Yanaco, Japan).

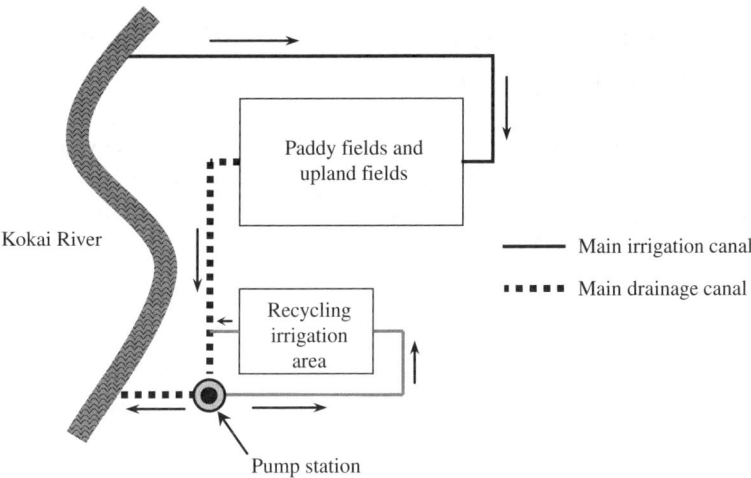

Figure 2 Outline of the recycling irrigation system

Table 2 Farming schedule of study paddy field in 2003

Dates	Farming practices
April 19	Basal fertilizer application (N 3.2 g m^{-2})
April 20	Start of irrigation
April 27	Puddling and weeding
April 29	Transplanting
May 17	Weedicide application
June 10–June 25	Mid-summer drainage
July 13	Topdressing (N 2.2 g m^{-2})
August 3	Pesticide application
August 30	End of irrigation
September 14	Harvesting

Results and discussion

Characteristic of N removal in the paddy field

Water balance in paddy field. The water balance in a paddy field is expressed by the following equation:

$$H_i = H_{i-1} + I_i + R_i - E_i - P_i - S_i,$$

where H_i = ponding water depth (mm), I_i = amount of irrigation water (mm), R_i = amount of rainfall (mm), E_i = amount of evapotranspiration (mm), P_i = amount of percolation (mm) and S_i = amount of surface drainage (mm) at time i (d).

We did not measure the amount of surface drainage or percolation. So we calculated the water balance on the basis of the observation that the maximum value of the sum of percolation and evapotranspiration was $10 \, \text{mm} \, \text{d}^{-1}$.

1. The amount of rainfall (R_i) was estimated from AMeDAS data.
2. The amount of evapotranspiration (E_i) was calculated by the Makkink method.
3. The amount of surface drainage on the ith day (S_i) was calculated as:

$S_i = 0 \, \text{mm}$ when $(H_{i-1} - H_i) + I_i + R_i < 10 \, \text{mm}$;

$S_i = (H_{i-1} - H_i) + I_i + R_i - E_i - P_i$ when $(H_{i-1} - H_i) + I_i + R_i \geq 10 \, \text{mm}$.

4. The amount of percolation on the ith day (P_i) was calculated as:

$P_i = 0 \, \text{mm}$ when $(H_{i-1} - H_i) + I_i + R_i < 0 \, \text{mm}$;

$P_i = (H_{i-1} - H_i) + I_i + R_i - E_i$ when $0 \, \text{mm} < (H_{i-1} - H_i) + I_i + R_i < 10 \, \text{mm}$;

$P_i = 10 - E_i \, \text{mm}$ when $(H_{i-1} - H_i) + I_i + R_i \geq 10 \, \text{mm}$.

Figure 3 shows the seasonal variations of the inflowing and outflowing water. The water requirement (evapotranspiration + percolation) during the irrigation period was $8.8 \, \text{mm} \, \text{d}^{-1}$. The optimum water requirement rate in Japan is $20-25 \, \text{mm} \, \text{d}^{-1}$. Thus, the study paddy field saved irrigation water.

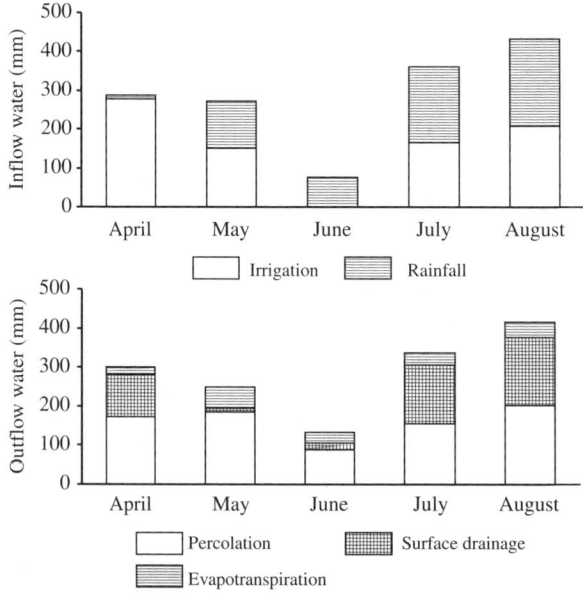

Figure 3 Seasonal variations in inflowing and outflowing water

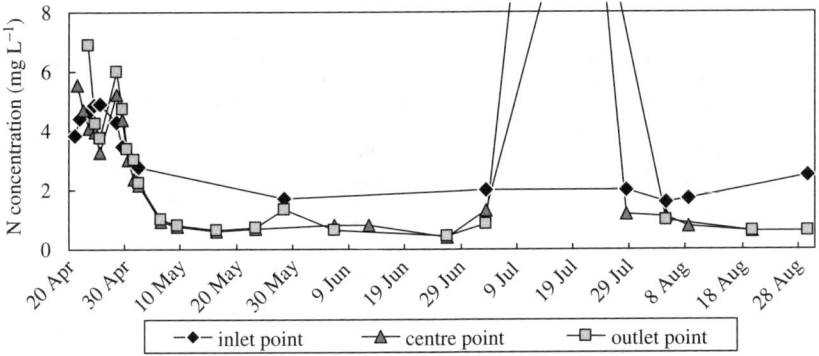

Figure 4 N concentrations of ponded water at inlet and outlet

N balance in paddy field. Figure 4 shows the N concentrations of ponded water at the inlet, centre and outlet points during the irrigation period. Most of the time, the N concentration at the centre and outlet was lower than that at the inlets, except during the start of irrigation, puddling and fertiliser topdressing.

Figure 5 shows the seasonal variations in the inflowing and outflowing N loads. These loads were calculated by multiplying the N concentration by water volume. The total N loads during the irrigation period were 30.7 kg ha^{-1} inflowing and 27.8 kg ha^{-1} outflowing (Table 3). The net outflow load of N (outflow minus inflow) was -2.9 kg ha^{-1}. The negative value means that this paddy field removed N.

N removal rate constant. In paddy fields, N is removed by denitrification, sedimentation and absorption by plants and algae. In general, this N removal is expressed by the following exponential formula:

$$N = N_0 \exp(-at/h),$$

where N = N concentration in the ponded water (mg L^{-1}), N_0 = initial N concentration (mg L^{-1}), h = ponded water depth (m), a = N removal rate constant (m d^{-1}) and t = time (d).

N removal rate constant means the potential for N removal. Figure 6 shows the decrease in N concentrations in the ponded water between two periods. In the first period, from 23 to 25 April, water with a high N concentration (4.73 mg L^{-1}) was supplied, and nutrients were dissolved from basal fertiliser and soil. In the second period, from 16 to 22 July, nutrients were dissolved from the topdressing. To calculate the N removal rate constant, we assumed that the ponding water depth was 0.05 m, the usual depth in Japanese paddy fields, and applied the above formula. The N removal rate constants were, respectively, 0.017 and 0.024 m d^{-1}. These values are close to previous results in the same field, 0.016 to 0.021 m d^{-1} (Yoshinaga et al., 2003).

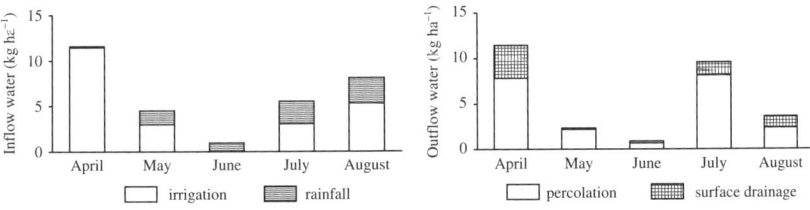

Figure 5 Seasonal variations in the inflow and outflow loads of N

Table 3 N loads of inflowing and outflowing water

	Inflow		Outflow		Net outflow
	Irrigation	Rainfall	Percolation	Surface drainage	
April	11.46	0.14	7.80	3.65	−0.16
May	2.95	1.55	2.21	0.13	−2.16
June	0.00	0.96	0.68	0.16	−0.12
July	3.05	2.46	8.13	1.45	4.08
August	5.30	2.82	2.31	1.27	−4.53
Total	22.76	7.92	21.13	6.66	−2.90

Tabuchi (2001) calculated N removal rate constants of some soil types at constant temperature in the dark, and found values of 0.007 to 0.014 m d^{-1}. In actual paddy fields, N is removed by absorption by vegetation and algae in the light as well, so actual values would be higher. Yamaguchi and Hata (1993) clarified the relationship between the N concentration of irrigation water and N removal. The N concentration dropped from an initial 20 mg L^{-1} to 1.0–2.0 mg L^{-1} by 7 d. Thus, the removal rate constant was around 0.03 m d^{-1}. Judging from these results, our study paddy field had a comparable potential for N removal.

N removal functions in catchment equipped with recycling irrigation system

Figure 7 shows the N concentrations of the river and main drainage canal. During most of the irrigation period, the N concentration at the outflow was lower than that at the inflow. During late April and late June to early July, however, the N concentration at the outflow was higher. Those results were due to nutrient enrichment in the paddy fields. Generally, during late April, the beginning of the irrigation period, a large amount of water is applied to paddy fields, and puddling and transplanting are carried out. These actions mix the water, basal fertiliser and soil, allowing nutrient-rich water to drain from the fields. In June, fields are drained to dry the soil as standard farming practice in Japan to maintain the condition of the rice roots. There was no water in our study paddy field during 10 to 25 June. In these aerobic conditions, organic matter breaks down quickly, and the nutrient concentrations in the soil become high. Irrigation began again in late June, and nutrient-rich water consequently drains from the paddy fields. Thus, the high N concentrations at the outflow were due to agricultural practices, and the paddy fields became a pollutant source.

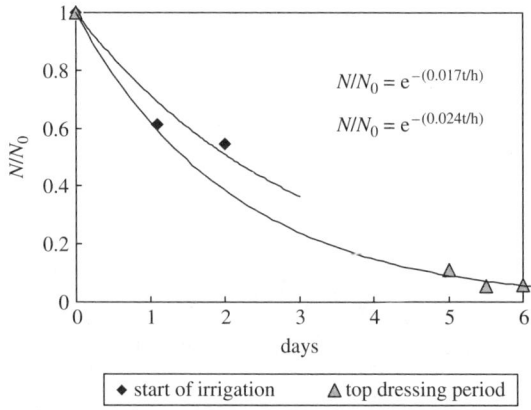

Figure 6 Decrease of N concentrations in the ponded water

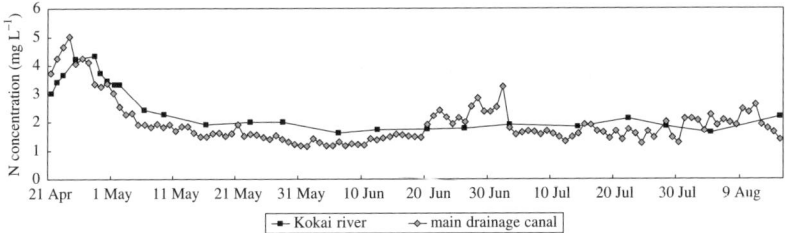

Figure 7 N concentrations in Kokai River and main drainage canal

Factors of N removal in recycling irrigation system. The recycling irrigation system removes N in two ways. First, reusing drainage water increases the retention time. Retention time has a great influence on nutrient removal. Retention time is calculated as follows:

$$RT = V/(Q_{in} + P),$$

where RT = retention time (d), V = water volume retained in the catchment (m^3), Q_{in} = flow of irrigation water to the catchment (m^3 d^{-1}) and P = precipitation (m^3 d^{-1}).

Takeda *et al.* (1997) found that nutrients were removed when retention time was 5 to 7 days. A higher recycling ratio requires less irrigation water to increase retention time. Feng *et al.* (2005) reported a retention time of 5–9 days (except in April) in the same paddy field as ours.

Second, a recycling irrigation system accumulates nutrients. To achieve this, a regulating reservoir is needed from which water is pumped up to the fields. The reservoir collects water and nutrients from the whole catchment and accumulates the nutrients in the recycling irrigation system by preventing the outflow of nutrient-rich water. It could be suggested that sedimentation of particle nitrogen in the catchment plays a great role in reducing nitrogen effluent load, because average turbidity was comparatively high (98.4 mg/L at the main drainage canal). In addition, paddy fields remove nutrients from nutrient-rich irrigation water (Kunimatsu, 1983). In our study catchment, the main drainage canal played the role of reservoir, and nutrient-rich water was reused on the upper fields. The average N concentration of the water in the main drainage canal, which was the source of recycled irrigation water, was 2.0 mg L^{-1}, a comparatively high concentration.

Relationship between recycling ratio and N removal. Our results show that a recycling irrigation system is effective at removing N (Figure 8). To evaluate the efficiency of our study site, we calculated the recycling ratio and difference in N concentration. The recycling ratio was calculated as follows:

$$R = 100 V_p/(V_p + V_s),$$

where R = recycling ratio (%), V_p = volume of reused water supplied through the recycling irrigation system (m^3), and V_s = volume of runoff from the catchment (m^3).

We estimated the N removal function from the difference in N concentration:

$$D = -\Delta N/t,$$

$$\Delta N = (N_o - N_i),$$

where D = difference in N concentration (mg L^{-1} d^{-1}), N_o = N concentration of outflow water in the main drainage canal (mg L^{-1}), N_i = N concentration of inflow water in Kokai River (mg L^{-1}), and t = time (d).

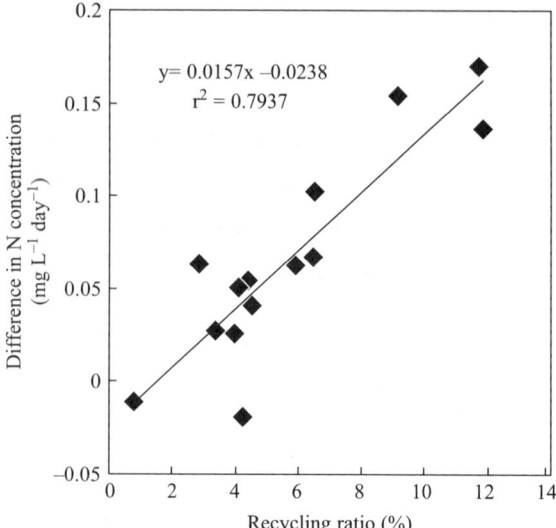

Figure 8 Relationship between recycling ratio and difference in N concentration

To reveal the effect of the recycling irrigation system on N removal, we calculated the recycling ratio and difference in N concentration for the period excluding when the paddy field was a pollutant source. Figure 8 shows the relationship between the two. A positive difference means that the N concentration at the outflow was lower than that at the inflow. This figure indicates that the N removal efficiency is proportional to the recycling ratio. Consequently, it is possible to raise the N removal effect of this recycling irrigation system by keeping the recycling ratio high. The mean recycling ratio was 5.7%, and the mean difference in N concentration was $0.066\,\mathrm{mg\,L^{-1}\,d^{-1}}$.

Conclusions

Recycling irrigation systems can remove N. Thus we found the system plays an effective role for nutrient load management in catchments. In this study, we evaluated the effect of the recycling irrigation system on N removal. We came to the following conclusions.

1. The total inflow and outflow loads during the irrigation period in the study paddy field were 30.7 and $27.8\,\mathrm{kg\,ha^{-1}}$ respectively. Therefore, the net outflow load was $-2.9\,\mathrm{kg\,ha^{-1}}$, indicating that this paddy field removed N.
2. The N removal rate constant was $0.017\,\mathrm{m\,d^{-1}}$ at the start of irrigation and $0.024\,\mathrm{m\,d^{-1}}$ during topdressing.
3. The difference in N concentration (inflow minus outflow) increased in proportion to recycling ratio.

The recycling irrigation system purifies water by accumulating N in the system. This function has a significant role to play in reducing the N load from rice-growing areas.

References

Feng, Y.W., Yoshinaga, I., Shiratani, E., Hitomi, T. and Hasebe, H., (2005). Nutrient balance in a paddy field with a recycling irrigation system. *Wat. Sci. Tech.*, **51**(3–4), 151–157.

Kaneki, R. (1991). Water quality of return flow and purification by using paddy fields. *J. Jpn. Soc. Irrig. Drain. Reclam. Eng.*, **59**(11), 31–36 (in Japanese).

Kudo, A., Kawagoe, N. and Sasanabe, S. (1995). Characteristics of water management and outflow load from a paddy field in a return flow irrigation area. *J. Jpn. Soc. Irrig. Drain. Reclam. Eng.*, **63**(2), 49–54 (in Japanese).

Kunimatsu, T. (1983). Crop land – recycling of nutrients and purification function in paddy fields. Research Report. Lake Biwa Research Institute, Shiga, Japan, pp. 28–35 (in Japanese).

Nagai, A. (1993). Estimation of pan evaporation by Makkink equation. *J. Jpn. Soc. Hydrol. & Wat. Resour*, **6**(3), 238–243 (in Japanese with English abstract).

Shiratani, E., Yoshinaga, I., Feng, Y.W. and Hasebe, H. (2004). Scenario analysis for reduction of effluent load from an agricultural area by recycling the run-off water. *Wat. Sci. Technol.*, **49**(3), 55–62.

Tabuchi, T. (2001). Nitrate removal in the flooded paddy field. *Proc. Int. Workshop on Efficiency of Purification Process in Riparian Buffer Zones* (edited by the organizing committee for international workshop of riparian buffer zones), pp. 81–90.

Tabuchi, T. and Takamura, Y. (1985). *Nitrogen and Phosphorus Outflow from Catchment Area*. Tokyo University Press, Japan.

Takeda, I., Fukushima, A. and Tanaka, R. (1997). Non-point pollutant reduction in a paddy-field watershed using a circular irrigation system. *Wat. Res.*, **31**, 2685–2692.

Yamaguchi, Y. and Hata, K. (1993). Change of water quality on nitrogen and phosphorus in surface of fallow paddy fields. *J. Jpn. Soc. Irrig. Drain. Reclam. Eng*, **61**(10), 7–12 (in Japanese).

Yoshinaga, I., Feng, Y.W., Hasebe, H. and Shiratani, E. (2003). Nitrogen removal function of paddy field in a circular irrigation system. *Proceedings of 7th International Conference on Diffuse Pollution*, 17–21 August 2003, Dublin, Ireland, pp. 56–61.

Recycling mineral nutrients to farmland via compost application

Y.Y. Liu, M. Ukita, T. Imai and T. Higuchi

Department of Civil and Environmental Engineering, Yamaguchi University, Tokiwadai 2-16-1, Ube, Yamaguchi 755-8611, Japan (E-mail: *yuyu_liu2004@hotmail.com*; *imai@yamaguchi-u.ac.jp*; *mukita@yamaguchi-u.ac.jp*; *takaya@yamaguchi-u.ac.jp*)

Abstract Increased cultivation of farmland has resulted in nutrient deficiency and consequently fertility degradation of soils. This research examined the application of composted wastes in terms of the feasibility and effectiveness of recycling plant essential minerals. Minerals in composts (derived from sewage sludge, livestock excrement, and municipal solid wastes, respectively) and in amended soils were observed. Ca/Mg ratios in amended soils and the effect of compost applications (mineral nutrients and heavy metals) on plant uptake were also studied. Results showed that composts, especially those made from sewage sludge and livestock excrement, were richer in mineral nutrients but also contained more heavy metals than untreated soil. The increase in some elements and plant-growth-essential Ca/Mg ratios were found in amended farmlands, implying that compost applications have made up for the nutrient deficiency and have adjusted chemical conditions of the soil. The soil contamination from heavy metals was noticeable. However, some results showed that the large existence of mineral nutrients and heavy metals in soils has caused no significant increase in the plant uptake of elements. The controlled composting process and farmland uses are believed necessary for reducing the heavy metal accumulation in agricultural plants.
Keywords Ca/Mg ratio; compost; farmland; heavy metal accumulation; mineral nutrient; transfer factor

Introduction

Mineral nutrients are essential for plant growth and human health, so the issues of nutrient deficiency and the long-term fertility degradation of cultivated lands are of great concern. On the other hand, solid wastes, mainly composed of C, H, O and N, and those abundant in inorganic nutrients, have been increasingly incinerated for landfill. All of these mineral nutrients originated from our earth, namely from farmlands. Therefore, the goal of our work is to assess the feasibility of recycling mineral nutrients to farmlands via a compost application. Finally, the heavy metal accumulation in amended soils and agricultural plants is well considered.

Methods

Site description and sampling

During a three-year period 2001–2003, a series of samplings were conducted in Yamaguchi, Nagano, and Miyazaki prefectures of Japan. In these regions, the farmland applications of composts had been carried out to different extents. Generally, composts were surface broadcast and mainly incorporated to a 15-cm depth after application. At first, five kinds of composting products [i.e. Garbage Compost (GC, derived from kitchen garbage and sawdust), Sewage Sludge Compost (SSC, from sewage sludge, 80%, and waste water sludge from food industry, 20%), Hen Excreta Compost (HEC, from hen excreta and sawdust), Swine Manure Compost (SMC, from swine manure, sawdust, tree rubbish and coffee dregs) and Cattle Excreta Compost (CEC, from cattle excreta, bark and sawdust, 1:1:1)] and three compost-amended soils (Table 1) were collected for

Table 1 Composting-amended soils (the 1st stage of experiment)

Compost amended soil			Compost application	Plants
Garbage compost amended soil (a)	GS_a	Sandy loam	GC, 10 t/ha/yr, greenhouse, 11 yrs	White small turnip, radish
Garbage compost amended soil (b)	GS_b	Sandy loam	GC, 30 t/ha/yr, greenhouse, 11 yrs	White small turnip, radish
Garbage compost amended soil (c)	GS_c	Sandy loam	GC, 100 t/ha/yr, greenhouse, 11 yrs	White small turnip, radish
Background soil corresponding to GS	GSB	Sandy loam	Fertilizer, greenhouse, 11 yrs	White small turnip, radish
Swine manure compost-amended soil (1)	SMS-1	Andosol	SMC, 40 t/ha/yr, paddy field, 6–7 yrs	Rice
Swine manure compost-amended soil (2)	SMS-2	Andosol	SMC, 40 t/ha/yr, paddy field, 1 yr	Rice
Background soil corresponding to SMS	SMSB	Andosol	Fertilizer, upland field	Rice
Sewage sludge compost amended soil	SSS	Andosol	SSC, 60 t/ha/yr, upland field, 7–8 yrs	Pasture
Background soil corresponding to SSS	SSSB	Andosol	Nothing, upland field	–

observing the elements in composts and amended soils. Soil sampling sites were located in the four corners and central parts of selected farmlands, and only the portions in the topsoil (0 ~ 15 cm) were taken. And then, the soil-plant transfers of elements were further examined. The samples taken were seven agricultural plants (i.e. carrot, Chinese cabbage, snap pea, cucumber, rice, broccoli and cabbage) and the soils near plant roots (Table 2). "Control soil" means the farmlands where relatively lower amounts of composts were applied. Impurities (e.g. stone and glass) were removed from the soil samples. Edible parts of agricultural plants were washed and then cut into small pieces. All compost, soil and plant samples were dried for one day at 105 °C, smashed to powder, and then stored at 4 °C.

Pretreatment and chemical analysis

The contents of eleven kinds of elements including Ca, Mg, Fe, Mn, Cu, Zn, Co, Ni, Cd, Cr and Pb in composts, soils and agricultural plants were measured. The EPA standard procedure was adopted in the sample digestion (USEPA, 1996). Atomic Absorption Flame Spectrophotometer (AAS, Shimadzu AA-66GPC) and Inductively Coupled

Table 2 Farmlands where agricultural plants were collected (the 2nd stage of experiment)

Plant	Compost amended soil		Control soil	
Carrot	CEC, 20 t/ha; rice bran, 0.750 t/ha; other additives, 3.6 t/ha. Upland field	Andosol	CEC, 6 t/ha. Upland field	Andosol
Chinese cabbage	CEC, 20 t/ha; other additives, 600 kg/ha. Upland field	Andosol	CEC, 6 t/ha. Upland field	Andosol
Snap bean	SMC, 20 t/ha; Ca fertilizer, 100 kg; and other additives. Upland field	Andosol	Additives, 1.33 t/ha. Upland field	Andosol
Cucumber	CEC, 20 t/ha; additives, 6 t/ha. Greenhouse	Andosol	Additives, 3.1 t /ha. Greenhouse	Andosol
Rice	HEC, 10 t/ha. Paddy field	Gley soil	Additives. Paddy field	Gley soil
Broccoli	HEC, 10 t/ha; additives, 6 t/ha. Upland field	Gley soil	Fertilizers, ~5 t/ha. Upland field	Gley soil
Cabbage	HEC, 10 t/ha; additives, 6 t/ha. Upland field	Gley soil	Additives. Upland field	Gley soil

*Additives: general designation of various other additives in farmland applications

Plasma-Atomic Emission Spectrometry (ICP-AES, PerkinElmer Optima 3300DV) were used in the measurement.

Results and discussion

Nutrients in composts and amended soils

A variety of elements, in differing amounts, were found in various composts and amended soils.

Except for Ca, Mg and Pb, it was found that Cu, Fe, Mn, Zn, Cd, Cr, Ni and Co were more abundant in SSC and SMC than in CEC, HEC and GC. Results of Cu and Ca are shown in Figures 1 and 2, respectively. It is estimated that both calcium and magnesium are two chemical elements, commonly existing in the environment and hence abundant in GC. The same abundance of lead in GC, as in other composts, could be attributed to the geological soil condition, fertilizer and the use of lead-containing products.

Comparisons of chemical elements in composted wastes and background soils are shown in Table 3. Results indicate that SSC, HEC, and SMC are rich in Ca, Mg, Cu, Zn and Cd, whereas CEC is relatively poorer in all elements except for Ca, Zn and Cd, as is GC for the element Ca. From the viewpoint on recycling of plant essential nutrients in organic solid wastes to farmlands, the feasibility of using composted wastes, mainly SSC, HEC, SMC, as well as CEC, in soil amendment can be well substantiated. Other composts, such as GC, can be reused as "the carbon additive". Reuse of compost to enrich farmland appears to be an economic and environmentally friendly alternative to the present practice of incineration and land filling.

In many countries today, various Max Permissible (Allowable) Concentrations, or Contaminant Concentration Limits for the biosolid application to land (USEPA, 1999)

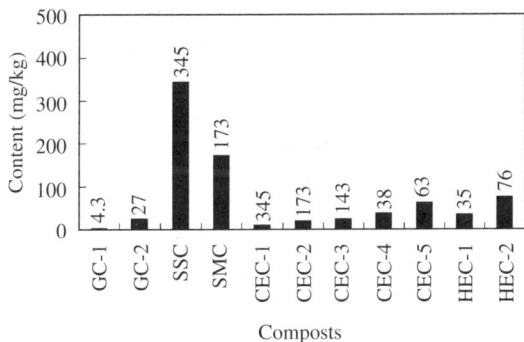

Figure 1 Cu in various composts

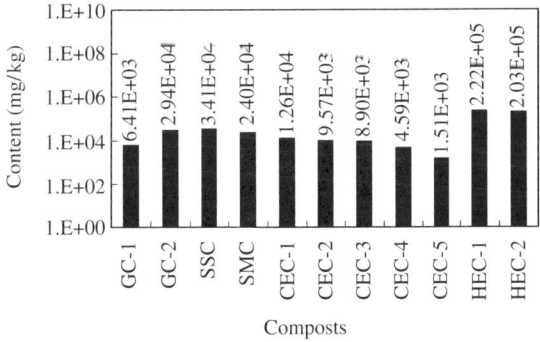

Figure 2 Ca in various composts

Table 3 Minerals in background soil and comparisons with composts

	Contents of background soil (control) mg/kg	Metal content of compost / Element content of soil (dry wt./dry wt.)				
		SSC	HEC	SMC	CEC	GC
Ca	3.13×10^3 (109–6.33×10^3)	10.9	68.0	7.68	2.38	5.73
Mg	5.99×10^3 (164–1.85×10^4)	1.87	1.69	2.10	0.80	0.43
Fe	4.29×10^4 (1.93×10^4–7.04×10^4)	0.50	0.03	0.14	0.07	0.17
Mn	676 (177–1.11×10^3)	1.07	0.50	0.56	0.51	0.29
Cu	41.3 (12.8–82.5)	8.35	1.34	4.20	0.76	0.37
Zn	115 (51.5–187)	8.06	4.27	3.61	2.72	0.46
Pb	23.0 (10.6–49.3)	0.26	1.11	0.25	0.25	0.15
Cd	0.16 (0.037–0.47)	10.6	2.39	1.42	3.16	0.48
Co	13.7 (1.78–29.0)	1.36	0.02	0.08	0.04	0.09
Ni	37.8 (5.87–84.7)	3.85	0.13	0.21	0.34	0.26
Cr	41.5 (7.68–81.3)	0.01	0.21	0.00	0.36	0.39

are being applied. Here, "Ceiling Concentrations of Inorganic Pollutants in Sewage Sludge" (suggested in "EPA CFR 40 Part 503") were used for evaluating chemical elements in composts. In this regulation, elements Cd, Cu, Pb, Ni and Zn are limited to below 85, 4,300, 840, 420 and 7,500 mg/kg, respectively. Experimental results indicated that all maximum values (Cd: 1.85 mg/kg CEC-5; Cu: 345 SSC; Pb: 49.6 HEC-1; Ni: 146 SSC; Zn: 1,050 CEC-5) of composts were far below those limits. It appears conclusive, therefore, that the environmental risk of compost application with inorganic pollutants is limited or at least controllable.

Practical land applications, with different loading rates, of GC, SMC and SSC were observed here. The elements in GC-amended soils (GS) contained quite similar levels to that in the control soil (GSB) (Figure 3(a)). The elements had scarcely accumulated in amended farmlands, that is, the GC application brought about no negative effect on the soil as to the heavy metal contamination. On the contrary, that in SMC-amended (SMS) and SSC-amended soils (SSS) increased to different degrees with compost application. Figure 3(b) shows that although seven-year applications of SMC have led to copper accumulation in soils, the values were still lower than Soil Boundary Value for

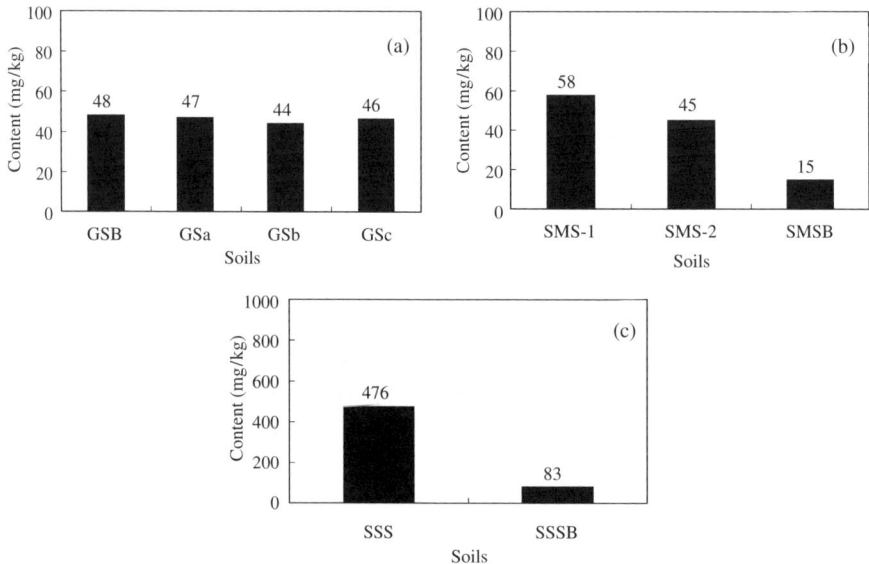

Figure 3 Cu accumulation in farmlands amended with GC (a), SMC (b) and SSC (c)

Cu, 50–140 mg/kg, recommended by the European Union (EU, 86/278/EC). Comparatively, the copper accumulation in SSC-amended farmland (476 mg/kg) was, remarkably, several times higher than the Soil Boundary Value (Figure 3(c)). Excessive compost application was considered to be one of the major causes. Another contributing factor may be that most of inorganic copper containing salts are often poorly soluble and even insoluble and therefore easily accumulate in soils. Evident accumulations of elements Mn, Zn, Cd, Ni, Co have been found in the SSC-amended farmland, while only a few appeared in SMC-amended soils. Heavy metal contamination remains a critical issue in the sewage sludge application in soil improvement. Controlled application of composted sewage sludge is essential for sustainable reuses of biosolid wastes in farmland amendment.

The optimum Ca/Mg ratio ($R_{Ca/Mg}$) in farmlands is often recommended for good crop production. Available evidence suggests that plants grow well and meet their Ca and Mg needs in soils with $R_{Ca/Mg}$ anywhere from 1:1 to 15:1 (Brady and Weil, 2001). Although the plant health and growth is not affected by soil $R_{Ca/Mg}$ in this range, $R_{Ca/Mg}$ with this range in the plant tissue may influence the mineral nutrition of grazing animals.

Results of soil samples in the three areas mentioned above (Table 1) have also demonstrated that the $R_{Ca/Mg}$ values were adequate by applying SSC, CEC and GC (Figure 4). Before composts were applied, values of $R_{Ca/Mg}$ in background soils (SSSB, SMSB and GSB) were far beyond the range from 1:1 to 15:1. In SSSB, $R_{Ca/Mg}$ went as high as 500, while those in SMSB and GSB were low, down to near 0 and 0.2, respectively. With the addition of composts, the $R_{Ca/Mg}$ gradually changed. In SSS, $R_{Ca/Mg}$ rapidly decreased down to nearly 10. Such a significant change could be attributed to both the long-term continuous application (6–7 yr) at a large loading rate of SSC (60 t/ha.yr), and considerably high contents of mineral elements in SSC. As for the application of SMC, $R_{Ca/Mg}$ was raised from 0.04 to 0.52 in the first year and then to 7.99 in the 6th to 7th year. Although the increase in $R_{Ca/Mg}$ for SMS was lower than that in SSS, due to a relatively low loading rate (40 t/ha) and fewer mineral elements in SMC, the imbalance of calcium and magnesium was largely improved. With respect to that of GC, slight but regular changes were observed: $R_{Ca/Mg}$ was gradually changed from 0.17 to 0.06

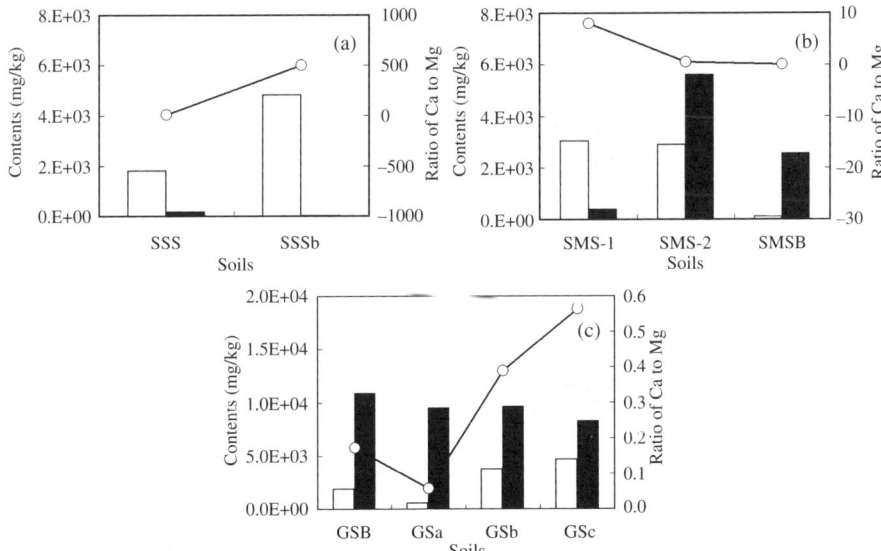

Figure 4 Contents and ratios of Ca and Mg in soils amended with SSC (a), SMC (b) and GC (c) at different degrees. □: Ca, ■: Mg, ○: Ratio of Ca to Mg

(although slightly down here), 0.39 and then 0.56, respectively. The extents of increase in $R_{Ca/Mg}$ can hardly be compared with what happened in SSS and SMS, but the improvement is still important, since the main contribution of the GC application is considered to be that of returning the chemically steady organic matters to the soil. By adjusting the ratio of calcium to magnesium (an essential plant-growth range from 1:1 to 15:1), compost applications to farmland can be very effective for soil improvement.

Plant uptake in amended soils

Minerals were even higher in organically grown potatoes and sweet corn than in conventionally grown ones (Warman and Havard, 1998). In the 2nd stage of this study, we examined seven other agricultural plants (Table 2). With a few exceptions, similar tendencies were observed. Figure 5 presents some selected results concerning carrot, Chinese cabbage, snap bean, cabbage, rice, and corresponding amended soils. It shows the comparison of metal contents before and after the compost applications, in which CEC, SMC and HEC were applied to different extents as shown in Table 2. This figure emphasizes the content changes (increase and/or decrease) of elements Ca, Mg, Fe, Mn, Cu, Zn, Co and Cr in agricultural plants and amended soils.

In carrots, the contents of minerals except Mn and Zn increased with the increase in the soils. As CEC was added, the element accumulation grew accordingly. Two exceptions showed the Mn and Zn accumulations in soils while there was a decrease in plant uptake. Such a phenomenon has also been found in the CEC application in Chinese cabbage production. Except for Mg, all elements significantly accumulated in the CEC-amended farmland soil, while there were notable decreases in Mn, Cu, Zn and Cr in the Chinese cabbage.

In the snap bean production, the SMC application caused Mn, Cu and Cr accumulations in the soil while only elemental Cu was significantly more than before. As for the HEC application in cabbage production, we found rather consistent increases in all minerals in plants and in the corresponding land amended with HEC. Previous studies show that, quite different from CEC, HEC contains plenty of water-soluble and exchangeable

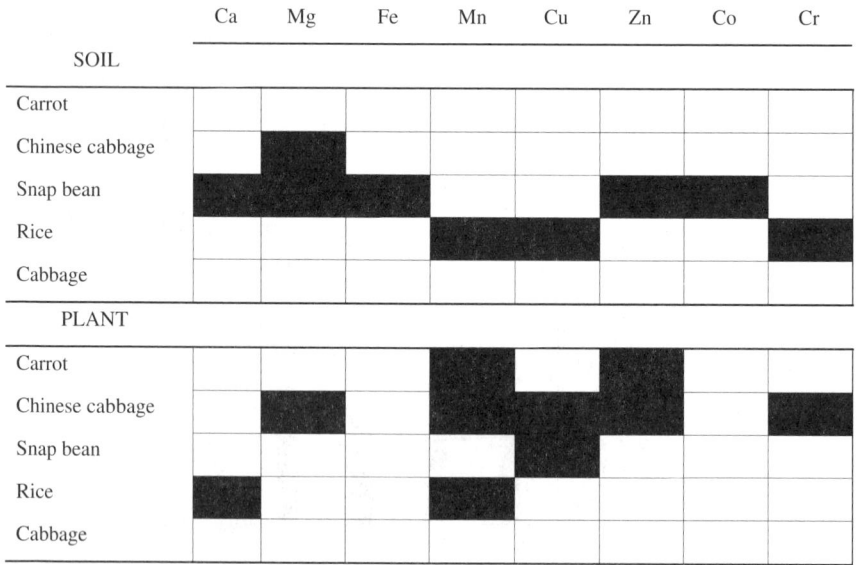

Figure 5 Content changes of elements in compost-amended soils and in vegetables: □: Increase, ■: Decrease of element contents

Table 4 Transfer factors of elements in agricultural plants grown in compost-amended farmlands

Agricultural plant	Compost	Elements							
		Ca	Mg	Fe	Mn	Cu	Zn	Co	Cr
Carrot	CEC	0.27	0.48	6.5E-4	1.2E-2	9.2E-2	0.19	3.7E-3	4.4E-3
Chinese cabbage	CEC	4.2	1.0	2.6E-3	3.8E-2	0.16	0.42	1.7E-2	1.8E-2
Snap bean	SMC	1.31	0.90	1.3E-3	0.15	0.17	0.41	2.5E-2	4.1E-3
Cucumber	CEC	1.60	0.73	1.7E-3	4.3E-2	0.96	1.1	2.9E-2	9.7E-3
Rice	CEC, HEC	6.8E-2	0.92	9.9E-4	0.10	0.30	1.0	8.3E-3	5.9E-3
Broccoli	HEC	1.4	1.8	6.6E-3	8.4E-2	0.56	1.9	5.1E-2	1.0E-2
Cabbage	HEC	1.6	1.4	4.0E-3	3.8E-2	0.26	0.55	3.3E-2	7.4E-3

(generally named as bioavailable, and sometimes EDTA and DTPA-extractable) forms of mineral elements (Liu, 2003), easily causing simple and rapid plant uptake. However, these two forms usually are far less in CEC (1–2% of total amounts of minerals) than in HEC (high up to 10%~20%) (Liu, 2003). Nevertheless, in another case of HEC application (rice production), the above-mentioned consistent increases didn't appear. Possibly in this case, irrigation water had reduced the influence of bioavailable minerals via efficient soil rinse. As a result, several exceptions were found for elements Ca, Mn, Cu and Cr.

Therefore, we shouldn't simply attribute the high element plant uptake to the element accumulation in soils. In practice, as reported in previous literature (Petruzzelli, 1989), the bioavailability of elements is the key factor affecting plant uptake. Further studies should focus on depressing the bioavailability of heavy metals and even the accumulation of toxic metals in edible portions of agricultural plants, effectively utilizing the plant-essential minerals. At least, the accumulation of designated components, e.g. toxic metals, in plants should be controllable.

In the studies of soil contamination and remediation involved in radioactive/toxic metals (Groudev et al., 2001a,b) and of food safety (McLaughlin et al., 1999), the soil-plant transfer of elements was an issue of great concern. It is often expressed by Transfer Factor (Ehlken and Kirchner, 2002; Cui et al., 2004). Transfer Factor is usually regarded as a constant for one sort of plants and can be expressed as follows:

Transfer Factor = Metal content of plant tissue / Metal content of the soil where the plant grew

Here, all contents are mg/kg, based on dry weight.

Here, we estimated the Transfer Factor of minerals to agricultural plants from farmlands amended with composts in different sorts and quantities (Table 2). The dry weight basis was measured for the metal concentration in plant tissue, although the fresh weight basis is also sometimes used (Cui et al., 2004). High Transfer Factors of mineral nutrients validate the compost application. Transfer of toxic elements is harmful for human health. The results (Table 4) show that:

- chinese cabbage, snap bean, cucumber, broccoli and cabbage can effectively accumulate mineral nutrients such as Ca, Mg, while all plants absorb less Fe and Mn;
- plant uptakes are low for heavy metals Cu and Zn, and especially low for Co and Cr.

Conclusions

Sewage sludge compost contains considerably more mineral nutrients and heavy metals than the composts derived from livestock excrement or from household garbage. Element contents of composts were all lower than the "Ceiling Concentrations of Inorganic Pollutants in Sewage Sludge" suggested in "EPA CFR 40 Part 503". The compost applications have brought out elemental accumulation in soil, mostly lower than the environmental criteria except for Cu, Zn and Pb. The nutrient deficiency was, to some extent, reduced. Moreover, the Ca/Mg ratios of several amended farmlands were adjusted to a range between 1:1 and 15:1, which is more suitable for promoting the healthy plant growth.

The soils with high element contents predictably generated the agricultural plants containing abundant mineral nutrients as well as heavy metals. However, there were a lot of exceptions. Some fields had a high plant uptake but low total element contents. It is probable that there was a high level available form but a low amount of elements in those fields. To some extent, the Transfer Factor was different from one element to another, i.e. generally low for heavy metals while relatively high for minerals such as Ca and Mg.

Finally, there is reason to believe that controlled farmland application of composted wastes is not only practical for soil production efficiency, but also essential for the safe and sustainable reuse of wastes. While still potentially problematic, prudent management can to some extent lessen heavy metal accumulation, and the farmland can be made more productive by the application of composted wastes.

References

Brady, N.C. and Weil, R.R. (2001). *The Nature and Properties of Soils*, 13th edn., Prentice Hall.

Cui, Y.J., Zhu, Y.G., Zhai, R.H., Chen, D.Y., Huang, Y.Z., Qiu, Y. and Liang, J.Z. (2004). Transfer of metals from soil to vegetables in an area near a smelter in Nanning, China. *Environ. Int.*, **30**(6), 785–791.

Ehlken, A. and Kirchner, G. (2002). Environmental processes affecting plant root uptake of radioactive trace elements and variability of transfer factor data: a review. *J. Environ. Radioactiv.*, **58**(2–3), 97–112.

Groudev, S.N., Georgiev, P.S., Spasova, I.I. and Komnitsas, K. (2001a). Bioremediation of a soil contaminated with radioactive elements. *Hydrometallurgy*, **59**, 311–318.

Groudev, S.N., Spasova, I.I. and Georgiev, P.S. (2001b). In situ bioremediation of soils contaminated with radioactive elements and toxic heavy metals. *Int. J. Miner. Process*, **62**(1–4), 301–308.

Liu, Y.Y. (2003). *Study on the biosolid waste disposal and mineral resource recycle*. Doctoral Dissertation, Yamaguchi University, Japan.

McLaughlin, M.J., Parker, D.R. and Clarke, J.M. (1999). Metals and micronutrients - food safety issues. *Field Crop Res.*, **60**(1–2), 143–163.

Petruzzelli, G. (1989). Recycling wastes in agriculture: heavy metal bioavailability. *Agr. Ecosyst. Environ.*, **27**(1–4), 493–503.

USEPA (1996). *Acid digestion of sediments, sludges, and soils*. USEPA SW-846; Method 3050B, 1996 Revise; USEPA: Washington, DC, 1996.

USEPA (1999). *Background report on fertilizer use, contaminants and regulations*. EPA 747-R-98-003, January 1999.

Warman, P.R. and Harvard, K.A. (1998). Yield, vitamin and mineral contents of organically and conventionally grown potatoes and sweet corn. *Agr. Ecosyst. Environ.*, **68**, 207–216.

Study on the potential of farmland soils as non-point sources of nitrogen and phosphorus in Japan

M. Ukita, X. Shi, T. Higuchi, Y. Arkin and M. Fukada*

Department of Civil and Environmental Engineering, Faculty of Engineering, Yamaguchi University, Tokiwadai Ube 755-8611, Japan
*Faculty of Agriculture, Yamaguchi University (E-mail: *mukita@yamaguchi-u.ac.jp*; *sxh10@hotmail.com*; *takaya@yamaguchi-u.ac.jp*)

Abstract The amounts of N and P accumulated in farmland soils of 50 cm depth were equivalent to the amount of chemical fertilizer supplied for 50–70 years. The values of N/P of surface soils in farmlands were 1.0–4.3, lower than expected. The median diameter of soil particles in run-off waters was generally less than 10 µm. The mean values of particulate fractions over 1 µm and over 0.22 µm were 19% for N, 27% for P, and 39% for N, 64% for P respectively. Fine particles of soil containing concentrated phosphorus should be carefully monitored as potential sources related to eutrophication.
Keywords Eutrophication; farmland; nitrogen; paddy field; phosphorus; run-off

Introduction

Eutrophication is still a worldwide problem and it is largely related to the modernized agriculture relying on chemical fertilizer. It is reported that phosphorus (P) concentration of farmland soils has been still increasing in Japan, although that of nitrogen (N) has been at steady state. It is easily understood if we consider the run-off rate of P from farmland is estimated to be 2–3% of the supply of fertilizer (Ukita and Nakanishi, 1999; Ukita and Prasertsan, 2002). This study aimed at confirming such situations in nearby areas around us, through the investigation of soil profiles in various farmlands and surveys on run-off waters from various farmland areas. In particular we focused on the accumulation of N and P in farmland soils and the run-off of soil particles containing highly concentrated phosphorus. As for the run-off of P from farmlands, many reports were already published (Sharpley, 1995; Weld *et al.*, 2001; Beckmann *et al.*, 2003).

Materials and methods

Survey on soil profiles

Representative fields were selected from various categories of farmlands and also from forests in Yamaguchi Prefecture near E131° 15′, N34° 05′ as summarized in Table 1. Samplings were conducted mainly in April 2000. Surface soils were sampled with a sampler of 5 cm depth and 100 ml volume at three points in each field. Composite samples were prepared for chemical analyses. Samples for soil profile were taken partly among those fields using a sprit-type core sampler of 8 cmφ, 50 cm long and a screw-drill-type sampler of 120 cm. Grass and fallen leaves were sampled in the quadrate of 30 × 30 cm square. These samples were analyzed for moisture, ignition loss (IL), Kjeldahl nitrogen (Kj-N), ammonium nitrogen (NH_4-N), nitrite and nitrate nitrogen ($NO_{2,3}$-N), total phosphorus (TP), phosphate phosphorus (PO_4-P). Water soluble N and P were analyzed for 4 g of wet soil extracted with 100 ml of distilled water after shaking for 30 min and then filtrated with GA100 (pore size 1 µm) and measured for inorganic N and P. For available

doi: 10.2166/wst.2006.045

Table 1 Sampling points for the survey on soil profiles in Yamaguchi Prefecture in Japan sampled in Apr. 2000

Categories		Grass & fallen leaves	Surface soil	Soil profile
Paddy field	Reductive wet	1 field * 3 points	1 field * 3 points	1 point
	Oxidative dry	1 field * 3 points	1 field * 3 points	1 point
	Before improvement works	1 field * 3 points	1 field * 3 points	1 point
	After improvement works	1 field * 3 points	1 field * 3 points	1 point
	Organic culture	–	2 fields * 3 points	2 points
	Ordinary culture	–	4 fields * 3 points	4 points
	Ibid. with barley culture	–	1 field * 3 points	1 point
Dry field	Vegetables	–	5 fields * 3 points	2 points
	Fruits orchard	3 fields * 3 points	6 fields * 3 points	5 points
	Pasture	3 fields * 3 points	3 fields * 3 points	3 points
	Tea garden	–	3 fields * 3 points	2 points
Forest	Deciduous broad leaf	2 fields * 3 points	2 fields * 3 points	1 point
	Evergreen broad leaf	6 fields * 3 points	6 fields * 3 points	1 point
	Evergreen needle leaf	3 fields * 3 points	4 fields * 3 points	3 points
	Bamboo	–	2 fields * 3 points	–

N and P, 2 g of wet soil and 100 ml of $0.0005\,\mathrm{mol\,l^{-1}}$ H_2SO_4 were used instead of water for extraction.

Surveys on run-off water

Run-off surveys were conducted in 10 rain events from May to October 2001 at six fields shown in Table 2 and Table 3. Surface run-offs were taken and partly subsurface penetrated waters were also taken for dry field A1, paddy field A and the upper part of dry field F. The tributary areas of subsurface run-off were not necessarily clear.

Suspended solid (SS) was filtered by GA100. Distribution of particle size was measured by using an apparatus of Horiba LA-920. Kj-N, NH_4-N, $NO_{2,3}$-N, TP, PO_4-P, soluble Kj-N (SN) and soluble TP (SP) were measured. The latter two were measured for the filtrate of SS measurement. Surface soils of the top 1 cm were sampled and measured for TN and TP and used in the experiments of washed-out load.

Results and discussion
Survey on soil profiles and accumulation of N and P in soils

Table 4 shows the results of estimating the accumulation of N and P in various surface soils of the top 5 cm. In this table dry field includes orchard and tea garden too. Accumulation of N in the 5 cm layer of soil ranges from $0.95\,\mathrm{t\,ha^{-1}}$ in paddy fields to $2.12\,\mathrm{t\,ha^{-1}}$ in dry fields. Accumulation of P ranges from 0.22 in forests to $1.06\,\mathrm{t\,ha^{-1}}$ in dry fields. Accumulation of soluble inorg-N ranges from $0.72\,\mathrm{kg\,ha^{-1}}$ in paddy fields to $73\,\mathrm{kg\,ha^{-1}}$ in dry fields. Accumulation of PO_4-P ranges from $0.02\,\mathrm{kg\,ha^{-1}}$ in forests to $27\,\mathrm{kg\,ha^{-1}}$ in dry fields.

Table 2 Fields for the surveys on run-off water

Categories	Location	Area (ha)	Gradient	Crop	Soil
Paddy field A	Hikino Ajisu	0.4	almost flat	rice	
Paddy field F	Fujigochi Ube	0.25	almost flat	rice	
Dry field A1	Hikino Ajisu	0.15	almost flat	onion, tobacco	reddish clay
Dry field A2	Hikino Ajisu	0.03	5.5°	tobacco	reddish clay
Dry field F	Fujigochi Ube	1.41	5 °(4 terraces)	tobacco, garlic, radish	sandy clay
Tea garden	Fujigochi Ube	2.21	7.4–11°	tea	

Table 3 Rainfall data during run-off survey

	May 21	May 30	June 13	June 19	July 6	July 12	Aug. 3	Sep. 14	Sep. 30	Oct. 9
Total rainfall (mm)	33	43	39	158	75.5	85.5	30	68	48	53
Max. intensity (mm/h)	6.5	9.5	11	41	16.5	20.5	7.5	60	9.5	11.5
Mean intensity (mm/h)	3	3	4	9	5.5	6	3	17	5	5.5
Duration of rainfall (h)	12	14	10	18	14	14	11	4	9	10
Preceding fine weather (day)	12	2	6	0	0	0	3	6	14	7
Preceding rainfall (mm)	3	2.5	5.5	7	21.5	12	0.5	0.5	1.5	8.5
Sampling period	17:45–2:20	18:00–0:45	19:10–23:45	9:45–20:24	9:55–15:30	6:20–7:55	16:00–20:50	21:10–22:45	12:04–18:10	12:48–19:50
Intensity during survey	1.0–1.5	5.3–1.0	9.0–3.0	6.9–13.0	0.5–0.5	16.0–16.5	3.5–0.5	2.5–0.5	1.0–1.5	8.5–10.0

Table 4 Accumulation of nutrients and organic matter in various surface soils of 5 cm (Apr. 2000)

Category	Number of samples	IL %	TN kg·ha^{-1}	TP kg·ha^{-1}	Available N (kg·ha^{-1})			PO$_4$-P	Water soluble (kg·ha^{-1})			PO$_4$-P
					NH$_4$-N	NO$_{2,3}$-N	PO$_4$-P		NH$_4$-N	NO$_{2,3}$-N		
Paddy field	6 × 3	7.30	951	365	0.81	1.59	11.4		0.35	0.37		2.79
		10	22	20	95	92	80		61	99		50
Dry field	12 × 3	14.39	2,117	1,064	29.0	55.9	53.6		25.7	46.9		27.0
		86	79	61	197	220	89		200	244		165
Pasture	3 × 3	12.28	1,426	831	0.85	8.93	4.4		0.36	2.63		3.58
		36	14	54	82	80	87		110	124		108
Forest	11 × 3	13.80	1,107	216	1.88	1.35	1.10		0.49	0.65		0.02
		31	27	87	72	96	95		107	127		332

* Lower lines are coefficients of variance (%)

The values of N/P are smaller than expected: 2.6, 1.0, 1.4, 4.3 and 1.7 for paddy fields, dry fields, orchard and pastures respectively. Forest soils contain a high amount of N and N/P ratio is 12.7.

Table 5 shows the standing stock of grass and fallen leaves on the surface of farmlands in April. The amount of biomass as dry matter ranges from $3.1\,t\,ha^{-1}$ for dry fields and 9.1 for forests. The values versus the accumulation in the 5 cm layer of soil range from 5.2 to 20%. Similarly, the values of N accumulation were $41-84\,kg\,ha^{-1}$ and correspond to 3.8–8%. Similarly, the values for P range from $19\,kg\,ha^{-1}$ for paddy fields to $71\,kg\,ha^{-1}$ for pastures and correspond to 4–37%. As these values seem to fluctuate seasonally, and are not so large, further discussions are omitted.

Table 6 shows the accumulation of I.L., N and P in the soil of 0–50 cm. Density distributions were considered in the estimation (Shi et al., 2002). I.L. accumulation ranged from $291\,t\,ha^{-1}$ in paddy fields to $848\,t\,ha^{-1}$ in tea gardens. Kj-N accumulation ranged from $13\,t\,ha^{-1}$ in paddy fields to $20\,t\,ha^{-1}$ in tea gardens. P accumulation ranged from $1.0\,t\,ha^{-1}$ in forests to $11\,t\,ha^{-1}$ in tea gardens. It should be noted that forest soils accumulate rather large amounts of N of $9.9\,t\,ha^{-1}$. The percentage of 5 cm layer versus 50 cm layer was 10–14% for I.L., 13–26% for Kj-N and 9–34% for TP. Figure 1 shows the vertical distribution of TP, available and water soluble PO_4-P in various soils. Figure 2 shows the vertical distribution of N as well. As general tendencies, the soluble fractions are larger for P than for N in paddy fields, dry fields, orchards and pastures. Available and soluble P distribute more highly near the surface, whereas those for N sometimes distribute more highly in deeper layers.

Survey on run-off water from farmlands

Table 7 shows the results of run-off surveys under rainy weather. PP or PN is the difference between T-P or T-N and S-TP or S-TN respectively. As soluble fractions were measured for the filtrate through GA-100 of 1 μm, the particulate fractions of P and N

Table 5 Standing stock of grass and fallen leaves (biomass) in various fields

Category	Number of samples	Biomass		TN		TP	
		t·ha^{-1}	%	kg·ha^{-1}	%	kg·ha^{-1}	%
Paddy field	4 × 3	5.2	14.6	41	4.0	19	6
Farm land	3 × 3	3.1	5.2	66	4.5	24	4
Pasture	3 × 3	4.1	8.1	76	3.8	71	6
Forest	11 × 3	9.1	20.3	84	8.0	32	37

*Percentage of grass and fallen leaves versus stock in surface soil of 5 cm

Table 6 Accumulation of organic matter, nitrogen and phosphorus in various lands in 0–50 cm soil

	Number of samples	IL		Kj-N		T-P	
		IL(t·ha^{-1})	%	Kj-N (t·ha^{-1})	%	TP (t·ha^{-1})	%
Paddy fields	6	291	10	5.7	13	2.5	12
Dry fields	2	313	11	6.7	23	6.6	34
Orchard	4	439	12	8.4	17	5.9	21
Tea garden	2	848	11	20	26	10.6	9
Pasture	3	530	14	12	20	5.9	25
Forest	3	619	13	9.9	16	1.0	13

The values of %: the part of accumulation in 0–5 cm versus accumulation in 0–50 cm

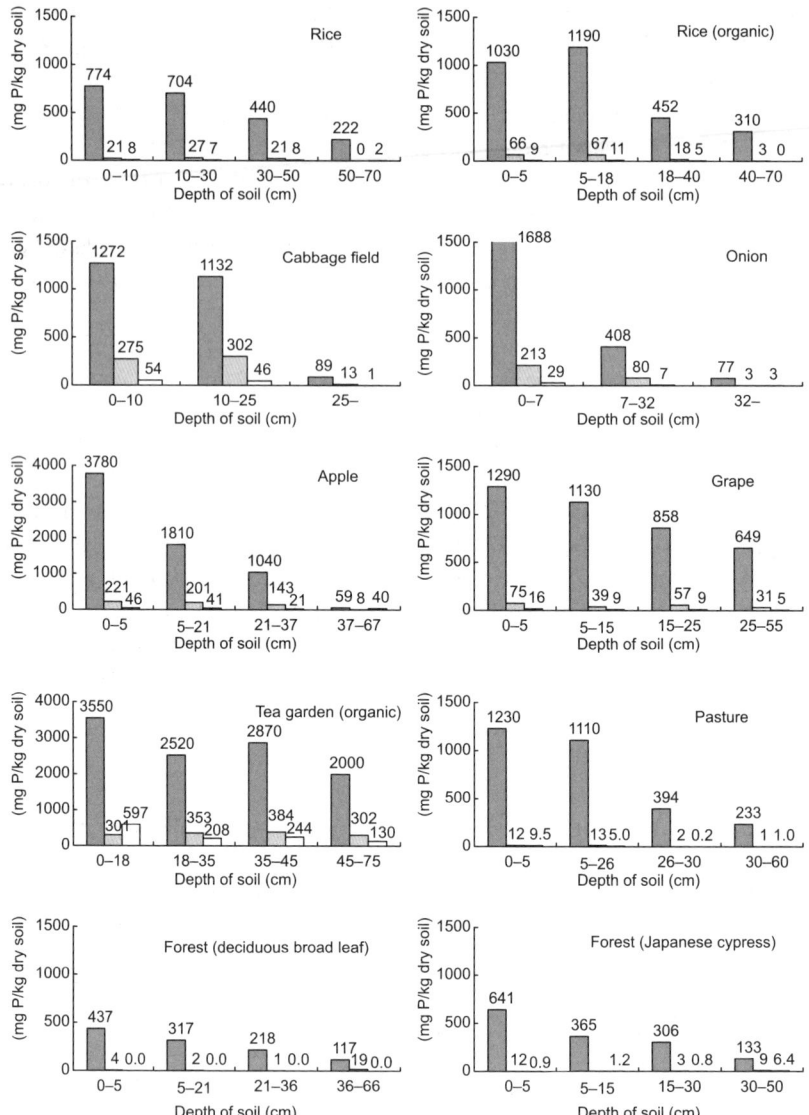

Figure 1 Vertical distribution of TP, available P and water soluble PO$_4$-P in various soils (Bars from left to right side represent TP, available P and soluble P respectively)

were relatively small, 10–50% for P and 10–62% for N. Median diameters of SS ranged from 1.2 to 7.3 μm; generally less than 10 μm.

Table 8 shows regression equations for various relationships among flow rate, suspended solids and size of particulate matter in run-off from farmland. SS loading is generally proportional to specific flow rate q_s or q_s^2. Median diameter tends to be larger as SS concentration becomes larger, and the larger the diameter of particulate matter, the larger becomes the P content in soils.

As the fraction of PP was smaller than expected, we tried to use filters with smaller pore size. Table 9 shows the particulate fraction through the filter of pore size 0.22 μm. As mean values, the fractions increase from 27(15–45) to 64(49–77)% or from 36(4–66) to 54(12–93)% for P. In the second case, the fractions of particulate matter increase from 19(1–48)% to 39(17–74)% for N.

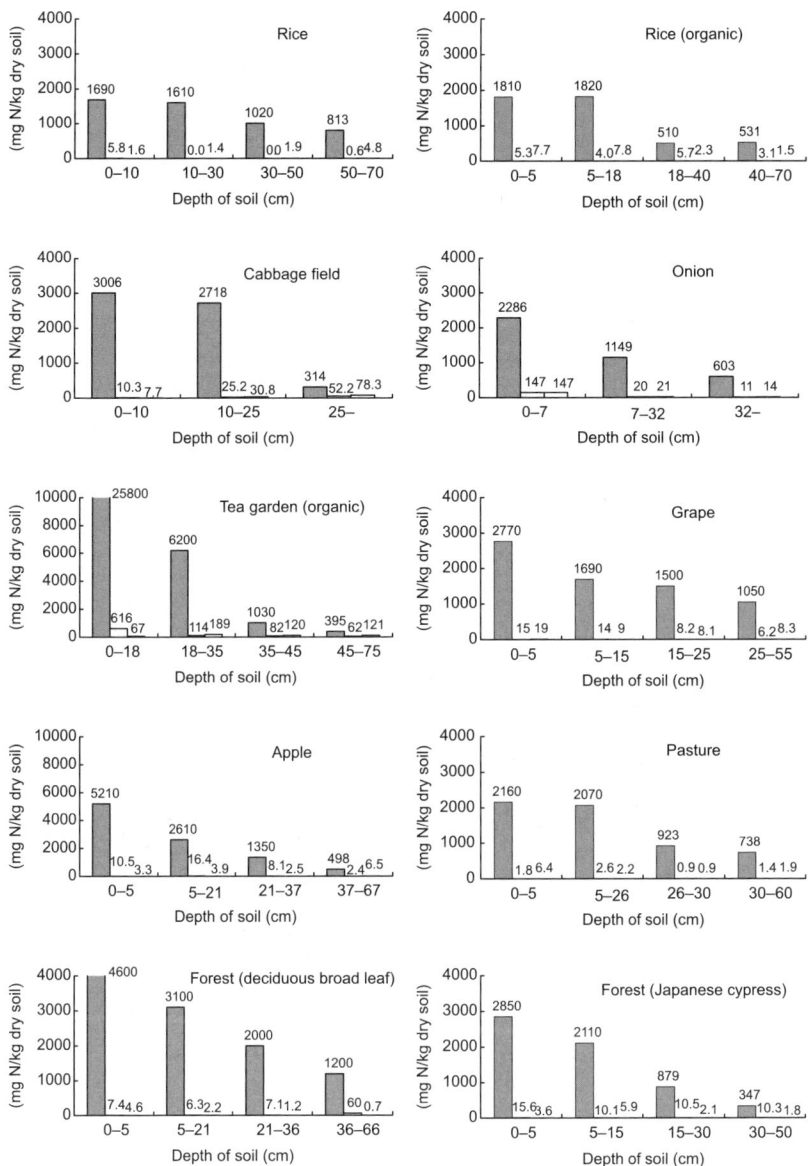

Figure 2 Vertical distribution of Kj-N, available and water-soluble inorg-N in various soils (Bars from left to right side represent Kj-N, available inorg-N and soluble inorg-N)

Figure 3 shows the comparison of run-off events at the beginning of rice cultivation with overlying water and before harvest without overlying water in a paddy field in the campus of Yamaguchi University. It should be noted that turbidity and TP concentration of run-off water on Oct.16 were clearly larger than those on Jun.19; nevertheless the rainfall intensity was rather smaller. This explains the environmental friendliness of paddy fields with overlying water.

We tried to measure the easily-washed-out fraction in soils, considering the importance of the potential source of P. 2 g of surface soil of 1 cm was taken to a vessel with a lid and 1L of distilled water added. After mixing, and leaving for 10 or 35 min, water in the upper 3 cm depth was sampled, by considering precipitation velocity using Stokes' law. Table 10 shows the result of such experiments. Comparing the particle size of

Table 7 Results of survey on run-off from farmland under rainy weather from May to Oct. 2001

	Specific flow (mm h^{-1})	Median diam. (μm)	SS (mg l^{-1})	PO$_4$-P (mg l^{-1})	S-TP (mg l^{-1})	TP (mg l^{-1})	PP/TP (%)	NO$_{2,3}$-N (mg l^{-1})	S-TN (mg l^{-1})	TN (mg l^{-1})	PN/TN (%)
Paddy field A (surface)	1.60	2.3	26	0.29	0.33	0.47	27	0.3	0.6	0.9	32
	2.33	1.7	30	0.22	0.23	0.30	19	0.5	0.5	0.5	18
Paddy field A (submerged)	0.17	1.2	2	0.03	0.04	0.05	17	0.9	1.0	1.1	10
	0.13	0.3	3	0.02	0.02	0.03	18	0.6	0.6	0.7	13
Paddy field F (surface)	5.98	1.3	3	0.14	0.18	0.20	16	2.5	2.7	2.9	12
	4.26	0.8	3	0.19	0.21	0.21	9	5.6	5.2	5.5	11
Dry field A1 (surface)	1.38	3.7	89	6.09	6.45	7.22	10	14.4	16.4	18.8	13
	1.17	1.9	75	1.87	1.95	2.23	9	15.6	17.8	21.9	12
Dry field A1 (submerged)	3.85	3.7	130	4.51	4.80	6.38	24	4.6	6.0	7.8	29
	5.09	1.4	77	2.59	2.78	3.05	19	6.3	5.3	5.2	18
Dry field A2 (surface)	4.38	7.3	1,175	3.59	4.22	7.85	50	2.9	3.9	9.1	62
	2.93	2.2	989	3.64	3.79	4.63	20	3.2	3.7	3.6	30
Dry field F (surface)	1.98	6.0	267	0.85	0.92	1.75	35	2.7	3.4	5.9	36
	2.62	4.1	380	0.33	0.32	1.00	28	3.1	2.9	4.6	29
Dry field F (submerged)	0.31	1.9	17	0.24	0.32	0.38	16	2.7	3.1	3.5	16
	0.23	2.0	19	0.20	0.20	0.21	9	5.5	4.7	4.9	13
Tea garden F (surface)	0.47	4.5	80	0.40	0.47	0.66	30	5.5	7.6	8.9	18
	0.59	5.6	119	0.45	0.53	0.55	25	4.7	5.7	5.6	18

A: Ajisu area, A1: plain field, A2 inclined field, F:Fujigochi area, terrace field; Upper lines: mean values, Lower lines: standard deviation

Table 8 Relationship among flowrate, suspended solids and size of particulate matter in run-off from farmland

	Run-off	Sample	q_s (mm hr^{-1})–SS load (g m^2 hr^{-1})		SS (mg/l)–median diameter (μm)		Median diameter (μm)–P contents (%)	
			Equation	R^2	Equation	R^2	Equation	R^2
Paddy field A	surface	21	$y = 0.0037e^{0.63x}$	0.82	$y = 0.054x + 0.93$	0.89	$y = -0.525x + 7.93$	0.06
	submerged	15	$y = 0.0002x^2 + 0.0018x$	0.25	$y = 0.036x + 1.13$	0.09		
Paddy field F	surface	18	$y = 0.0016x^{070}$	0.50	$y = 0.068x + 1.16$	0.07	$y = 9.2x^{-0.217}$	0.05
Dry field A1	surface	7	$y = 0.163x$	0.99	$y = 0.59 \ln(x) + 0.92$	0.10		
	submerged	16	$y = 0.044x^2 + 0.036x$	0.92	$y = 0.87 \ln(x) + 0.061$	0.12		
Dry field A2	surface	7	$y = 0.227x^{1.95}$	0.58	$y = 0.61x - 0.053$	0.22	$y = 9.0x^{-0.45}$	0.09
Dry field F	surface	30	$y = 0.334x$	0.41	$y = 1.89 \ln(x) - 1.96$	0.79	$y = 6.46x^{-0.356}$	0.39
	submerged	26	$y = 0.0177x^{1.40}$	0.63	$y = 0.050x + 0.217$	0.20	$y = 5.00x^{-0.388}$	0.25
Tea garden F	surface	22	$y = 0.150x^2 + 0.0247x$	0.97	$y = 0.036x + 1.65$	0.58	$y = 5.82x^{-0.359}$	0.35

A: Ajisu area, A1: plain field, A2 inclined field, F: Fujigochi area, terrace field

Table 9 Particulate fraction through different sizes of filter

Experiment	Filter size	Paddy A	Paddy B	Dry field A1	Dry field A2	Dry field B	Tea garden F	Average (%)
1st P	1 μm	15	20	22	45	30	29	27
	0.22 μm	70	55	49	70	61	77	64
2nd P	1 μm	9	51	66	47	4	42	36
	0.22 μm	12	61	93	76	30	51	54
2nd N	1 μm	1	2	26	10	26	48	19
	0.22 μm	74	17	32	19	36	58	39

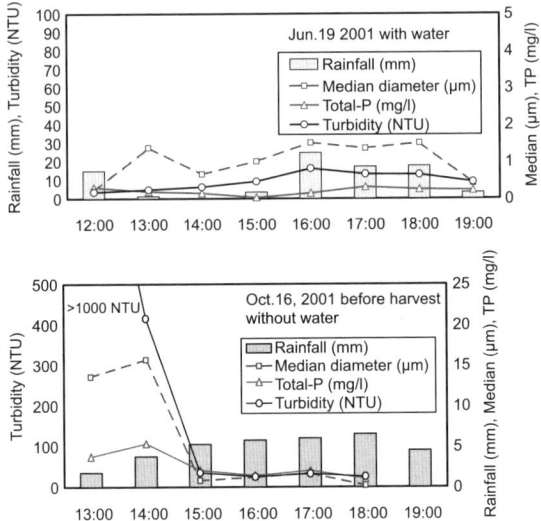

Figure 3 Comparison of two run-off events from paddy field in Yamaguchi Univ.

easily-washed-out fractions targeting less than 5 or 10 μm with those of SS in run-off waters as shown in Table 9, big differences were not found among them except the tea garden, which contained a high amount of organic matter. Moreover, pooling water on the surface of farmland under rain, median diameters of SS were around 5 μm and SS contain 3–4 mg P/g. These fine particles of soil less than 10 μm are incapable of precipitating soon after re-suspension by the turbulence of rain drops. Easily-washed out fractions occupy 3–20% of farmland soil. Anyhow, these fine particles concentrating P by 3–5 mg/g should be monitored carefully hereafter.

Conclusions

The main results obtained in this research are as follows.
- The accumulation of N and P in surface (5 cm) layers of farmland soils was 0.95–2.12 t N ha^{-1} and 0.22–1.06 t P ha^{-1}. Ratios of N/P were rather low at 2.6, 1.0, 1.4, 4.3 and 1.7 respectively for paddy fields, dry fields, orchards, tea gardens and pastures. The value for forests was as high as 12.7.
- Available fractions in surface soil are 0.4–4% for P, 0–0.6% for N in the case of paddy fields, and 2–8.5% for P, 0.1–22.7% for N in the case of dry fields. Generally, the fraction of available P tends to be larger than that of N. Available and soluble P distribute more highly near the surface, whereas those for N sometimes distribute more highly in deeper layers.
- The accumulation of P and N in 0–50 cm layers is equivalent to the amount of chemical fertilizer used for 56 years and 72 years, assuming fertilizer supply to be 44 kg P

Table 10 Comparison of P, easily-washed-out fraction in soils and SS in run-off water

	TP in soil of top 1 cm (mg g^{-1})	Exp.1 expecting less than 10 μm			Exp.2 expecting less than 5 μm			Run-off water	
		Median diameter (μm)	PP/SS (mg g^{-1})	Easily-washed-out fraction (%)	Median diameter (μm)	PP/SS (mg g^{-1})	Easily-washed-out fraction (%)	PP/SS (mg g^{-1})	Number of samples
Paddy field A	0.81	7.1	3.4	20	4.7	3.7	16	5.5	34
Dry field F	0.50	4.5	2.7	16	5.3	2.9	13	3.9	38
Dry field A2	0.81	6.1	2.1	5	5.5	3.0	3	3.2	8
Tea garden F	1.39	4.9	20.4	15	1.4	8.0	3	2.3	22

and 85 kg N ha^{-1}year^{-1} in the case of paddy fields, and 56 years and 51 years assuming 80 kg P and 175 kg N ha^{-1}year^{-1} in the case of dry fields respectively.
- Median diameter of soil particles in run-off water was less than 10 μm. The values are 1.2–2.3 μm for paddy fields and 1.9–7.3 μm for dry fields, and tend to be smaller for subsurface run-off and larger for inclined dry fields.
- There were seen certain relationships between specific flow rate and SS loading, SS concentration and median diameter, and median diameter and P contents in SS. Therefore, it is possible to estimate particulate P loading from specific flow rate in each farmland. The overlying water on paddy fields was very effective to decrease SS loading in rainy weather.
- A simple method of measuring the easily-washed out fraction of P was proposed. Targeting the particulate matter less than 10 μm, median diameter ranged from 4.5 to 7.1 μm, and the fractions were estimated to be 5–20% for the surface soils in various farmlands. Fine particles of soil containing concentrated phosphorus should be carefully attended because of the eutrophication factor.

Acknowledgement

The authors thank Foundation of River & Watershed Environment Management Japan for their financial support.

References

Beckmann, M.E., Krogstad, T. and Sharpley, A.N. (2003). Diffuse Pollution Conference Dublin 3, 163–169.

Sharpley, A. (1995). Fate and transport of nutrients: Phosphorus, RCA Publication Archive, http://www.nrcs.usda.gov/technical/land/pubs/wp08text.html.

Shi, X., Ukita, M., Higuchi, T., Imai, T. and Sekine, M. (2002). Evaluation of eutrophication potentials of soils representing non-point sources. *J. of Japan Society on Water Environment*, **25**(2), 112–118.

Ukita, M. and Nakanishi, H. (1999). Pollutant load analysis for the environmental management of enclosed sea in Japan. In *Proc. Joint Conf. MEDCOAST '99 and EMECS '99 in Antalya Turkey*, 1227–1238.

Ukita, M. and Prasertsan, P. (2002). Present state of food and feed cycle and accompanying issues around Japan. *Water Science and Technology*, **45**(12), 13–21.

Weld, J.L., Sharpley, A.N., Beegle, D.B. and Gburek, W.J. (2001). Identifying critical sources of phosphorus export from agricultural watersheds. *Nutrient Cycling in Agroecosystems*, **59**, 29–38.

Decrease in herbicide concentrations and affected factors in lagoons located around Lake Biwa

M. Sudo*, M. Nishino** and T. Okubo**

*School of Environmental Science, The University of Shiga Prefecture, 2500 Hassaka Hikone, 522-8533, Japan (E-mail: sudo@ses.usp.ac.jp)
**Lake Biwa Environmental Research Institute, 5-34 Yanagihira 522-0022, Japan (E-mail: nishino@lberi.jp; okubo-t@lberi.jp)

Abstract The contamination levels and changes in the concentrations in four lagoons around Lake Biwa of paddy-use herbicide were studied. Four lagoons, Sone-numa (52 days of HRT (hydraulic residence time) estimated from the lagoon volume and the average discharge at the outlet, 21 ha area), Yanagihira-ko (40 days, 5.0 ha), Noda-numa (11 days, 6.0 ha), and Iba-naiko (2 days, 55.5 ha), were selected as monitoring sites. Intensive water sampling was carried out once a week from May to June at the outlet of each lagoon. Although twelve of the monitored herbicides were detected, the maximum concentrations did not exceed the guidelines for water-supply law in Japan. The relation between half-lives in herbicide concentrations and characteristics of a lagoon such as HRT and chlorophyll-a concentrations were examined. The shorter half-lives of herbicide concentrations in lagoons with shorter HRT means that replacement by influent water effectively decreased the pesticide concentrations. Shorter half-lives in lagoons with high chlorophyll-a concentrations between the lagoons with similar HRT suggest that biological degradation during the residence time worked more efficiently in the lagoon with high chlorophyll-a concentrations.
Keywords Half-life; herbicide; lagoons; Lake Biwa; water contamination

Introduction

Located on central Honshu Island, Lake Biwa has the largest surface area in Japan (674 km^2) and is used by 14 million people in the Kinki district as potable water and for industrial, recreational, and agricultural purposes. Although more than 30 lagoons have existed around Lake Biwa (1 ha – 1,150 ha), half of them have been altered entirely or partially, being established as drainage areas for paddy-fields or upland fields between 1944–1971 (LBRI, 1983). Recently, depression of agriculture and reevaluation of the environmentally beneficial aspects of lagoons have resulted in the development of several lagoon regeneration projects. Lagoons provide habitat for aquatic plants, microorganisms, insects, birds, and fishes, including some endemic species. Lagoons have also been recognized as active ecosystem components and are thought to be potential sites for the removal of nutrients before their flow into lakes (Fukushima et al., 1989; Toda et al., 1994; Okamoto et al., 1997; Okubo, 1998; Okubo et al., 2000). Pesticides, which may have a detrimental impact on human health and natural ecosystems, are a major source of non-point pollution in the Lake Biwa basin (Sudo et al., 2002a,b) and are expected to be mitigated in lagoon areas during their retention. There are several studies reporting removal of pesticides used in upland fields (Kadlec and Hey, 1994; Detenbeck et al., 1996; Schulz and Peall, 2001; Moore et al., 2001; Kao et al., 2001, 2002; Stearman et al., 2003); however, there have been few published studies of paddy use pesticides in lagoons around Lake Biwa (Sudo et al., 2003; Hinokio et al., 1996). The objective of this study was to evaluate the actual environmental risk of pesticides and their potential for removal

doi: 10.2166/wst.2006.046

in lagoons under realistic environmental conditions. Although the pesticide budget based on long-term monitoring is essential to evaluating the accurate mass removal rate and mass removal efficiency, this study focused on pesticide concentrations at lagoon outlets. The comparison of pesticide concentrations among lagoons provides the relative contamination levels. The influence of residence time and biosphere activity as indicated by chlorophyll-a concentrations on half-lives in pesticide concentrations were examined to evaluate the lagoon buffer function and the pesticide-removing capabilities of a site.

Materials and methods

Study area

Prior to this study, pesticide concentrations were monitored in 20 lagoons around Lake Biwa once a month from May to August in 2001 (LBRI, 2002). Four of these 20 lagoons, Sone-numa, Yanagihira-ko, Noda-numa, and Iba-naiko, were selected as monitoring sites in this study on the basis of the maximum concentrations and the contaminations two or three months after the application period. Maps of the lagoons and the locations of the sampling sites are shown in Figure 1. Excluding the Noda-numa, water in the lagoons drains from a single outlet, although some rivers or canals are inflows to the lagoons. The areas and average water depths are also shown in Figure 1. The hydraulic residence time (HRT) estimated from the lagoon volume and the average discharge at the outlet, assuming no short-circuiting, are shown in Table 2.

Rice is a major crop in the catchment areas of lagoons, and related pesticides are drained through a discharge river and canal. In the majority of paddy fields, rice seedlings are transplanted between the end of April and the beginning of May, and are harvested until late September. Pre-emergence herbicides are applied by sequential treatments or a one-shot treatment up to 3 weeks after transplantation. In a sequential treatment, a first-stage herbicide is applied up to 5 days after transplanting, followed by second-stage herbicide application between 2 and 3 weeks after that. In a one-shot treatment, a one-shot herbicide is applied once between 3 and 14 days after

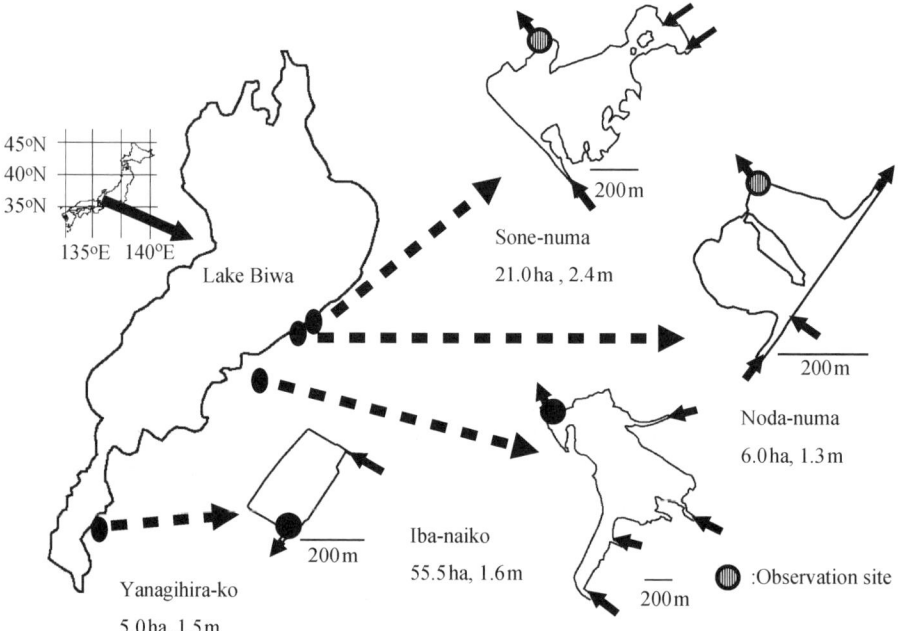

Figure 1 The locations, areas, and average water depths of observation sites in the lagoons

transplanting. A second-stage herbicide is often applied additionally when the one-shot herbicide does not work efficiently.

Sample collection and sample analysis

Water samples were collected at the single outlet of three lagoons and the outlet at the bottom of the Noda-numa once a week from May to early July. Sampling was also carried out every 2 weeks at the beginning of April and in the latter half of July. Water temperature, pH, conductivity, and discharge at the sampling site were also measured on-site. Water samples were analyzed for pesticide and chlorophyll-a concentrations. Twelve herbicides were analyzed (Table 1) according to our previously described method (Sudo et al., 2004). Chlorophyll-a concentrations were measured using 90% acetone as the extract solvent according to the method of SCOR/Unesco (1966).

Results and discussion

Pesticide contamination in lagoons

The maximum concentrations of pesticides in the lagoons are shown in Table 1. The detection frequencies, calculated via 12 observations, are shown in the same table. Although the maximum concentrations and detection frequencies may be affected by the differences in application amounts in the catchment area and by the timing between sampling and the pesticide application period, all of the herbicides monitored in this study were detected more or less in four lagoons. Among the four lagoons, the concentrations in the Sone-numa and the Noda-numa were relatively high, suggesting that these lagoons may receive more agricultural non-point runoff loads and/or be less diluted with discharge originating from non-treated areas.

Among the pesticides detected in the four lagoons, the maximum concentrations of daimuron, mefenacet, and simetryn exceeded 5 μg/L. The water-supply law, which is designed to supply secure drinking water by securing water sources, was amended in 2004 and established guidelines for 101 pesticides. The values specified in the guideline are in general lower than the reported LC_{50} value for aquatic species such as carp,

Table 1 The maximum concentrations of herbicides and detection frequencies in lagoons in 2003

	Sone-numa		Yanagihira-ko		Noda-numa		Iba-naiko		Guideline*** μg/L
	Max.* μg/L	Freq.** %	Max. μg/L	Freq. %	Max. μg/L	Freq. %	Max. μg/L	Freq. %	
First-stage herbicide									
Bromobutide	4.72	83	1.12	83	2.90	83	1.29	75	40
Pretirachlor	0.64	58	0.91	67	1.49	58	1.94	58	40
Tenylchlor	0.61	92	0.19	75	0.59	83	0.17	100	200
One-shot herbicide									
Daimuron	8.13	92	5.11	92	7.27	92	5.82	92	800
Mefenacet	3.84	75	0.59	83	5.25	83	3.02	83	0
Thiobencarb	0.29	58	0.02	17	0.83	75	0.52	92	20
Esprocarb	0.13	83	0.08	33	0.78	83	0.77	83	10
Dimepiperate	0.14	17	0.05	17	0.54	75	0.36	33	3
Pyributicarb	0.12	75	0.12	67	0.09	50	0.13	67	20
Second-stage herbicide									
Simetryn	4.42	83	0.76	92	7.98	92	1.54	75	30
Benfresate	0.51	67	0.14	58	3.90	67	0.68	83	–
Molinate	0.32	75	0.11	50	1.28	92	0.18	50	5

*Maximum concentration
**Detection frequency
***Guideline for water-supply law in Japan

rainbow trout, and daphnia (Kanazawa, 1996). The maximum concentrations of mefenacet, bromobutide, pretirachlor, dimepiperate, simetryn, and molinate were within one order of magnitude lower than the guideline values. Those of other herbicides were beyond two orders of magnitude lower. Herbicide concentrations in influent water would be higher than that in lagoons; lagoons were effective for decreasing peak concentrations in contaminated water and decreasing toxicity for aquatic species by dilution and degradation.

Seasonal variations in herbicide concentrations

The following results and discussion focus on the concentrations of bromobutide, daimuron, mefenacet, and simetryn because of their high concentrations in the four lagoons. Bromobutide is applied between late April and mid May, as it is contained in both first-stage herbicides and one-shot herbicides. The application periods for daimuron and mefenacet are between early and late May, while that for simetryn is between late May and early June. Seasonal variations in the four lagoons are shown in Figure 2. In the lagoons with relatively short HRT, the Iba-naiko and the Noda-numa, the contamination levels of herbicides increased with almost direct correspondence to the application timing. The fall of concentrations occurred following the peak and remained at less than 1 μg/L for daimuron and 0.2 μg/L for the three herbicides in July.

In the Sone-numa and the Yanagihira-ko, the appearance of detectable contaminations tended to be delayed by approximately 1–2 weeks compared with the Iba-naiko and the Noda-numa, although the herbicides were applied at almost identical times in the catchment area. An exception with regard to the appearance of bromobutide concentrations may be due to the longer sampling interval in April. The high contamination periods also tended to be delayed and prolonged in these lagoons. The lag time may have been due to their long HRT. Following the peak concentrations, the concentrations gradually decreased and remained at 2 μg/L for daimuron and at more than 0.5 μg/L for the three herbicides in July.

Seasonal variations in chlorophyll-a concentrations

Seasonal variations in chlorophyll-a concentrations are shown in Figure 3. Concentrations in the Sone-numa ranged between 5 and 15 μg/L from April to early June, and increased to more than 30 μg/L in mid- and late June. A temporal increase was also observed at the Noda-numa in the middle of June. The concentrations in the Yanagihira-ko and the Iba-naiko remained relatively stable from April to June, ranging from 5–15 μg/L and 10–20 μg/L. In all lagoons, the concentrations decreased to less than 5 μg/L after mid-July.

Elimination rate of pesticides at the lagoon outlets

The lumped first-order decay function is used to relate the decrease in concentrations at the outlet in a lagoon as follows:

$$dC/dt = -kC$$

where C is the pesticide concentration and k is the elimination rate constant. Because herbicides are applied almost at the same time during a fixed application period, pesticide inputs from rivers or drains are limited at the period. Thus, the sampling date on which the maximum concentrations were determined was substituted as $t = 0$. The data set was analyzed up to early July. Table 2 shows the half-lives of pesticide concentrations calculated from gradients of the regression curve. Other parameters such as chlorophyll-a concentration, water temperature, pH and conductivity are also shown in

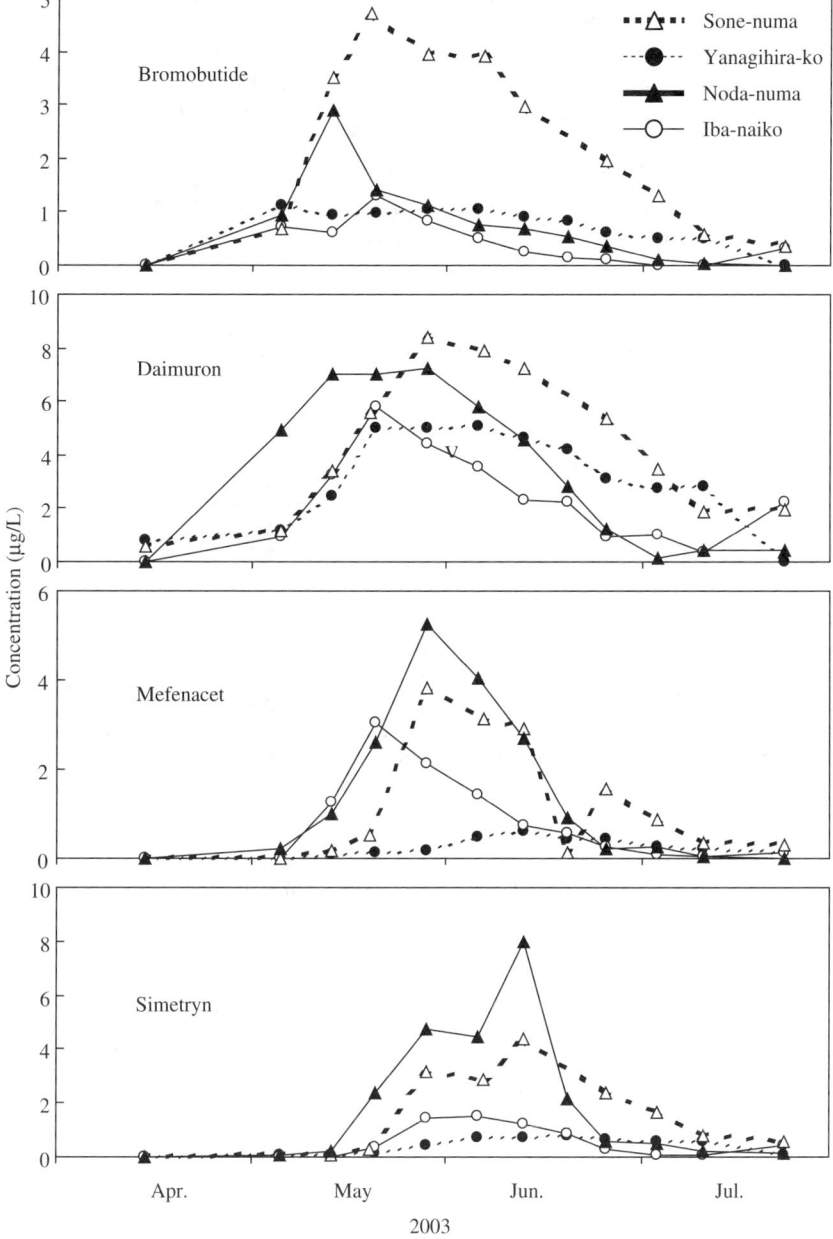

Figure 2 Seasonal variations in herbicide concentrations in the four lagoons

the same table. The chlorophyll-a concentrations were relatively high in the lagoons with higher conductivity, suggesting these lagoons may receive more nutrients from their catchment area. There are no significant differences in water temperature and pH among lagoons.

As shown in Table 2, the half-lives of herbicides in lagoons with relatively shorter HRT, the Noda-numa and the Iba-naiko, were estimated to be 5–10 days, while in those with longer HRT, the Sone-numa and the Yanagihira-ko, the half-lives were 20–30 days. The shorter half-lives in lagoons with shorter HRT means that replacement by influent water efficiently promotes pesticide disappearance.

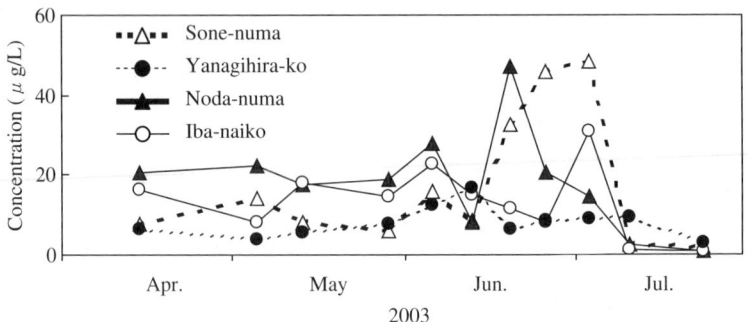

Figure 3 Seasonal changes in chlorophyll-a concentrations in the four lagoons

Table 2 HRT, half-life of herbicides and chlorophyll-a concentration in lagoons

			Sone-numa	Yanagihira-ko	Noda-numa	Iba-naiko
Half-life	Bromobutide	days	18.0	33.8	10.9	9.5
	Daimuron		24.5	28.9	10.2	14.5
	Mefenacet		14.7	18.9	11.7	8.6
	Simetryn		35.0	46.3	4.8	6.5
HRT		days	52	40	11	2
Chlorophyll-a*		µg/L	16.3	7.7	20.4	12.3
Water Temp.*		°C	24.6	22.8	23.0	22.1
pH*			7.75	7.70	7.60	7.96
Conductivity*		mS/m	19.3	18.2	23.6	13.8

*Average between May and June

If the half-lives of pesticides depend only on the HRT, half-lives in the Noda-numa could be expected to be longer than those of the Iba-naiko. But there was no significant difference in the half-lives between the lagoons. The fate of pesticide flow into a lagoon is degradation, outflow, internal storage and vaporization. Among them, internal storage can be considered negligible as the herbicides monitored in this study exist mainly in the aqueous phase due to their high water solubility and relatively low octanol-water partition coefficients. Vaporization can also be negligible because of their low Henry's law constant. Thus, the fate contributing to pesticide elimination during the residence time is degradation. Although each degradation process such as biological degradation and photo-degradation was not examined, they were lumped together into the chlorophyll-a concentration as an indicator of degradation. Chlorophyll-a concentrations represent the phytoplankton biomass, and the abundance of phytoplankton may depend on the abundance of aquatic biomass. Higher chlorophyll-a concentrations may bring advantages for the biodegradation process; however, they may reduce the ability of photo-degradation because of a decrease in transparency. Previous studies suggest that atrazine (McKinlay and Kasperek, 1999; Kao et al., 2001; Kao et al., 2002; Anderson et al., 2002), azinphosmethyl, chlorpyriphos and endosulfan (Schulz and Peall, 2001) are degraded in wetland by biodegradation. As shown in Table 2, the shorter half-lives in the Noda-numa than in the Iba-naiko may be due to higher chlorophyll-a concentrations, suggesting that biological degradation during the residence time in the Noda-numa worked more efficiently to reduce pesticide concentrations than in the Iba-naiko. The difference between expected half-lives based on HRT and the observed ones in the Sone-numa and Yanagihira-ko could be explained in the same manner.

Conclusions

Herbicide concentrations were monitored at four lagoons around Lake Biwa during a 4-month period in 2003. Although the concentrations of herbicide were low, long-term effects on the aquatic environment are of concern in lagoons with long HRT because of the herbicide residue. Our results also suggest that lagoons with longer HRT and higher chlorophyll-a concentrations may reduce pesticide inputs to the lake. These conditions, however, are favorable to primary production, and they may lead to adverse effects on the water quality of a lagoon. It is therefore necessary to determine the most effective conditions and design best management practices for lagoons.

Acknowledgements

The authors express our appreciation to Tatsuhito Ogata and Misachi Shikimoto for their assistance in field sampling and laboratory analyses.

References

Anderson, K.L., Wheeler, K.A., Robinson, J.B. and Touvinen, O.H. (2002). Atrazine mineralization potential in two wetlands. *Water Res.*, **36**, 4785–4794.

Detenbeck, N.E., Hermanuts, R., Allen, K. and Swift, M.C. (1996). Fate and effects of the herbicide atrazine in flow-through wetland mesocosms. *Environ. Toxicol. Chem.*, **15**, 937–946.

Fukushima, T. and Iwata, Y. (1994). On water purification by aquatic plant (lotus) within a farm pond where domestic sewage is mixed. *Trans. Japanese Society Irrigation, Drainage Reclamations Engineering*, **142**, 99–105 (in Japanese).

Hinokio, R., Jiku, F. and Yokota, K. (1996). Fundamental study on pesticide storage and behavior in a lagoon around Lake Biwa. *Env. Conservation engineering*, **25**(8), 434–437 (in Japanese).

Kadlec, R.H. and Hey, D.L. (1994). Constructed wetlands for river water quality improvement. *Wat. Sci. Tech.*, **29**(4), 159–168.

Kanazawa, J. (1996). *Environmental Science of Pesticides*, Godo shuppan, Tokyo (in Japanese).

Kao, C.M., Wang, J.Y. and Wu, M.J. (2001). Evaluation of atrazine removal processes in a wetland. *Wat. Sci. Tech.*, **44**(11–12), 539–544.

Kao, C.M., Wang, J.Y., Dhen, K.F., Lee, H.Y. and Wu, M.J. (2002). Non-point source pesticide removal by a mountainous wetland. *Wat. Sci. Tech.*, **46**(6–7), 199–206.

LBRI (1983). Lagoons – its ecological function. *Lake Biwa Res. Inst. Bull.*, **2**, 46–54 (in Japanese).

LBRI (2002). *Characteristics of nutrient and pesticide concentration in lagoons around Lake Biwa*, Research Report 2002-01, Lake Biwa Research Institute, Otsu, Japan (in Japanese).

McKinlay, R.G. and Kasperek, K. (1999). Observations on decontamination of herbicide-polluted water by marsh plant systems. *Water Res.*, **33**, 505–511.

Moore, M.T., Rodgers, J.H. Jr., Cooper, C.M. and Smith, S. Jr. (2001). Mitigation of metolachlor-associated agricultural runoff using constructed wetlands in Mississippi, USA. *Agric. Eco. Environ.*, **84**, 169–176.

Okamoto, Y., Nakamura, K. and Kobayashi, H. (1997). Experiment for direct purifying on irrigation pond utilizing fallow paddy fields that were not expected to produce rice profitably. *Trans. Japanese Society Irrigation, Drainage Reclamations Engineering*, **187**, 151–160 (in Japanese).

Okubo, T. (1998). Water purification in wetlands and lagoons. *J. Water and Waste*, **40**(10), 35–45 (in Japanese).

Okubo, T., Tsujimura, S. and Sudo, M. (2000). Removal of pollutant loading in a lagoon. *Lake Biwa Res. Inst. Bull.*, **18**, 36–48.

SCOR/UNESCO Working Group 17 (1966). *Determination of photosynthetic pigments in sea water*, UNESCO, Paris, p. 69.

Schulz, R. and Peall, S.K.C. (2001). Effectiveness of a constructed wetland for retention of nonpoint-source pesticide pollution in the Lourens River Catchment, South Africa. *Environ. Sci. Technol.*, **35**, 422–426.

Stearman, G.K., George, D.B., Carlson, K. and Lansford, S. (2003). Pesticide removal from container nursery runoff in constructed wetland cells. *J. Environ. Qual.*, **32**, 1548–1556.

Sudo, M., Kunimatsu, T. and Okubo, T. (2002a). Concentration and loading of pesticide residues in Lake Biwa basin (Japan). *Water Research*, **36**(1), 315–329.

Sudo, M., Okubo, T., Kunimatsu, T., Ebise, S., Nakamura, M. and Kaneki, R. (2002b). Inflow and outflow of agricultural chemicals in Lake Biwa. *Lakes and Reservoirs*, **7**, 301–308.

Sudo, M., Okubo, T. and Kunimatsu, T. (2003). Herbicide loadings from paddy fields and purification effect in lagoon. *J. of Japanese society of Irrigation, Drainage and Reclamation Engineering*, **223**, 71–77 (in Japanese).

Sudo, M., Kawachi, T., Hida, Y. and Kunimatsu, T. (2004). Spatial distribution and seasonal changes of pesticides in Lake Biwa, Japan. *Limnology*, **5**, 77–86.

Toda, H., Matsumoto, E., Miyazaki, T., Shibano, K. and Kawashima, H. (1994). Nitrate disappearance in an irrigation pond. *Soil Science and Plant Nutrition*, **65**, 266–273 (in Japanese).

Estimation of pesticide runoff from paddy fields to rural rivers

A. Numabe and S. Nagahora

Hokkaido Institute of Environmental Sciences, Kita-19 Nishi-12, Kita-ku, Sapporo, Hokkaido, 060-0819, Japan

Abstract The runoff characteristics of pesticides from paddy fields to rural rivers were investigated over a period of three years in Hokkaido Prefecture, Japan. High pesticide concentrations were usually observed in rivers during pesticide application periods. In one year, the growth of rice seedlings slowed down after transplantation owing to low temperatures and lack of sunshine, and many farmers delayed herbicide application. In that year, high-concentration runoff of herbicides in rivers was observed 1–3 weeks later than in average years. The pesticide runoff rates ranged from 0.3% for fenthion to 42% for benfuresate. The runoff rates of pesticides applied post-flood were large. Furthermore, the larger the water solubility of the pesticide, the larger the runoff rate. The highest concentrations of herbicides in paddy water were observed on the day of application or 1–2 days later, and the concentrations decreased exponentially afterwards. The half-lives of the herbicides ranged from 1.2 days for pretilachlor and esprocarb to 5 days for simetryn; the concentrations of the herbicides in paddy water had decreased to 1/10 of their initial concentrations by about 7 days after application. Therefore, the runoff amounts of pesticides from paddy fields could be decreased by improving irrigation-water management.

Keywords Herbicide concentrations; irrigation water management; paddy fields; paddy water; pesticide runoff; runoff rate

Introduction

Approximately 500 chemicals are currently used as pesticides in Japan. The pesticides, both singly and in combination, are diluted and formulated as dusts, granules, emulsions, wettable powders, and so on. About 350,000 tons of pesticide formulations are purchased every year. Ninety-eight per cent of all the formulations sold are for agriculture and forestry uses, and 50% of those are applied to paddy fields. Pesticides applied to paddy fields run off into environmental waters more easily than pesticides applied to upland fields, because paddy water is discharged directly into rivers and lakes.

Irrigation water for paddy fields is usually managed artificially. Therefore, if irrigation discharges are carefully timed after pesticide applications, high-concentration runoff of pesticides can be controlled. However, most previous studies have indicated that high concentrations and large loadings of pesticides in river and lake waters are observed immediately after pesticide application and during rainfall-runoff events (Numabe et al., 1992; Ebise et al., 1993; Nagafuchi et al., 1994; Sasagawa et al., 1996). In agricultural regions where most farmers have office or factory jobs, pesticides are applied to paddy fields during weekends. Therefore, high concentrations and large loadings of pesticides are observed in drainage rivers at the beginning of the week (Numabe et al., 1992). These findings clearly indicate that water management in paddy fields in these regions is inappropriate with regard to pesticide runoff.

Many kinds of pesticides are used for rice plantings, and the pesticides used and the application periods differ from region to region, because of the significant climatic differences between the northern and southern regions of Japan. Although many studies of pesticide runoff from paddy fields have been conducted, results from cold regions are limited.

In this report, we describe the runoff characteristics of pesticides as determined from weekly river observations in typical rice-planting regions in Hokkaido Prefecture, which is located in the northernmost area of Japan. We also report on changes in herbicide concentrations in paddy waters over time.

Observation areas and methods

River water samples were collected on Monday or Tuesday from early May to late August from 1997 to 1999 from the downstream areas of two tributaries of the Chitose River: the Old-Yubari River (St. 1) and the Horomui-Unga River (St. 2) (Figure 1). The Chitose River is one of the main tributaries of the Ishikari River, which is the largest river in Hokkaido, and supplies tap water to the city of Ebetsu and the town of Nanporo at sites 1 and 2. The Old-Yubari River and the Horomui-Unga River are 18.5 and 9 km in stream length and have 100 and 34.8 km^2 watershed areas, respectively. Rice cultivation areas ranged from 3,690 ha (1997) to 3,220 ha (1999) at St. 1 and from 1,430 ha (1997) to 1,200 ha (1999) at St. 2. In these watersheds, rice seedlings were transplanted from mid to late May, and in particular between 20 and 25 May. Because most farmers in this region are full-time farmers, their farm work is not concentrated at the weekends.

When river water samples were collected, pH, water temperature, and river flow were also measured. River flow was estimated by measuring the cross-sections of the rivers and the water velocity with an electromagnetic current meter.

Changes in the herbicide concentration of paddy water were investigated in three paddies (Figure 2). Paddies A and B had the same dimensions, 176 m × 28.5 m (50 a area); paddy C was 58 m × 29.5 m (17 a). Rice was cultivated by means of a standard method in these paddies. Paddy water samples were collected daily for several days after herbicide applications and thereafter at 2- to 7-day intervals. There were three sampling sites in paddies A and B (A/B-1–3) and four in C paddy (C-1–4).

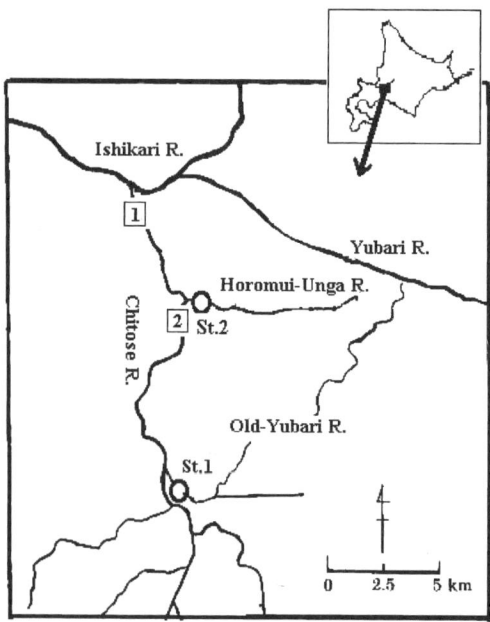

Figure 1 Location of studied watershed

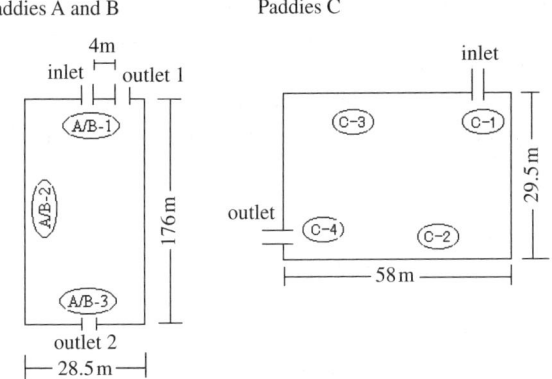

Figure 2 Schematic of observation paddies

The pesticides were analyzed using the methods described previously (Inoue *et al.*, 2002). However, Sep-Pak Plus PS-2 or tC-18 cartridges (Waters) for solid-phase extraction were used, and only the filtrates were analyzed. Twelve herbicides, three insecticides, four fungicides and three oxidation metabolites of pyributicarb (Pyr) and fenthion (MPP) were analyzed. Herbicides comprised pretilachlor (Pre), Pyr, esprocarb (Esp), mefenacet (Mfn), thenylchlor (Tnl), benfuresate (Bnf), thiobencarb (Thio), molinate (Mol), simetryn (Sim), daimuron (Dmr), bensulfuron-methyl (Bnm), and bentazone (Btz); insecticides comprised fenobucarb (BPMC), fenitrothion (MEP), and MPP; fungicides comprised edifenphos (EDDP), fthalide (Ftl), pyroquilon (Prq), and metalaxyl(Mtl); and the oxidation metabolites of Pyr and MPP comprised pyributicarb-oxon (Pry-ox), fenthion-sulfoxide (MPP-sfox), and fenthion-sulfone (MPP-sfon). Three of the herbicides, Dmr, Bnm, and Btz, were analyzed by high-performance liquid chromatography (2690 Separations Module equipped with a 2487 Dual λ Absorbance Detector, Waters). The others were analyzed on a gas chromatograph (HP-6890, Hewlett-Packard) equipped with a mass spectrometer (HP-5973, Hewlett-Packard).

Pesticide application methods

There are two application methods for herbicides: the sequential treatment method and the one-shot treatment method. In the sequential treatment method, the herbicides are applied sequentially in three stages, depending on the growth of weeds.

(1) In the first-stage, the initial herbicides Pre and Pyr are applied until either 4 days before or 7 days after transplantation.
(2) In the second-stage, the mid-term herbicides Mol, Sim, and Thio are applied from 15 to 30 days after transplantation.
(3) In the third-stage, if necessary, the later-term herbicide Btz is applied beginning 30 days after transplantation.

In the one-shot treatment method, a one-shot herbicide is applied once during the period from 3 to 15 days after transplantation. In these watersheds, nine herbicides (not including Mol, Sim, and Btz) of twelve herbicides mentioned above were used as one-shot herbicides.

Pre and Pyr are usually applied both before and after transplantation, and Thio is usually applied as a one-shot herbicide and a mid-term herbicide. However, in these watersheds, Pyr was predominantly applied as an initial herbicide before transplantation, and Thio was applied as a mid-term herbicide.

Insecticides and fungicides were also applied to the paddy water from mid June to early July and to the rice vegetation after the paddy water was discharged from late July to mid August, depending on the forecast and the occurrence of diseases and insect pests. However, one fungicide, Mtl, was applied to the rice seedling box before transplantation and was not directly applied to paddy fields.

The amounts of pesticides applied on the studied watersheds were estimated based on the amounts sold by the agricultural cooperatives, which are the main pesticide distributors in these areas.

Results and discussion

The concentration changes of the main pesticides at St. 2 in 1998 are shown in Figure 3, and the highest concentrations and the detection dates of 19 pesticides and three metabolites are shown in Table 1.

High-concentration runoffs of Pyr and Pre were observed in late May, during the rice-transplantation period. The concentration of Pre then decreased but increased again in early June, and a second concentration peak was observed during early to mid June. In this watershed, less Pre was applied as an initial herbicide before transplantation than was applied as a one-shot herbicide (for example, initial, 128 g/ha; one-shot, 229 g/ha, in 1998). But the highest concentration was observed in late May, before transplantation. The one-shot herbicides Esp and Bnf were detected beginning in late May, with declining concentrations after mid June (similar to the declines in Pre during the same period). The high-concentration runoff period of the mid-term herbicide Mol was found to be mid June to early July, and the late-term herbicide Btz was observed after July.

The runoff periods of the herbicides shifted slightly over the years. The runoff patterns of Pre, Mfn, and Mol after three years' investigation are shown in Figure 4. The highest concentrations of all the pesticides monitored in 1997 were observed later than in the other two years. In particular, the runoff period of Mol in 1997 was about 3 weeks later than in 1998 and 1999. In 1997, the growth of rice seedlings after transplantation was reduced due to low temperatures and lack of sunshine; as a result, it seems that many farmers delayed their herbicide applications.

Ebise *et al.* reported that the largest loadings of one-shot herbicides in Ibaraki Prefecture were found between 1 and 2 weeks after transplantation (Ebise *et al.*, 1993). In our observations, the high-concentration period was observed between 2 and 3 weeks after transplantation. It appears that herbicides in our observation areas were applied later than in Ibaraki Prefecture as a result of reduced rice seedling growth after transplantation, due to low (10–15 °C) irrigation water temperature.

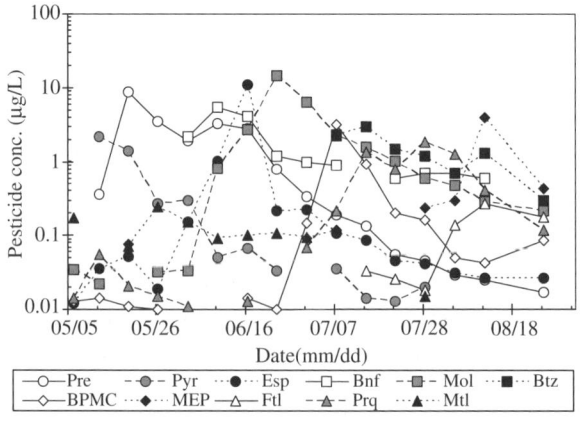

Figure 3 Changes in pesticide concentration at St. 2 in 1998

Table 1 Maximum pesticide concentrations (ng/ml) in river water and detection dates (in parentheses)

	Year	Herbicides										
		Pre	Pyr	Pyr-ox	Esp	Mfn	Dmr	Tnl	Bnf	Bnm	Thio	Mol	Sim	Btz
St. 1	1998	6.84 (5/19)	1.46 (5/19)	0.17 (5/19)	0.35 (6/16)	0.57 (6/16)	3.4 (6/16)	0.17 (6/09)	7.6 (6/16)	0.6 (6/16)	0.08 (6/23-30)	2.78 (6/23)	0.46 (6/23)	6.7 (8/04)
	1999	7.75 (5/25)	0.59 (5/25)	0.24 (5/25)	0.34 (6/15)	0.35 (6/22)		0.26 (6/15)	5.65 (6/15)		0.05 (6/22)	1.62 (6/22)	0.55 (6/08)	
St. 2	1997	6.73 (5/26)	0.62 (5/19)	0.23 (5/19)	1.10 (6/16)	1.04 (6/23-30)	1.6 (6/23)		5.0 (6/16)	0.2 (6/16-23-00)	1.48 (6/30)	11.3 (6/30)	2.66 (6/30)	
	1998	8.77 (5/19)	2.17 (5/12)	0.27 (5/19)	11.1 (6/16)	0.46 (6/16)	0.7 (6/02)	0.07 (6/02)	5.4 (6/09)	0.3 (6/02)	0.27 (6/23)	14.6 (6/23)	1.53 (6/23)	9.5 (7/14)
	1999	5.48 (5/25)	0.56 (5/13)	0.19 (5/25)	0.53 (6/08)	0.24 (6/15)		0.13 (6/01)	4.55 (6/15)		0.27 (6/22)	8.30 (6/15)	1.61 (6/22)	3.0 (7/14)

		Insecticides						Fungicides			
		BPMC	MEP	MPP	MPP-sfon	MPP-sfox		EDDP	Ftl	Prg	Mtl
St. 1	1998	2.58 (7/07)	0.34 (8/04)	0.31 (8/04)	0.71 (8/25)	1.45 (8/04)		1.52 (8/04)	0.17 (8/04)	1.37 (7/14)	0.18 (5/26)
	1999	1.96 (7/06)	0.19 (8/10)	0.75 (7/21)	0.09 (7/27)	0.43 (8/10)		0.93 (7/27)	0.16 (8/03)	0.77 (7/27)	0.08 (6/08)
St. 2	1997	6.10 (7/07)	1.22 (8/11)	0.22 (8/11)				0.63 (8/11)	0.23 (8/11)	7.77 (7/14)	0.15 (5/26)
	1998	3.21 (7/C7)	4.00 (8/11)	0.15 (8/11)	0.20 (8/25)	1.66 (8/25)		0.73 (8/25)	0.27 (8/11)	1.87 (7/28)	0.24 (5/26)
	1999	3.36 (7/C6)	0/19 (8/10)	0.25 (7/27)	0.10 (7/27)	0.31 (7/27)		0.25 (8/10)	0.29 (8/10)	2.59 (7/21)	0.11 (5/25-6/01)

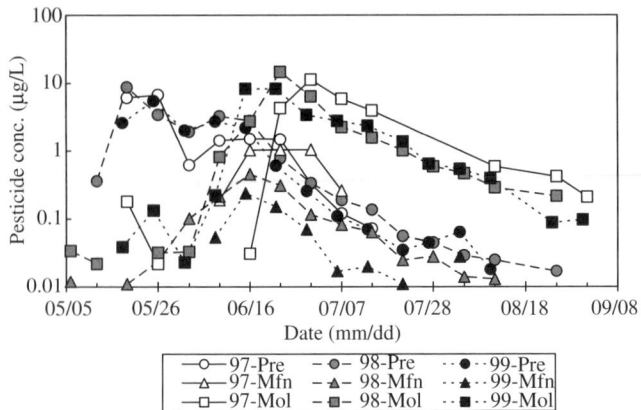

Figure 4 Comparison of herbicide runoff periods, 1997–1999

The insecticide BPMC began to appear in mid June, with a runoff peak in early July. MEP was detected beginning in late July, with a runoff peak observed in August.

Initial detection of the fungicide Prq occurred 1 week later than that of BPMC; the high-concentration period for Prq was from mid July to mid August. The runoff behavior of Ftl was similar to that of MEP. The fungicide Mtl, which was applied to the seedling boxes, was detected from the transplantation period until late June. It appears that Mtl was transported to the paddy field through transplantation of the rice seedlings. The detection of the investigated metabolites was delayed with respect to the detection of the parent compounds.

The relationship between solubility and the runoff rate is shown in Figure 5. The runoff rates of herbicides, insecticides, and fungicides ranged from 1.8% for Esp to 42% for Bnf, 0.3% for MPP to 9% for BPMC and 0.4% for Ftl to 25% for Prq, respectively. Large variations were recorded in the range of runoff rates of each pesticide; greater water solubility resulted in a larger runoff rate. Among the pesticides shown in Figure 5,

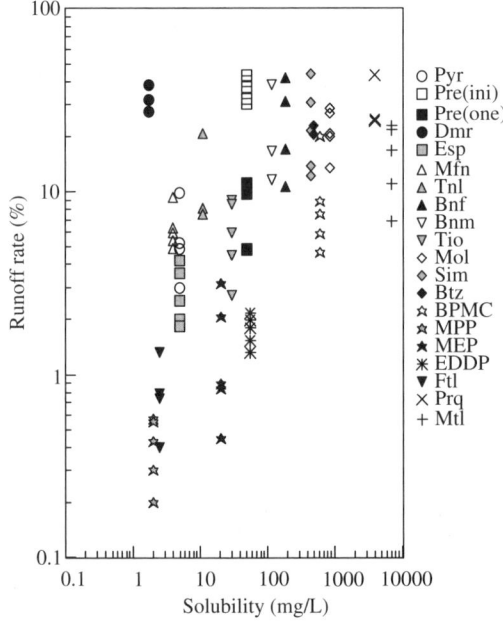

Figure 5 Relationship between solubility and runoff rate. ini: initial herbicide; one: one-shot herbicide

all herbicides, the insecticide BPMC, and the fungicide Prq were applied to paddy water. The other insecticides and fungicides (except for Mtl) were applied to the rice vegetation. The runoff rates of pesticides applied to paddy water were larger than those of insecticides and fungicides applied to the rice vegetation after the paddy water was discharged.

The application dates, amounts, and forms of herbicides; the maximum concentrations; the initial concentrations (extrapolated values); the half-lives and 1/10-lives (the number of days required for the concentrations of herbicides applied to paddy water to decrease to 1/10 of initial concentration) are shown in Table 2. During the observation period, paddy waters were not artificially discharged except before transplantation.

The highest concentrations of herbicides in paddy water were detected on the day of application or 1–2 days later, and herbicide concentrations in paddy water subsequently decreased according to the following equation:

$$C = C_0 \cdot e^{-kt}$$

where C is the pesticide concentration in paddy water; k is a constant; t is the elapsed time in days; and C_0 is the initial concentration of pesticide.

The half-lives of herbicides in paddy water ranged from 1.2 days for Pre and Esp to 5 days for Sim, and the concentrations of most pesticides in the paddy water had decreased to less than 1/10 of their initial concentrations by about 7 days after application.

The flooding water in the paddy fields should not be usually discharged for 4–5 days after pesticide application. Therefore, if paddy water discharge is managed rationally, the amount of pesticide runoff from paddy fields can probably be controlled. High concentrations of many pesticides were observed in river and lake water immediately after application, however, and the runoff rates of pesticides applied to paddy water obtained in our study were also large. These results suggested that the discharge of irrigation water after pesticide application is not being managed appropriately.

Moreover, initial herbicides, applied before rice is transplanted, were detected at relatively high concentrations in rivers, and the runoff rates of these pesticides were large

Table 2 Concentrations and persistence of herbicides in paddy water

Herbicide	Date:Paddy	Form	Amount g/a	Max-conc. mg/ml	Ini-conc. mg/ml	1/2-life days	1/10-life days
Pyr	97/5/19:B	E	360	89.1 (0)	143	1.81	6.02
	97/6/16:B	F	600	366 (0)	290	1.38	4.57
	98/5/18:C	E	350	123 (0)	125	1.91	6.33
Pre	97/5/19:B	E	240	168 (1)	200	4.32	14.4
	98/5/18:C	E	240	167 (0)	146	3.44	11.4
	98/6/16:A	G	240	104 (1)	176	1.23	4.08
	99/6/15:A	G	u	82.6 (0)	66	1.21	4.03
Esp	97/6/16:A	G	2,100	1,650 (1)	2,075	1.21	4.02
Bnf	98/6/16:A	G	240	390 (1)	499	2.20	7.30
Bnm	97/6/16:A	G	75	127 (1)	145	4.07	13.5
	97/6/16:B	F	70	190 (0)	174	2.01	6.67
Mol	98/6/17:C	G	282	1,940 (1)	2,312	2.47	8.20
	99/6/15:A	G	240	1,950 (1)	2,541	2.05	6.80
	99/6/15:C	G	282	2,650 (1)	4,656	1.64	5.46
Sim	98/6/17:C	G	530	154 (2)	168	4.93	16.39
	99/6/15:A	G	450	437 (0)	647	1.86	6.17
	99/6/15:C	G	530	720 (1)	1,094	2.32	7.69
Mfn	97/6/16:A	G	u	11.2 (1)	18	2.18	7.25

Max-conc. = detected maximum concentration of herbicide (duration of maximum concentration in days)
Ini-conc. = extrapolated initial concentration of herbicide
E = emulsion, F = flowable, G = granule, u = unknown

because paddy waters were discharged immediately before transplanting the rice. For example, the half-life of the initial herbicide Pre is about 4 days. If paddy waters are discharged 4 days after Pre application, the high-concentration runoff of Pre cannot be prevented. Therefore, the application of an initial herbicide having a long half-life before transplanting rice should be discontinued, or the herbicide should be applied 7 days or more before transplantation. Our results indicate that appropriate irrigation-water management is necessary for the conservation of an optimum water environment with regard to pesticide contamination.

Conclusion

Pesticides applied to paddy fields were observed at high concentrations in rivers during pesticide application periods. In particular, initial herbicides, applied before rice was transplanted, were detected at high concentrations in rivers during the transplantation period; runoff rates were large as well. When rice seedling growth after transplantation was reduced, due to low temperatures and lack of sunshine, many farmers delayed herbicide application. As a result, the high-concentration runoff periods of herbicides in rivers were 1–3 weeks later in the year with poorer weather than in average years.

The runoff rates of pesticides applied to paddy fields post-flood were larger, and runoff rates of insecticides and fungicides applied to rice vegetation after paddy water was discharged were smaller. In addition, the larger the water solubility of the pesticide, the larger the runoff rate.

The highest concentrations of herbicides in paddy water were detected on the day of application or 1–2 days later, and the concentrations decreased exponentially after that time. The concentrations of many herbicides in paddy waters had decreased to less than 1/10 of the initial concentrations by about 7 days after application. Therefore, high-concentration runoffs can be controlled by improving the management of irrigation, and appropriate irrigation-water management is necessary to conserve an optimum water environment.

References

Ebise, S., Inoue, T. and Numabe, A. (1993). Runoff characteristics and observation methods of pesticides and nutrients in rural river. *Wat. Sci. Tech.*, **28**(3–5), 589–593.

Inoue, T., Ebise, S., Numabe, A., Nagafuchi, O. and Matsui, Y. (2002). Runoff characteristics of particulate pesticides in a river from paddy fields. *Wat. Sci., Tech.*, **45**(9), 121–126.

Nagafuchi, O., Inoue, T. and Ebise, S. (1994). Runoff pattern of pesticides from paddy fields in the catchment area of Rikimaru Reservoir, Japan. *Wat. Sci. Tech.*, **30**(7), 137–144.

Numabe, A., Ebise, S. and Inoue, T. (1992). Estimation on the runoff amounts of pesticides applied after transplanting of rice plant by drainage river (in Japanese). *J. Japan Society on Water Environ.*, **15**(10), 662–671.

Sasagawa, Y., Matsui, S. and Yamada, H. (1996). Run-off of herbicides from paddy fields around the southern basin of Lake Biwa (in Japanese). *J. Japan Society on Water Environ.*, **19**(7), 547–556.

Economic valuation of reduction in nitrogen outflow from a paddy field area equipped with a recycling irrigation facility

Y.W. Feng*, E. Shiratani**, I. Yoshinaga** and T. Hitomi**

*Japan Society for the Promotion of Science, 6 Ichibancho, Chiyoda-ku, Tokyo, 102-8471, Japan
(E-mail: *ab-feng@criepi.denken.or.jp*)
**National Institute for Rural Engineering, 2-1-6 Kannondai, Tsukuba City, Ibaraki 305-8609, Japan
(E-mail: *eisaku-shiratani@nm.maff.go.jp; yoshi190@nkk.affrc.go.jp; thitomi@nkk.affrc.go.jp*)

Abstract We estimated the reduction in nitrogen outflow load from a paddy field that had a recycling irrigation facility and, by using a replacement cost method, evaluated the economic effect of nitrogen removal by the paddy field during the irrigation period in the Yoshinuma region of Tsukuba City, Japan. The recycling ratio of outflow water (proportion of outflow reused) was 13.5%. The nitrogen (N) outflow load was reduced by about 45 kg ha^{-1} by the N removal function of the paddy field and by about 39 kg ha^{-1} by the recycling irrigation facility. The paddy field equipped with a recycling irrigation facility as an N removal facility was valued at 32.6 million Japanese yen (JPY) ha^{-1} and 0.72 million JPY ha^{-1} per year, which compare it with the construction and maintenance costs, respectively, of a water quality improvement facility. The recycling irrigation facility was costed at 17.3 million JPY ha^{-1} for construction and 0.21 million JPY ha^{-1} for maintenance per year. The cost for constructing and maintaining a recycling irrigation facility was 53% of the value of the paddy field area equipped with a recycling irrigation facility as an N removal facility.
Keywords Effective; nitrogen; outflow load; replacement cost method

Introduction

In Japan, paddy fields cover 55% of all land used for agriculture. They require fertilizers with an N at 70 kg ha^{-1} y^{-1} and abundant amounts of water, accounting for 95% of the total agricultural water demand (Tabuchi and Hasegawa, 1995). Most paddy fields use streams and lakes as their main water sources and discharge the outflow back into them. Therefore, outflow water containing nitrogen from fertilizer application is a causative factor in the deterioration of water quality of streams and lakes. To reduce the nitrogen outflow from paddy fields, many approaches are called for, including improvement of soil properties and of methods of fertilizer application, and the introduction of a recycling irrigation facility. A recycling irrigation facility is often constructed in some paddy fields along the lower parts of rivers; river water is reused within the paddy to make better use of limited amounts of water. The introduction of a recycling irrigation facility has received much attention, because it may not only save irrigation water, but also reduce nutrient outflow loads from agricultural areas (Misawa, 1987; Kudo *et al.*, 1995). Some investigations have shown that a paddy field area equipped with a recycling irrigation facility reduces the nutrient outflow loads (Takeda *et al.*, 1997; Feng *et al.*, 2004; Shiratani *et al.*, 2004a), but the economic effect has not been evaluated.

The objectives of our study were to elucidate the reduction in nitrogen outflow load in a paddy field equipped with a recycling irrigation facility and to evaluate the economic effect by using the replacement cost method.

doi: 10.2166/wst.2006.048

Methods

Study site

This investigation was carried out during the irrigation period from April to September 2002 in the Yoshinuma region (36°8′N, 140°0′E), which is located about halfway down the Kokai River, Tsukuba City, Ibaraki Prefecture (Figure 1). The paddy field area we studied belongs to a larger area of paddies and is equipped with a system to recycle the irrigation water. Figure 2 shows the irrigation and drainage systems and water sampling points. Water from the Kokai River is pumped into the upper paddy field (43.2 ha) via pump station 2. The outflow water from that field drains into the main drainage canal. Water from this canal is pumped into the study paddy (7.3 ha) via pump station 1. The outflowing water from the study paddy field also discharges into the main drainage canal, which drains into the Kokai River. The soil is a Gray Lowland soil, which has good permeability.

From farmers' records, we found that the usual amounts of fertilizer applied to paddy fields in this region are 32 and 30 kg N in late April and mid July, respectively.

For this study, we used hydrological data and concentrations of N in irrigation water, outflow water, and precipitation from Feng *et al.* (2004) to calculate the amounts of reduction in nitrogen outflow load.

Economic valuation method

The replacement cost method (RCM) is often used to evaluate the multifunctional roles of paddy fields. In RCM, goods and services traded on the market are substituted for the functions to be valued. The functions are then evaluated according to the market prices of these goods and services. This method has two advantages. First, it is possible to evaluate each function separately, and second, the valuation is easy to understand, since goods and services are used instead of functions. Several studies using RCM have evaluated the nutrient reduction by tidal flats or wetlands on the basis of the costs of construction and maintenance of sewage treatment plants (Barbier *et al.*, 1997; Sasaki, 1998). However, it is difficult to apply this method to paddy field areas because sewage treatment plants treat water containing a high concentration of nitrogen (N: 15–40 mg L^{-1}), while paddy field areas equipped with a recycling irrigation facility use water containing a low concentration of nitrogen (N: < 3 mg L^{-1}). In this study, we evaluated the economic effect of reduction in nitrogen outflow load from a paddy field equipped with a recycling irrigation facility on the basis of the costs of construction and maintenance of

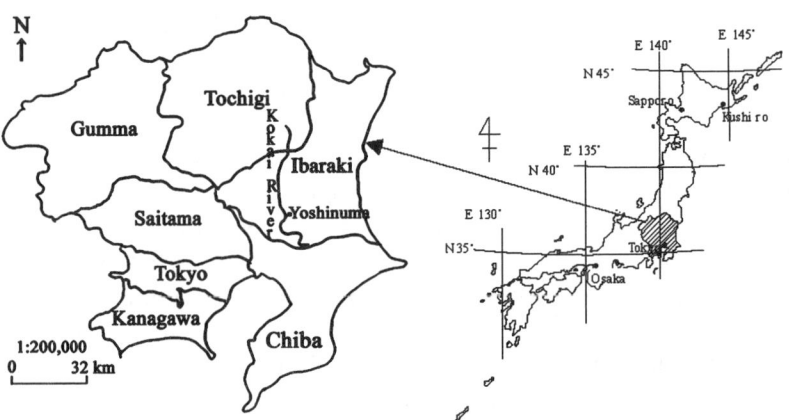

Figure 1 Location of the paddy field

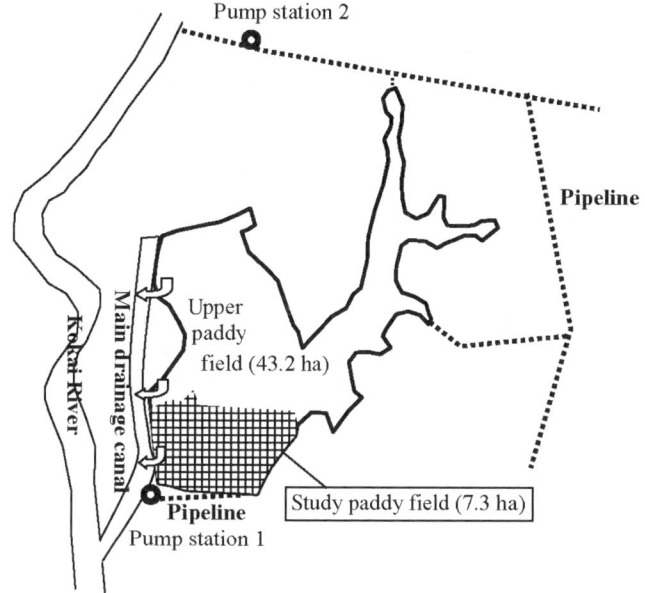

Figure 2 Schematic representation of irrigation and drainage system

a water quality improvement facility, that has been constructed throughout Japan to directly purify river water, irrigation water, and field outflow water.

Results

Reduction in nitrogen outflow load

The reduction in N outflow load from a paddy field equipped with a recycling irrigation facility is caused not only by the reuse of N for irrigating a paddy field by a recycling irrigation facility, but also by the paddy field's removal functions, e.g. the denitrification of NO_x–N and sedimentation of particulate N.

Kunimatsu (1983) reported the possible removal functions of paddy fields irrigated with water containing N at concentrations higher than $2\,mg\,L^{-1}$. In our study area, a recycling irrigation facility was implemented to reuse the outflow water from the upper paddy field for irrigation. The N concentration in the irrigation water averaged $2.31\,mg\,L^{-1}$ and reached $4.91\,mg\,L^{-1}$. This paddy field can remove N because it uses irrigation water with a relatively high concentration of N and has a good retention time, both of which factors facilitate N removal (Feng et al., 2004). The N concentration decrease in the ponded water can be expressed by the following first-order-kinetic reaction formula according to Shiratani et al. (2004a).

$$C = C_0 \exp^{-\frac{\alpha}{h}t} \tag{1}$$

where C is the concentration in the ponded water ($mg\,L^{-1}$), C_0 is the initial N concentration in the ponded water ($mg\,L^{-1}$), α is the rate constant for N removal ($m\,d^{-1}$), h is the depth of the ponded water (m), and t is time (in days). Although ponded depth in the paddy field was not always constant, h was taken to be 0.05 m, in accordance with the standard paddy field conditions in Japan. α was $0.016\,m\,d^{-1}$, similar to the findings of another investigation in the same paddy field area (Yoshinaga et al., 2003).

The N removal rate in our study area was calculated according to Tabuchi et al. (1993):

$$R = 10{,}000\alpha C_{irrigation} \tag{2}$$

where R is the nitrogen removal rate (g ha^{-1} d^{-1}), $C_{irrigation}$ is the N concentration in the ponded water (mg L^{-1}), α is the rate constant for N removal (m d^{-1}). The resulting N removal rate was 370 g ha^{-1} d^{-1}. The area of the study paddy field was 7.3 ha and there were 121 days of irrigation. The amount of N removed from our study paddy field area during the irrigation period was calculated to be 326 kg.

The amount of N reused for irrigating the paddy field by the recycling irrigation facility was calculated with the following formula.

$$M = 10{,}000 Q_i C_{irrigation} \tag{3}$$

where M is the amount of reused N (g ha^{-1} d^{-1}) and Q_i is the average daily water requirement (m). The results indicate that the amount of reused N was 323 g ha^{-1} d^{-1}, when the average daily water requirement was 0.014 m and the N concentration was 2.31 mg L^{-1}. About 285 kg N was reused for irrigating the paddy field during the irrigation period.

Economic valuation

Paddy field. The average daily rate of water (Q, m s^{-1}) flowing into a water quality improvement facility is required for calculating the construction and maintenance costs. This rate was calculated with the following equation:

$$Q = \frac{(M + R) \times A}{C_{in} \times r \times 86{,}400} \tag{4}$$

where M is the amount of reused N for irrigating the paddy field by the recycling irrigation facility (g ha^{-1} d^{-1}), R is the N removal rate by the paddy field removal function (g ha^{-1} d^{-1}), A is the area of the paddy field (7.3 ha), C_{in} is the N concentration of flowing water (mg L^{-1}), r is the N removal rate of a water quality improvement facility. The range of r is from 4% to 18.2%, depending on the different treatment methods used by the water quality improvement facility. In this study, r was 12.1% (the average value reported by the Ministry of Construction of Japan, 1994) and C_{in} was equated with $C_{irrigation}$ (2.31 mg L^{-1}). $Q = 0.21$ m s^{-1} when the water quality improvement facility removed the same N load in the paddy field.

The paddy field's valuation (E_c and E_m, JPY) as an N removal facility can be expressed by the following equations, which compare it with the construction and maintenance costs, respectively, of a water quality improvement facility:

$$E_c = Q \times W_c \tag{5}$$

$$E_m = Q \times W_m \tag{6}$$

where W_c and W_m are the construction and maintenance costs per unit flowing water of the water quality improvement facility, respectively, and Q is the average daily rate of flowing water (m s^{-1}). W_c and W_m were 1,132 million JPY m^{-3} and 25 million JPY m^{-3} per year (the Ministry of Construction of Japan, 1994), respectively. As an N removal facility, the paddy field was valued at 238 million JPY for the construction cost and 5.25 million JPY for the maintenance cost per year.

Recycling irrigation facility. The construction and maintenance costs of a recycling irrigation facility were calculated according to the Japan Sewage Works Association (1994) as:

$$E_p = 85.51 \times Q_p^{0.598} \times (113.2/90.1) \tag{7}$$

$$E_i = 1 \times Q_p^{0.690} \times (113.2/90.1) \tag{8}$$

where E_p and E_i are construction and maintenance costs, respectively, of a recycling irrigation facility and Q_p is pumpage (m^3 min^{-1}). The resulting costs were 126 million JPY for construction and 1.52 million JPY per year for maintenance. The cost for constructing and maintaining a recycling irrigation facility was 53% of the value of the paddy field area equipped with a recycling irrigation facility as an N removal facility. Shiratani et al. (2004b) calculated the average paddy field's valuation in Japan as an N removal facility as 6 to 10 million JPY ha^{-1} and 0.20 million JPY ha^{-1} per year, respectively, by RCM when compared with the costs of construction and maintenance of water quality improvement facilities. In our study, the paddy field as an N removal facility was valued at 112 million JPY for the construction cost and 3.73 million JPY for the maintenance cost per year, if we subtract the costs for constructing and maintaining a recycling irrigation facility, respectively. The studied paddy field area was 7.3 ha; therefore, the paddy field as an N removal facility was valued at 15.3 million JPY per unit area (ha) for the construction cost and 0.51 million JPY per unit area (ha) for the maintenance cost per year. These are 1.5 to 2.5 times and 2.6 times higher than the valuations reported by Shiratani et al. (2004b), when compared with costs of construction and maintenance, respectively. Because the paddy field reported by Shiratani et al. (2004b) was a normal irrigation paddy field area without a recycling irrigation facility, the reduction in N outflow loads appeared just as the paddy field's N removal function. The above results indicate that a recycling irrigation facility can be considered an economic and effective way to reduce nutrients flowing out from a paddy field area.

Recycling ratio

To reduce the N outflow load from a paddy field area, it is necessary to reuse as much of the outflow water as possible as irrigation water. The recycling ratio (R_{rp}) of outflow water is defined as the ratio of the volume of irrigation water applied to the paddy field (reused water by a recycling irrigation facility, V_p) to the sum of that of the upstream outflow (V_d) and the outflow from the paddy field (V_s).

$$R_{rp} = \frac{V_p}{V_d + V_s} \times 100 \tag{9}$$

$R_{rp} = 13.5\%$. The area of paddy field able to be irrigated with the total outflow water increased as the recycling ratio increased (Figure 3). This is because the volume of irrigation water increased, but the water requirement per unit paddy field remained the same.

The N concentration in ponded water at the ith day was simulated with Shiratani's model (2004a):

$$N_{paddy,i} = \frac{(h_{i-1} - O_i)N_{irrigation,i-1} \exp\left(-\frac{\alpha}{h_{i-1}}\right)\Delta t + W_i N_{irrigation,i} + R_i N_{rain,i} + L_i/10}{h_i} \tag{10}$$

where N_{paddy}, $N_{irrigation}$, and N_{rain} are the N concentrations (mg L^{-1}) in ponded water, irrigation water, and precipitation, respectively. L is the amount of applied N fertilizer (kg ha^{-1}), h is the depth of ponded water (m), O is the amount of outflow water (m), W is the amount of irrigation water (m), R is the amount of precipitation (m), α is the rate constant for N removal (m d^{-1}), and $\Delta t = 1$ day. The N concentration in ponded water appears to have a significant exponential correlation with the recycling ratio ($r^2 = 0.99$) (Figure 3). Hidaka (1990) reported that ponded water harmed rice growth when the N concentration was higher than 3 mg L^{-1}. The simulation results indicate that when the N concentration in ponded water is 3 mg L^{-1}, at a 30% recycling ratio the area of paddy field that can be irrigated with outflow water is 15.5 ha. The valuations of paddy field

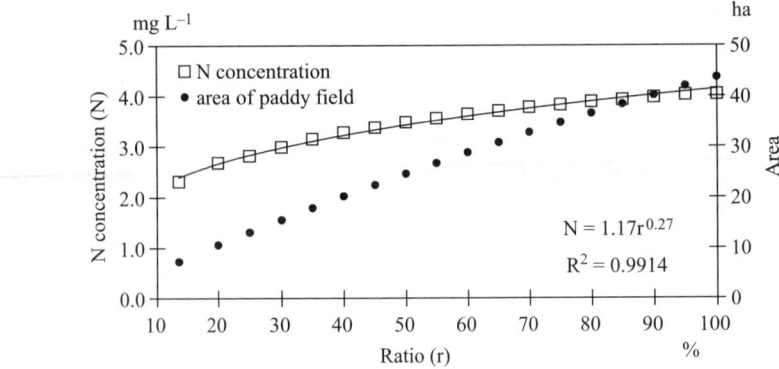

Figure 3 The relationship between the recycling ratio and nitrogen concentration

Table 1 The recycling ratio and construction and maintenance costs

Ratio %	Construction cost million JPY	Maintenance cost million JPY
13.5	238	5.2
20	344	7.6
25	426	9.4
30	504	11.1
35	581	12.8
40	656	14.5
45	728	16.1
50	799	17.6
55	867	19.2
60	936	20.7
65	1,001	22.1
70	1,065	23.5
75	1,127	24.9
80	1,189	26.3
85	1,249	27.5
90	1,306	28.8
95	1,364	30.1
100	1,420	31.4

areas equipped with a recycling irrigation facility as an N removal facility and recycling ratios are shown in Table 1. The results indicate that the most economic and effective recycling ratio in this study paddy field area is 30%, while the paddy field as an N removal facility is valued at 504 million JPY in comparison with the construction cost and at 11.1 million JPY per year in comparison with the maintenance cost of a water quality improvement facility. The costs of construction and maintenance of a recycling irrigation facility are 199 million JPY and 2.40 million JPY per year, respectively. The cost for constructing and maintaining a recycling irrigation facility is 39% of the value of a paddy field area equipped with a recycling irrigation facility as an N removal facility. The effective construction of a recycling irrigation facility not only saves abundant amounts of irrigation water, but also reduces nutrient outflow loads from paddy field areas. A recycling irrigation facility can be considered an economic and effective way to reduce non-point source nutrients.

Conclusions

We evaluated the reduction in nitrogen outflow load from a paddy field area equipped with a recycling irrigation facility and estimated the paddy field's valuation as an N removal facility. We calculated the recycling ratio of outflow water and proposed the most economic and effective recycling ratio. We drew the following conclusions.

1. The recycling ratio of outflow water was 13.5%, and the N outflow load was reduced by 326 kg by the study paddy field's removal function and by 285 kg by the recycling irrigation facility.
2. The value of the paddy field area equipped with a recycling irrigation facility as an N removal facility was 112 million JPY for construction and 3.73 million JPY for maintaining per year, subtracting the cost of the recycling irrigation facility, when replaced by a water quality improvement facility.
3. The N concentration in ponded water appeared to have a significant exponential correlation with the recycling ratio.
4. The most effective and economic recycling ratio of outflow in this study paddy field area would be 30%, the paddy field would have an economic value of 515 million JPY when replaced by the sum of the construction and maintenance costs of a water quality improvement facility.

The effective construction of a recycling irrigation facility not only saves abundant amounts of irrigation water and reduces nutrient outflow loads from paddy field area, but also has a huge economic effect. A recycling irrigation facility can be considered an economic and effective way to reduce non-point source nutrients.

References

Barbier, E.B., Acremam, M.C. and Knowler, D. (1997). *Economic valuation of wetlands: a Guide for policy markers and planners*, Ramsar Convention Bureau, Gland, Switzerland.

Feng, Y.W., Yoshinaga, I., Shiratani, E., Hitomi, T. and Hasebe, H. (2004). Characteristics and behavior of nutrients in a paddy field area equipped with a recycling irrigation system. *Agri. Wat. Mgmt.*, **68**, 47–60.

Hidaka, S. (1990). Studies on effect of irrigation water quality upon the rice growth and soil and its usable marginal concentration. *Bulletin of Saitama Prefecture Agricultural Experiment Station*, **44**, 1–93.

Kudo, A., Kawagoe, N. and Sasanabe, S. (1995). Characteristics of water management and outflow load from a paddy field in a return flow irrigation area. *J. Jpn. Soc. Irrig. Drain. Reclam. Eng.*, **63**(2), 179–184 (in Japanese with English abstract).

Kunimatsu, T. (1983). *Crop land – recycling of nutrients and purification function in paddy fields*, Research report, Lake Biwa Research Institute, Shiga, Japan (in Japanese with English abstract).

Ministry of Construction of Japan (1994). Evaluation documents: The efficient water improvement facility in public water source such as river – A water improvement facility of the multi-tank oxidation reactor type, pp. 19–21.

Misawa, S. (1987). Mechanism of water quality change in paddy field. *Trans. Jpn. Soc. Irrig. Drain. Reclam. Eng.*, **127**, 69–79 (in Japanese with English abstract).

Sasaki, K. (1998). Material circulation and production in estuary and tidal flat. *Aquabiology*, **20**(2), 132–137.

Shiratani, E., Yoshinaga, I., Feng, Y.W. and Hasebe, H. (2004a). Scenario analysis for reduction of effluent load from an agricultural area by recycling the run-off water. *Wat. Sci. Tech.*, **49**(3), 55–62.

Shiratani, E., Feng, Y.W., Yoshinaga, I., Hitomi, T. and Yamaguchi, Y. (2004b). Economic valuation of N purification/pollution of cultivated land by a new replacement cost method. In *Proceedings of 4th World Water Congress*, 19 September 2004.

Tabuchi, T. and Hasegawa, S. (1995). *Paddy fields in the world*. The Japanese Society of Irrigation, Drainage and Reclamation Engineering, Japan.

Tabuchi, T., Shinoda, Y. and Koroda, H. (1993). Experiment on nitrogen removal in the flooded paddy field. *J. Jpn. Soc. Irrig. Drain. Reclam. Eng.*, **61**(12), 1123–1128.

Takeda, I., Fukushima, A. and Tanaka, R. (1997). Non-point pollutant reduction in a paddy field watershed using a circular irrigation system. *Water Res.*, **31**(11), 2685–2692.

Yoshinaga, I., Feng, Y.W., Hasebe, H. and Shiratani, E. (2003). Nitrogen removal function of paddy field in a circular irrigation system. In *Proceedings of 7th International Conference on Diffuse Pollution*, 17–21 August 2003, Dublin, Ireland, pp. 56–61.

On-site treatment of turbid river water using chitosan, a natural organic polymer coagulant

M. Sekine*, A. Takeshita**, N. Oda***, M. Ukita*, T. Imai* and T. Higuchi*

*Civil & Environmental Engineering, Yamaguchi University, 2-16-1 Tokiwadai, Ube, Yamaguchi 755-8611, JP
(E-mail: *ms@env.civil.yamaguchi-u.ac.jp*; *mukita@yamaguchi-u.ac.jp*; *imai@yamaguchi-u.ac.jp*; *takaya@yamaguchi-u.ac.jp*)
**WESCO, 1-101 Tokura, Shikama ward, Himeji, Hyogo 672-8046, JP (E-mail: *a-takesita@wesco.co.jp*)
***Ube Environmental Technology Center, 4-23 Bunkyouchou, Ube, Yamaguchi 755-0056, JP
(E-mail: *nozomuo@ukgc-eco.ac.jp*)

Abstract Chitosan, acetylate of chitin, is a biodegradable cationic polymer. The objective of this study is to assess the applicability of chitosan as an on-site treatment agent of turbid water caused by river construction works and other diffused pollutions. The results of jar-tests indicate that floc of chitosan is much larger than that of aluminium sulfate, and turbidity treated by chitosan under moving water conditions is much lower than that of aluminium sulfate. Chitosan is applied to Imou River in Yamaguchi prefecture, where river construction work is going on. St.1 is located just below the construction work, St.2 is located about 250 m downstream from St.1, and St.3 is located about 350 m downstream from St.2. Initial turbidity of each station is 1,100, 937 and 313 NTU, respectively. By applying chitosan at St.1, turbidity of each station is drastically reduced to 1,100, 12 and 0 NTU. Chitosan could be helpful to reduce problems caused by turbidity in rivers.
Keywords Chitosan; coagulation; fish toxicity; sedimentation; turbidity removal

Introduction

Although there are a lot of demands for on-site turbidity treatment methods in relation with river construction works, there is no effective method applicable to small rivers especially under a cost sensitive situation. In our former study, we proposed a chemical precipitation using aluminium sulfate combined with temporary dams (Takeshita *et al.*, 2002). In the field experiment, we could reduce turbidity of 250 NTU down to 50 NTU in a river in which construction work was going on. But the appearance of the water was still not clear enough, and our proposal was not adopted after comparing the effectiveness of the treatment and the risk of aluminium to living organisms in the river. In this study, we focus on chitosan, acetylate of chitin, as a coagulant (Huang *et al.*, 1996; Pan *et al.*, 1999; Huang *et al.*, 2000). Chitosan is a cationic polymer obtained from natural resources like shells of shrimps. Thinking the fact that chitosan is not only widely used for a coagulant in food industry but also sold as a health food, chitosan can be a safer coagulant than aluminium sulfate for applying directly in rivers. The objective of this study is to assess the applicability and safety of chitosan as an on-site treatment agent of turbid water caused by river construction works and other diffused pollutions.

Jar-tests of chitosan and aluminium sulfate

Determining optimum coagulant concentrations

A test solution of 210 NTU, pH 7.6 and 20 °C is prepared by using river water and mud. After two minutes pre-stirring at $181\,s^{-1}$ (G value (Camp, 1943)), coagulants are injected in various concentrations, then kept stirring for three minutes at $181\,s^{-1}$ and ten minutes at $49\,s^{-1}$. After five minutes of still standing conditions, the turbidity of the supernatant water is measured. For aluminium sulfate, we need to add sodium hydroxide to make it work as coagulant, whereas for making a solution of chitosan, we need to add the same amount of acetic acid. Although these additional agents are needed, pH of the test solution changes little after coagulant injection. Figures 1 and 2 show the results of the jar-tests. Based on these results, the optimum coagulant concentrations are determined as 1.0 mg-chitosan/L for chitosan and 4.1 mg-Al_2O_3/L for aluminium sulfate.

Stirring condition and treatment performance

A test solution of 213 NTU, pH 8.03 and 20 °C is prepared by using river water and mud. After two minutes pre-stirring at $181\,s^{-1}$, coagulants are injected in optimum concentrations, and then kept stirring for three minutes at $181\,s^{-1}$ and ten minutes with different G values. After five minutes of still standing conditions, the turbidity of supernatant water is measured. Figure 3 shows the result. Aluminium sulfate receives small influence from the stirring conditions, whereas chitosan shows higher performance in higher stirring conditions. In addition, much larger particles are observed in chitosan solution which seem almost to settle down even under stirring conditions.

Based on this observation, we test coagulant performance under stirring conditions in the next experiments. After two minutes pre-stirring at $181\,s^{-1}$, coagulants are injected in optimum concentrations, and then kept stirring for three minutes at $181\,s^{-1}$ and 160 minutes with various stirring conditions. During the 160 minutes stirring, the turbidity is measured several times at a depth of 2 cm (out of 12.7 cm total depth). Figure 4 shows the turbidity at the end of 160 minutes stirring. The figure also contains G values at pools and rapids in a river. Chitosan shows high performance under high G value compared to aluminium sulfate. Figure 5 shows the particle size created by chitosan and aluminium sulfate under stirring conditions. The particle size created by chitosan is ten times larger than that of aluminium sulfate. This characteristic might be another strong point of chitosan as an on-site river turbidity treatment coagulant. Easily we hit on an on-site turbidity treatment process of injecting chitosan before the rapids of a river then let particles precipitate in a pool of it.

Field experiment in Imou River

We conducted a field experiment on 18 November 2003 in Imou River in Yamaguchi prefecture. Figure 6 shows the location of observing stations. St.1 is a rapid right downstream from a construction site. Coagulant is injected at St.1 (Figure 7). The stirring condition of St.1 is: B = 0.95 m, v = 0.35 m, H = 11 cm, $354\,s^{-1}$ of G value and two minutes retention time. St.2 is a pool with B = 2.46 m, v = 0.07 m, H = 18.1 cm, $42\,s^{-1}$ of G value and seven minutes retention time. St.3 is a rapid with B = 1 m,

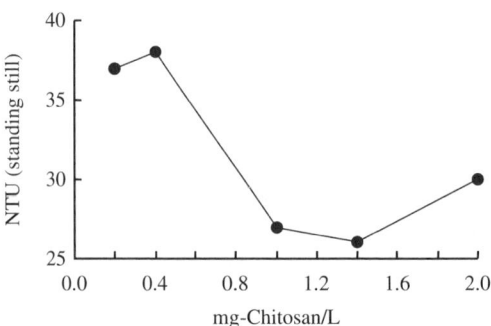

Figure 1 Chitosan concentration and turbidity under standing still conditions

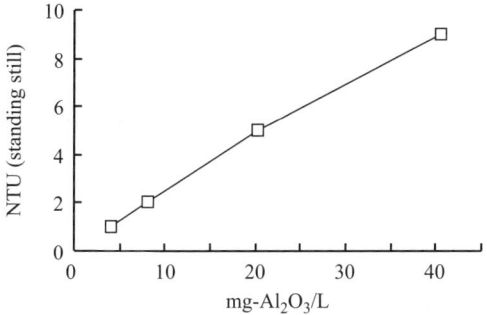

Figure 2 Aluminium sulfate concentration and turbidity under standing still conditions

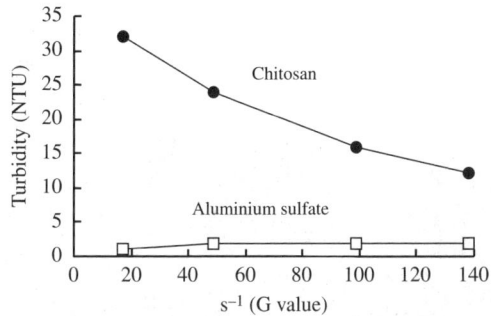

Figure 3 Turbidity under standing still conditions with different G values

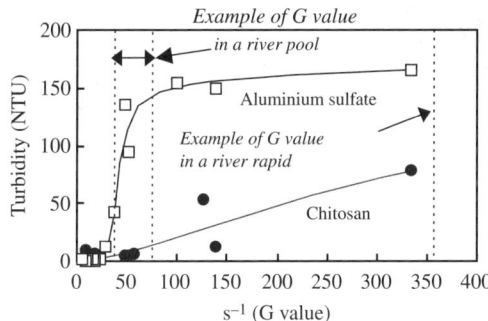

Figure 4 Turbidity under stirring conditions with different G values (after 160 min stirring)

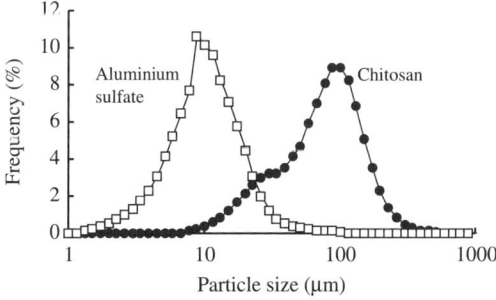

Figure 5 Particle size distribution created by chitosan and aluminium sulfate under stirring conditions

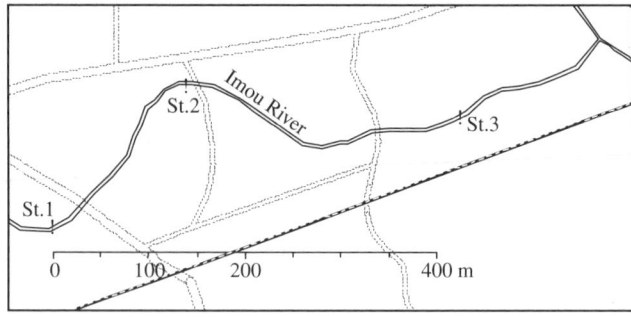

Figure 6 Location of observing stations in Imou River

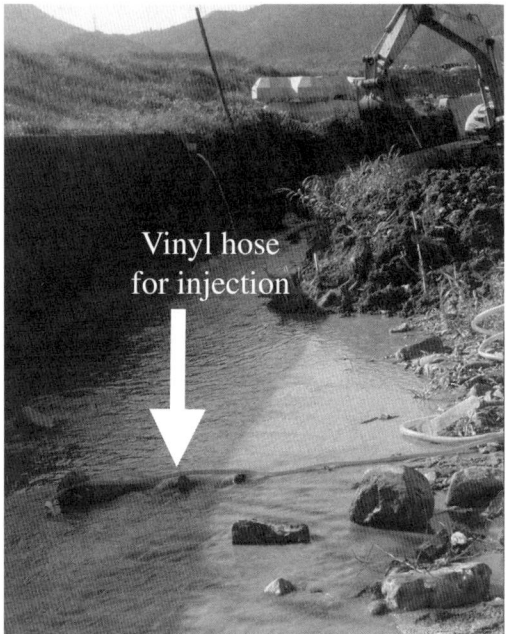

Figure 7 St.1

v = 0.38 m, H = 7.7 cm (Figures 8 and 9). In this experiment, pure chitosan is not used but FLONAC #250 is used, which is a trade name of 1:1 mixture flake of chitosan and acetic acid to increase solubility of chitosan. The optimum coagulant concentration is determined as 1.5 mg-FLONAC/L from a jar-test. The retention times are about 45 minutes from St.1 to St.2 and 60 minutes from St.2 to St.3. FLONAC solution is continuously injected at St.1 for two hours and turbidity and SS are measured at all stations. Figure 10 shows the time series of turbidity and Figure 11 shows SS concentrations taking retention time into account. As you can see from these figures and photos, chitosan shows quite high performance in removing turbidity compared to our experience with aluminium sulfate in former research. Precipitation effectively occurs in natural pools and we don't even need a temporary dam.

Safety tests of coagulants

Although we reported the excellent on-site treatment performance of chitosan in the previous chapter, we also observed the strange behavior of fish in the high turbidity section

Figure 8 St.3 before coagulant injection (You cannot see bottom materials)

Figure 9 St.3 after coagulant injection (You can see bottom materials clearly)

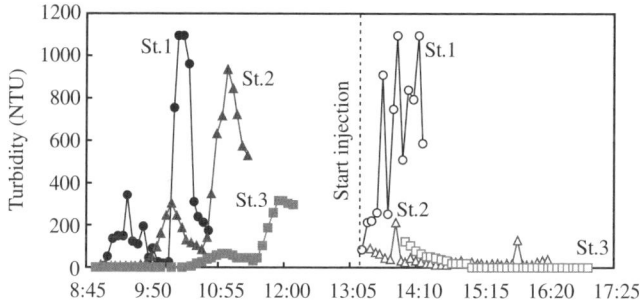

Figure 10 Time series of turbidity in Imou River

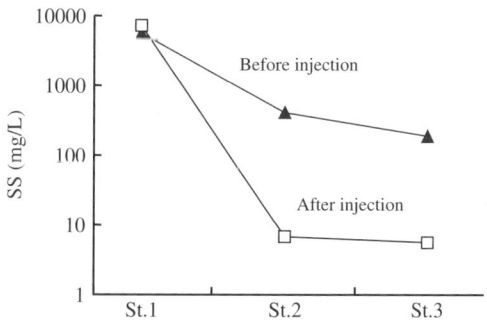

Figure 11 SS change caused by chitosan injection

which was about 50 m downstream from St.1. In the section, a bunch of fish exhibited "surfacing", but the behavior was somewhat different from the common behavior known as surfacing. They didn't take breath at the surface but just stayed calm and quiet right below the surface with heads slightly up (Figure 12). We didn't observe any dead fish or quickly moving fish in the section, nor any strange behavior of fish in the other downstream sections. Although we started this research based on the expectation that chitosan is safer than aluminium sulfate, now a toxicity test is required.

An acute toxicity test using larva of *Oryzias latipes* is employed. Ten larvae with 25 mL test water are put into a Petri dish (φ90 mm × 40 mm) and cultivated 48 hours under 25 °C. The number of dead larvae and abnormal larvae is counted at 1, 2, 3, 6, 12, 24, 48 hours. A control test is also conducted using water treated with activated carbon at the same time. In case the mortality rate of the control test exceeds ten per cent, the result is omitted. Figures 13 and 14 show the results of the toxicity tests.

Chitosan shows a low rate but constant abnormal behavior. From the close observation of the abnormal behavior, fins of the larva seem to cling to its body. In the other experiment, we observe that 96% of chitosan goes to the bottom sediment by precipitation very shortly. Supernatant water causes no abnormal larva behavior.

A fishery science researcher privately mentioned to us that the observed "surfacing" might be the common behavior for fish when they are in turbid water. He added that it was no wonder that no one had observed the behavior before because we couldn't see fish through turbid water. We haven't found any method to test his opinion, but it could be one reason for the "surfacing" behavior.

Although we couldn't clearly prove the safety of chitosan from these tests, the abnormal behavior of fish seems not to be caused by toxicity but to be caused by the physical viscosity of chitosan. High turbidity also causes the avoiding behavior of fish or deadly effects on bottom living organisms. One should compare the risk of turbidity over a long distance and the "toxicity" of chitosan over a short distance.

Cost analysis

The cost of FLONAC is ¥2,000/kg and aluminium sulfate is ¥326/kg. The optimum coagulant concentrations for 1,000 NTU water are 1.2 mg-FLONAC/L and 6.1 mg-Al_2O_3/L. Based on these conditions, the cost of chitosan is 1.2 times higher than aluminium sulfate. But you should remember that there is a big difference in on-site treatment performance. By using chitosan, you don't need any special settling pool.

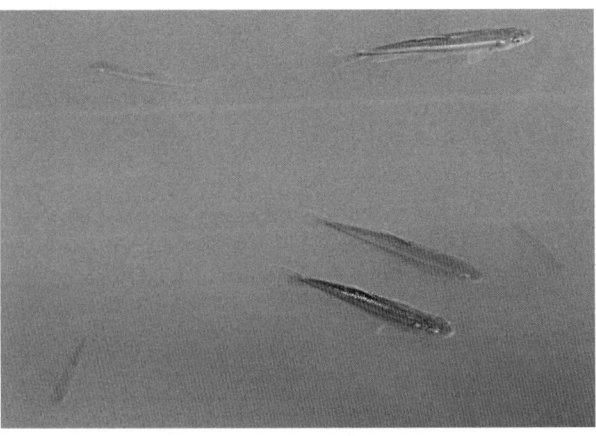

Figure 12 "Surfacing" fish 50 m downstream from St.1. They just stayed calm and quiet near the surface

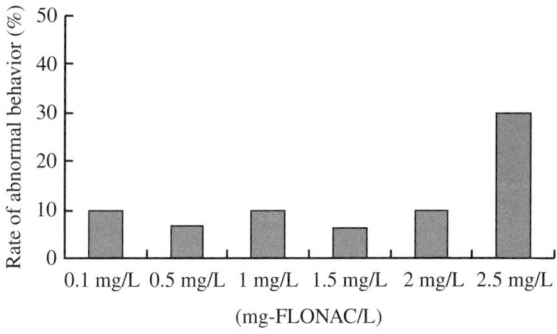

Figure 13 Rate of abnormal behavior of larvae for FLONAC

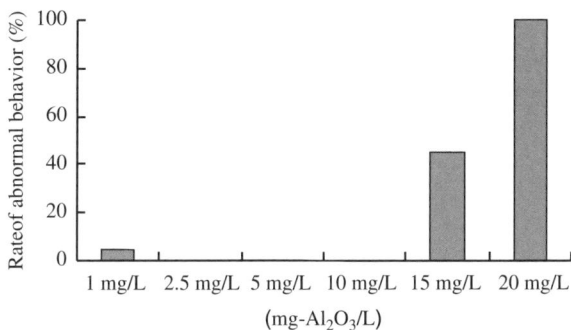

Figure 14 Rate of abnormal behavior of larvae for aluminium sulfate

Conclusion

In this research, chitosan shows excellent performance and usability as an on-site turbidity treatment agent. It is comparable in cost and much better in performance compared with aluminium sulfate. Although we still need to be careful about the effect of chitosan on living organisms, chitosan can greatly reduce the distance where turbidity reaches. The supernatant water treated by chitosan is quite safe. We hope this technique can reduce the impact of turbid water on living organisms.

References

Camp, T.R. (1943). Velocity gradient and internal work in fluid motion. *Journal of Boston Society of Civil Engineers*, **30**(4), 219–237.

Huang, C. and Chen, Y. (1996). Coagulation of colloidal particles in water by chitosan. *Journal of Chemical Technology and Biotechnology*, **66**(3), 227–232.

Huang, C., Chen, S. and Pan, J.R. (2000). Optimal condition for modification of chitosan: a biopolymer for coagulation of colloidal particles. *Water Research*, **34**(3), 1057–1062.

Pan, J.R., Huang, C., Chen, S. and Chung, Y.C. (1999). Evaluation of a modified chitosan biopolymer for coagulation of colloidal particles. *Colloids and Surfaces A: Physicochemical and Engineering Aspects*, **147**(3), 359–364.

Takeshita, A., Sekine, M., Ukita, M., Imai, T. and Higuchi, T. (2002). Field survey on the effect of turbidity caused by river construction works on living organisms and field experiment of on-site treatment (Japanese). *Proceedings of Environmental Engineering Research Forum*, **39**, 128–130.

Study on purification mechanism in soil penetration facility for effluents from urban area and control strategies

K. Yamada*, D. Ujiie* and K. Nishikawa**

*Department of Environmental Systems Engineering, Ritsumeikan University, 1-1-1 Nojihigashi, Kusatsu, Shiga 525-8577, Japan (E-mail: *yamada-k@se.ritsumei.ac.jp; daiki_ujiie@mhi.co.jp*)
**NJS Consultants Co. Ltd., 9-15 Kaigan 1 –Chome, Minato-ku, Tokyo 105-0022, Japan
(E-mail: *kouichi_nishikawa@njs.co.jp*)

Abstract In this study, demonstration experiments for removal of pollutants from road surface runoff during storm events were carried out under natural conditions in an outdoor pilot-scale soil penetration facility. In general, soil retains suspended matter and removes dissolved matter by adsorption. However, issues such as reduced purification capacity resulting from clogging and recovery of purification capacity during periods of intermittent supply of the storm water affect the removal efficiency of pollutants. Therefore, this study aimed at clarifying purification mechanisms during storm events and understanding how the structural characteristics of the soil penetration facility affect purification capacity based on long-term continuous measurements. In addition, modeling the purification mechanism under changing characteristics of rainfall in the long-term was undertaken.
Keywords Infiltration capacity; mass balance model; road surface runoff; soil penetration facility; soil purification mechanisms; time-series model

Introduction

Lake Biwa is an important source of water for about 14 million people in the Kansai Metropolitan Region. The lake has undergone tremendous pollution since the period of high economic growth in Japan that started in the 1950s. The water quality of the lake improved to some extent in the 1980s due to a number of measures taken such as formulation of the Basic Law for Environmental Pollution Control and the setting of Environmental and Effluent Standards. However, the water quality has not significantly improved and still does not meet the Environmental Standards (Figure 1).

The pollutant load entering the lake as estimated by Shiga Prefectural Government is shown in Table 1. The pollutant load from urban and industrial point sources decreased between 1995 and 2000. However, the pollutant load from the non-point sources shows an increasing trend. The non-point pollutant load from land, especially from urban areas, has increased due to increased development of urban areas. Therefore, measures for reduction of pollutant load in road surface runoff during storm events have recently been emphasized. As one of the specific measures, treatment of road surface runoff by soil penetration was investigated in this study. The characteristics of road surface runoff and soil penetration purification during storm events are shown in Table 2.

In this study, continuous pilot-scale demonstration experiments were carried out to investigate the removal of pollutant load from road surface runoff using a soil penetration facility. In earlier studies it was shown that characteristics of pollutant runoff from road surface were strongly influenced by amount of precipitation, rainfall intensity and antecedent days (Sugihara *et al.*, 2000; Nishikawa *et al.*, 2001; Nishikawa *et al.*, 2002). These studies reported average pollutant removal efficiencies of 80%, 50% and 40% for COD,

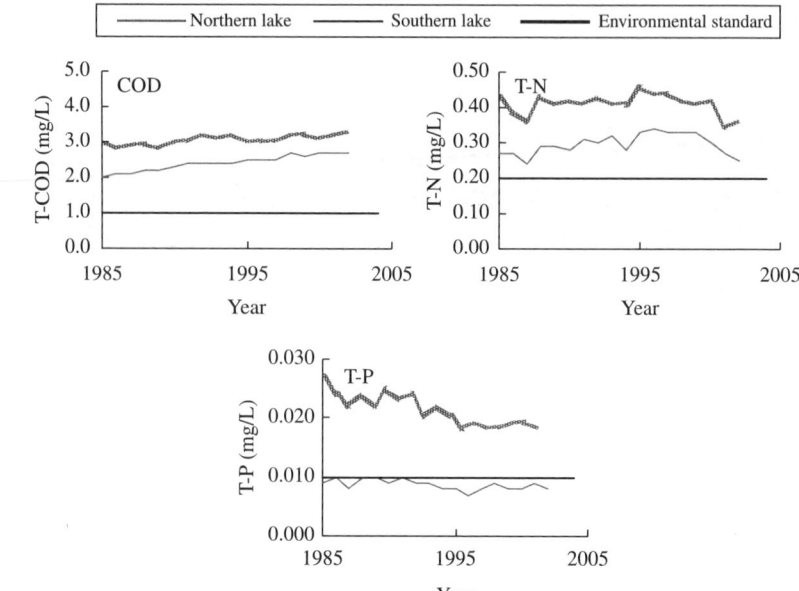

Figure 1 Trend of water quality of Lake Biwa, Shiga Prefectural Government (2004)

T-N and T-P, respectively. The studies further reported that clogging occurs in the 5 cm surface layer of soil and that the removal efficiencies of pollutants could be drastically increased by using an improved type of soil. As a continuation of the studies referred to above, the current study focused on understanding the mechanism of purification of soil and the effect of structural characteristics of the soil penetration facility on the purification capacity. The specific objectives of the study were: 1) to examine the purification mechanism of soil in relation to precipitation characteristics, 2) to model the purification mechanism of soil in the long term.

Outline of survey

Outline of soil penetration facility

A series of penetration experiments were continuously carried out at an experimental field located in Kusatsu City over a period of two years and two months. The outline of the soil penetration facility is shown in Figure 2 and Table 3. Runoff collected from 750

Table 1 Pollutant load going into Lake Biwa

		Point source		Non-Point source			
		Urban	Industrial	Agricultural	Natural	Land	Total
COD (t/day)	1995	16.0	11.5	7.5	15.8	5.4	56.2
	2000	11.5	7.1	7.0	15.8	5.8	47.2
	2005	9.3	6.6	7.0	15.7	6.1	44.7
T-N (t/day)	1995	5.7	3.4	3.4	7.7	1.4	21.6
	2000	5.5	2.5	3.2	7.7	1.5	20.4
	2005	5.4	2.3	3.1	7.7	1.6	20.1
T-P (t/day)	1995	0.52	0.37	0.18	0.21	0.07	1.35
	2000	0.43	0.25	0.16	0.21	0.08	1.13
	2005	0.38	0.23	0.16	0.20	0.09	1.06

Source: Shiga Prefectural Government (2004)

Table 2 Characteristics of purification by soil penetration for road surface runoff

Road surface runoff	Soil penetration
Irregular drainage	**Merit**
Intermittent drainage	Recovery of soil condition
Variation of flow	Pollutant removal
High concentration	Adsorption, Air exchange
SS, COD, T-P, T-N,	Microbe decompositon
Heavy metals and PAHs	Economical efficiency
	Low cost
Irregular pollutant load	**Demerit**
First flush	Sustainability
	Clogging

m² of road surface area was divided into two equal flows after which it flowed into each of the tanks in parallel. This water was called inflow water. Some of the inflow water penetrates the soil and is gathered by the drainage pipe at the bottom of the tank and then drained. This is called drained water. Some of the inflow water is retained in the soil and is referred to here as water held in soil. As the inflow water increases, some of the water flowing on the surface of the tank (surface flow) goes over the weir and is drained. This is called overflow water.

Outline of measurement

The facility was used continuously from September 2001 to November 2003. Over this period there were 253 rainfall events with a total precipitation of 3,500 mm. Measurement items were precipitation, flow rate and water quality. Precipitation was measured using a rain gauge installed near the experimental field. Runoff from the road surface was supplied to the field for all storm events. Effluent from the field was composed of surface flow and infiltrated flow. Water flow was gauged and water sampled for water quality analysis at three points for inflow, infiltrated flow and overflow. Precipitation was measured for all the 253 rainfall events that occurred over the period of the experiments. Water quality analyses were done for 14 of these rainfall events. The sampling interval for water quality analyses was 5 minutes during the early stage of runoff and peak discharge while a longer interval of about 30 minutes was used for other periods. Water quality was analyzed for SS, T-COD, D-COD, T-N, D-N, T-P, D-P and heavy metals.

Model

The basic concept of mass balance for runoff volume and pollutant load is shown in Figure 3. The mass balance of the facility was separated into two stages, namely, the

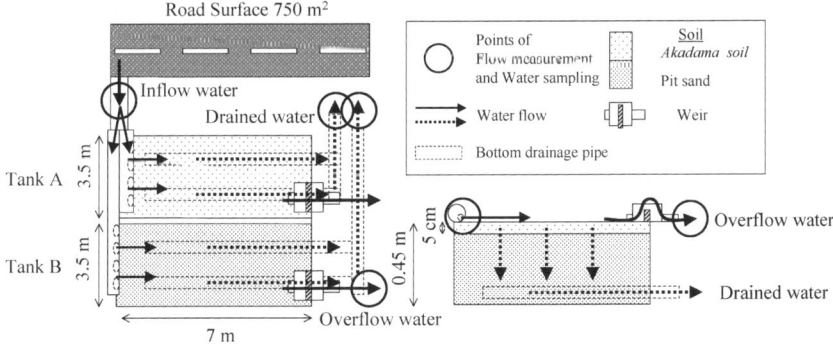

Figure 2 Soil penetration facility

Table 3 Outline of soil penetration facility

Tank	Surface area (m²)	Depth (m)	Kind of soil (depth, m) Upper layer	Kind of soil (depth, m) Lower layer	Height of weir (m)
A	25	0.45	AKADAMA soil (0.05)	pit sand (0.40)	0.03
B	25	0.45	pit sand (0.45)		0.03

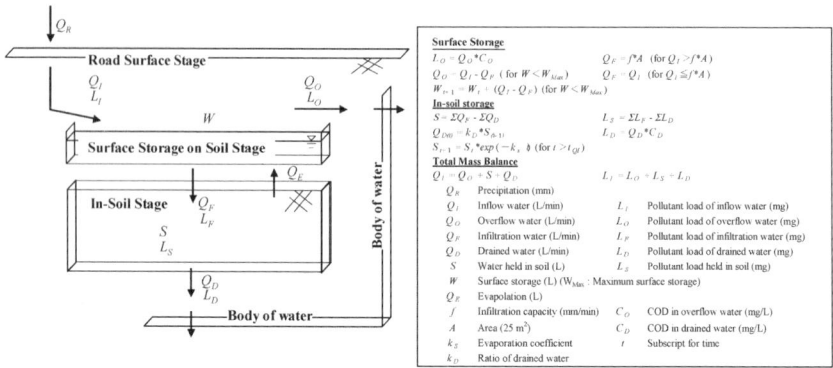

Figure 3 Basic concept of mass balance

surface storage stage representing storage of pollutants in water that flows on the surface of soil, and in-soil storage stage. The model parameters in Figure 3 were obtained statistically.

Results and discussion

Road surface stage

The characteristics of 50 rainfalls for which overflow occurred over the weir are shown in Table 4, while the characteristics of 14 rainfalls for which water quality was analyzed are given in Table 5.

Surface storage

Quantity of Water. The maximum amount of inflow water that penetrates through the soil corresponds to the infiltration capacity. When the inflow water exceeds the penetration capacity, the excess water is stored on the surface (this is referred to as surface storage). When the surface storage reaches the maximum surface storage capacity, excess water is discharged through the weir as overflow water.

The infiltration capacity was investigated during overflow conditions. Infiltration capacity was calculated by the following equation:

$$f = \Sigma Q_I / t / A \tag{1}$$

Table 4 Characteristics of rainfalls with overflow over the weir (50 rainfalls)

	Precipitation (mm)	Duration time (h)	Ave. rainfall intensity (mm/h)	Max. rainfall intensity (mm/h)	Antecedent days (day)
Ave.	25.5	9.47	4.69	24.48	2.83
Med.	20.0	8.42	3.00	18.00	1.34
Max.	131.0	41.83	24.00	111.00	17.49
Min.	3.5	0.17	0.41	3.00	0.25

Table 5 Characteristics of rainfalls with water quality analysis

Date	Precipitation (mm)	Duration time (h)	Ave. rainfall intensity (mm/h)	Max. rainfall intensity (mm/h)	Antecedent days (day)
2001					
9/10	24.5	18.55	1.32	15.00	2.39
10/9	52.5	14.17	3.71	39.00	5.25
11/29	7.5	10.33	0.73	12.00	22.43
2002					
3/5	52.0	15.67	3.32	18.00	5.18
7/13	6.0	3.00	2.00	21.00	2.32
9/6	14.5	8.83	1.64	15.00	8.64
9/26	5.5	3.50	1.57	6.00	9.28
10/15	5.5	0.50	11.00	18.00	7.17
11/1	33.5	10.50	3.19	6.00	8.06
2003					
5/30	19.5	8.00	2.44	9.00	4.10
6/16	20.0	6.00	3.33	18.00	0.61
7/29	31.5	17.00	1.85	18.00	5.91
10/13	15.5	8.67	1.79	15.00	17.49
11/3	20.5	20.00	1.03	21.00	11.80

The terms in Equation (1) are as defined in Figure 3. The variation of initial infiltration capacity (at 30 minutes from the start of overflow) and terminal infiltration capacity (at 30 minutes before overflow stopped) with time is shown in Figure 4.

From Figure 4, it was confirmed that the infiltration capacity decreased with time from the start of overflow and approached a constant value. This demonstrates the commencement of saturation of the soil. The relationship between the cumulative inflow water and infiltration capacity is shown in Figure 5. The following relationships were established:

Tank A $\quad f = -1.82 \ln(\Sigma Q_I) + 21.66 \; R^2 = 0.52$ (2)

Tank B $\quad f = -1.96 \ln(\Sigma Q_I) + 21.66 \; R^2 = 0.73$ (3)

The terms in Equations (2) and (3) are as defined in Figure 3.

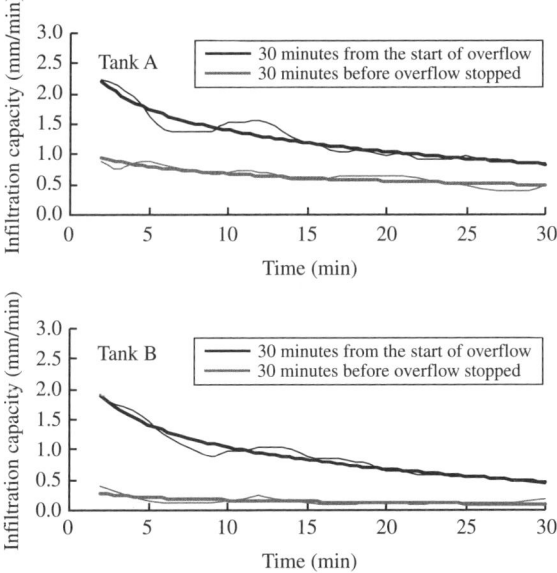

Figure 4 Variation of infiltration capacity with time

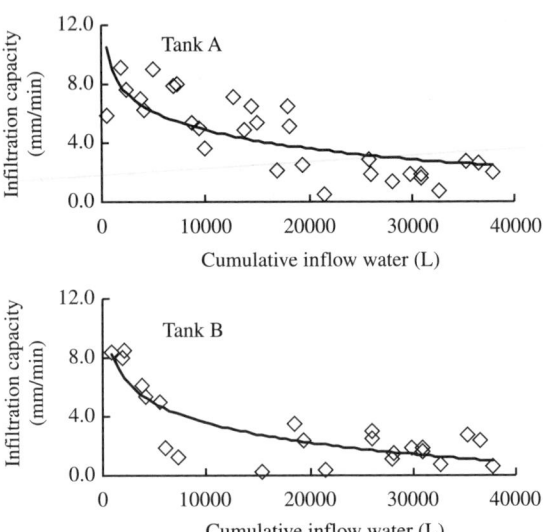

Figure 5 Variation of infiltration capacity with cumulative inflow water

Pollutant Load. Using results of actual measurement of the quality of overflow water, the pollutant load (expressed as COD) was simulated. The relationship between mean pollutant concentration of inflow water ($\Sigma L_I/\Sigma Q_I$) and the initial pollutant concentration in overflow water (C_{Fa}) and the terminal pollutant concentration in overflow water (C_{Ter}) is shown in Figure 6.

The following relationships were formulated:

Initial water quality

Tank A $C_{Fa} = 0.58x + 6.20 \ R^2 = 0.33$ (4)

Tank B $C_{Fa} = 0.63x + 6.34 \ R^2 = 0.38$ (5)

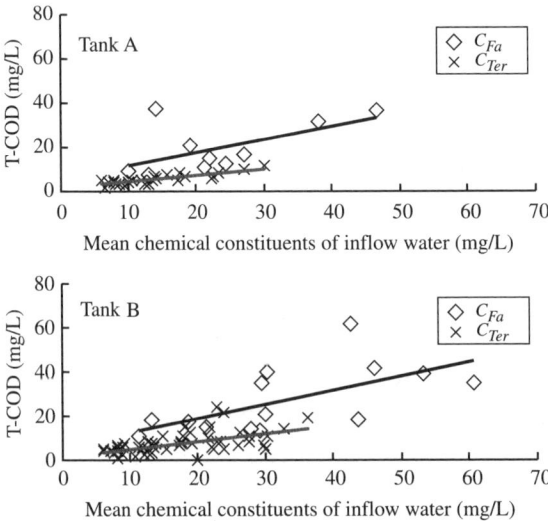

Figure 6 Concentration of pollutant in overflow water

Terminal water quality

Tank A $\quad C_{Ter} = 0.31x + 1.02 \ R^2 = 0.69$ (6)

Tank B $\quad C_{Ter} = 0.36x + 1.47 \ R^2 = 0.32$ (7)

In Equations (4)–(7), x is T-COD of inflow water (mg/L). Using the above results, the following relationship was established for COD in overflow water:

$$C_O = (C_{Fa} - C_{Ter}) * (1 - \exp(-k_{CO} * t) + C_{Ter} \quad (8)$$

In Equation (8), k_{CO} is a constant and other terms are as defined in Figure 3.

In-soil storage

Quantity of Water. As penetrated water infiltrates in the soil, drained water starts to appear depending on the amount of water held in the soil. Initially water held in soil increases with time. However, as penetrated water decreases, water held in the soil decreases. When inflow of water to the facility stops, water held in the soil continues to decrease due to water loss and drainage until water held in the soil reaches the amount of the water of adhesion on the soil. Using the mass balance for water volume (Figure 3), the maximum amount of water held in the soil for a given precipitation was determined. The relationship between cumulative penetrated water (at the time when maximum water held in soil occurs) and the maximum water held in soil is shown in Figure 7. The maximum water held in the soil (S_{Max}) for each of the tanks was determined as the maximum value of water held in the soil lying within the range of 2 standard deviations from the mean of the normal distribution curve. The following values of S_{Max} were obtained:

Tank A $\quad S_{Max} = 5,661 \, (L)$ (9)

Tank B $\quad S_{Max} = 3,980 \, (L)$ (10)

The soil condition in which water started to drain was determined from the relation between antecedent days (which affect the soil condition) and water held in the soil at the start of draining as shown in Figure 8.

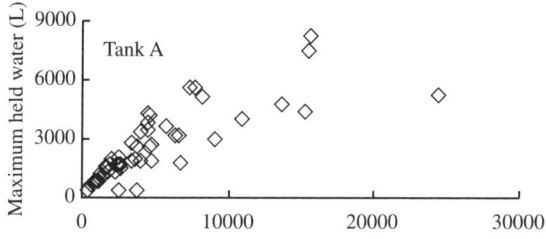

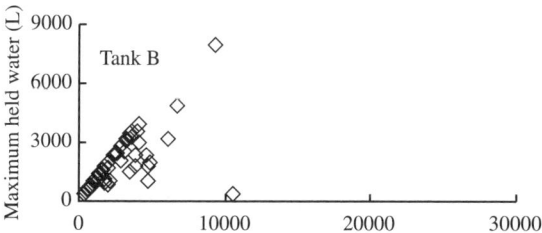

Figure 7 Maximum held water

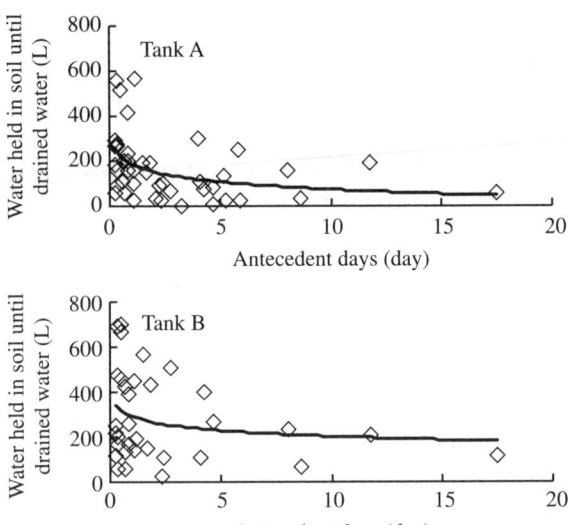

Figure 8 Water held in soil until drained water

Water held in soil at the start of draining decreased with increase in number of antecedent days. This is explained by the fact that the infiltration capacity recovers with increase in number of antecedent days due to increase in aeration of the soil. The following relationships were formulated:

Tank A $S_0 = -49.03 \ln(x) + 186.75 \quad R^2 = 0.17$ (11)

Tank B $S_0 = -38.34 \ln(x) + 290.51 \quad R^2 = 0.05$ (12)

In Equations (11) and (12), S_0 is water held in the soil at the stopping of water inflow and x is antecedent days (days).

There was seen to be a relation between drained water and water held in soil. At each survey, the relation between the ratio of drained water at maximum drained water (k_D) and water held in soil is shown in Figure 9.

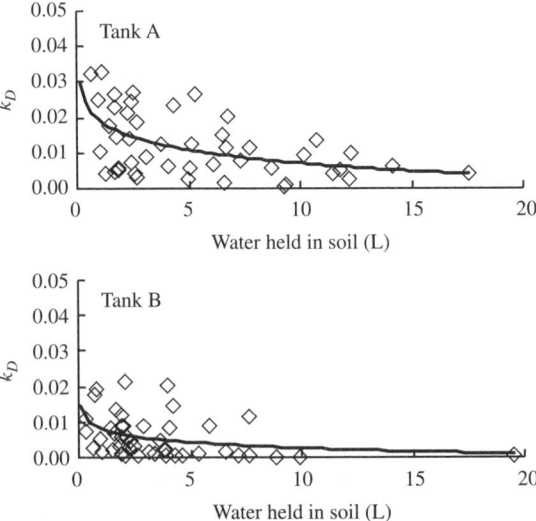

Figure 9 Ratio of drained water

The following relations were formulated:

Tank A $\quad k_D = -0.0054 \ln(S) + 0.052 \quad R^2 = 0.23$ (13)

Tank B $\quad k_D = -0.0024 \ln(S) + 0.023 \quad R^2 = 0.12$ (14)

In Equations (13) and (14), S is water held in soil. The above relations were used to determine k_D in time-series. From the stopping of water inflow, water held in the soil was lost through draining and evaporation. Therefore, the behavior of water held in the soil was investigated to determine drainage coefficient at the stopping of water inflow (ks_0). The relation between precipitation and ks_0 is shown in Figure 10.

The following relationships were formulated:

Tank A $\quad ks_0 = 0.0018 \ln(x) - 0.0083 \quad R^2 = 0.21$ (15)

Tank B $\quad ks_0 = 0.0008 \ln(x) - 0.0042 \quad R^2 = 0.11$ (16)

In Equations (15) and (16), x is precipitation (mm). The evaporation coefficient (k_s) was determined to converge to 0 with time from the stopping of water inflow and the following relationship was formulated:

Tank A, B $\quad k_s = m * \ln(t) - ks_0$ (17)

In Equation (17), m is a constant.

Pollutant Load. Using results of actual measurement of the quality of drained water, the pollutant load (expressed as COD) was simulated. At each tank, the relationship between mean pollutant concentration of inflow water ($\Sigma L_I/\Sigma Q_I$) and the initial pollutant concentration in drained water (C_{Fa}) and the terminal pollutant concentration in drained water (C_{Ter}) is shown in Figure 11.

The following relationships were formulated:
Initial water quality

Tank A $\quad C_{Fa} = 0.84x + 18.96 \quad R^2 = 0.90$ (18)

Tank B $\quad C_{Fa} = 1.30x + 31.34 \quad R^2 = 0.42$ (19)

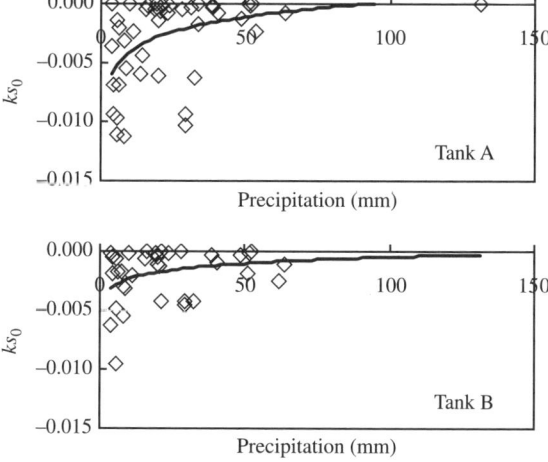

Figure 10 Variation of ks_0 with precipitation

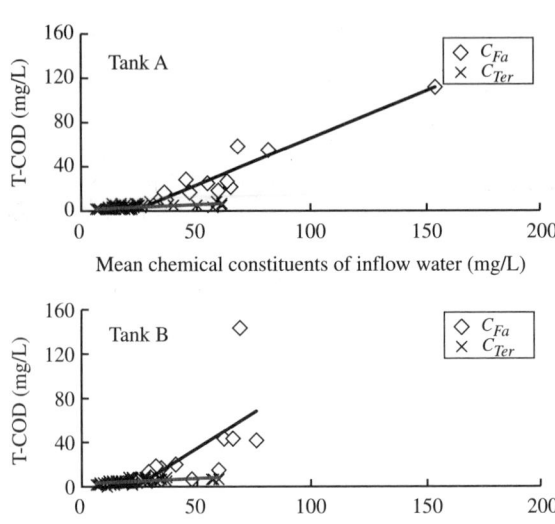

Figure 11 Quality of drained water

Terminal water quality

Tank A $C_{Ter} = 0.086x + 1.56 \; R^2 = 0.37$ (20)

Tank B $C_{Ter} = 0.106x + 1.96 \; R^2 = 0.44$ (21)

In Equations (18) – (21), x is T-COD of inflow water (mg/L). Using the above results, the following relationship was established:

$$C_D = (C_{Fa} - C_{Ter}) * (1 - \exp(-k_{CD} * t) + C_{Ter}$$ (22)

In Equation (22), k_{CD} is a constant.

Water flow simulation

Water flow simulation was done for six rainfalls from 2002/07/09 to 2002/07/19 among the 50 rainfalls that were analyzed in this study. Measured and estimated values for Tank

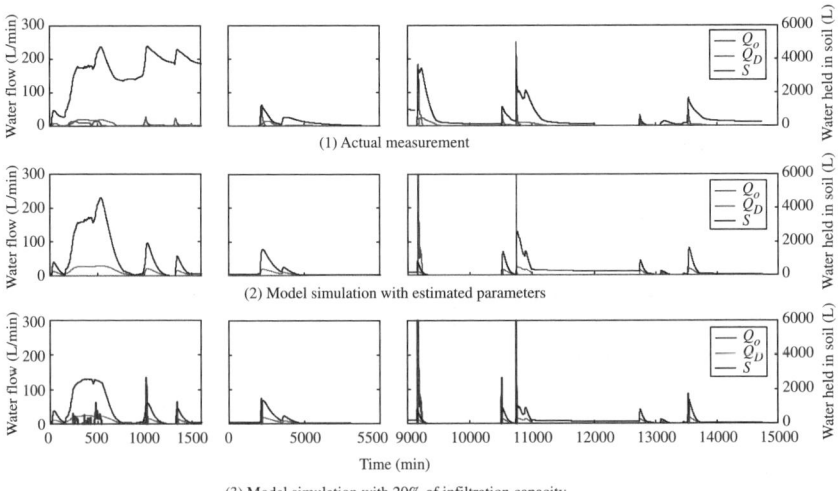

Figure 12 The result of water flow simulation

Table 6 The result of water flow simulation

Date	Actual measurement			Model simulation with estimated parameter							Model simulation with 20% of infiltration capacity						
				Estimate value				Relative error			Estimate value				Relative error		
	ΣQ_o (L)	ΣQ_D (L)	S(L)	ΣQ_o (L)	ΣQ_D (L)	S(L)		ΣQ_o (%)	ΣQ_D (%)	S(%)	ΣQ_o (L)	ΣQ_D (L)	S(L)		ΣQ_o (%)	ΣQ_D (%)	S(%)
2002/7/9	5,382	11,049	3,414	0	19,728	117		−100	79	−97	4,091	15,676	78		−24	42	−98
2002/7/13	738	1,621	–	0	2,390	48		−100	47	–	164	2,201	43		−78	36	–
2002/7/15	787	1,710	956	0	3,063	405		−100	79	−58	495	2,589	384		−37	51	−60
2002/7/16	5,372	4,605	126	9,365	927	8		74	−80	−94	9,365	917	7		74	−80	−94
2002/7/17	3,882	4,133	121	1,823	5,976	348		−53	44	187	4,035	3,923	188		4	−5	55
2002/7/18	1,610	1,850	232	0	3,839	78		−100	107	−66	278	3,481	54		−83	88	−77

A are shown in Figure 12 and the results of numerical analysis are shown in Table 6. The results show significant differences between the actual measurement and the estimated values of the amount of overflow. The discrepancy could be due to the fact that the estimated value of infiltration capacity was smaller than the actual measured value. The estimated values of infiltration capacity used in the model were reduced to 20%. As a result, the simulated value of flow approximated the actual measurement.

Conclusions

Time-series models were used to simulate the purification mechanism of road surface runoff in a soil penetration facility. The following results were obtained: 1) infiltration capacity was determined to be decreasing from the start of overflow, and to converge to a constant value at the end, 2) the start of flow of drained water was shown to be influenced by the recovery of infiltration capacity of the soil in antecedent days, 3) the decrease of water held in the soil from the time of stopping of inflow of water, due to drainage and water loss by evaporation, was illustrated, and 4) results of model simulation for flow did not fit well with measured values. Further analysis of the model parameters is needed.

References

Sugihara, M., Yamada, K. and Terada, A. (2000). Study on soil filtration treatment for pollutants from road surface during storm events. *JSWE, Proceedings, 3rd Annual Symposium* 95–96.

Nishikawa, K., Sugihara, M. and Yamada, K. (2001). Study on soil penetration facility for pollutants discharged from road surface. *Proceedings, 56th Annual Meeting JSCE*, 488–489.

Nishikawa, K., Sugihara, M. and Yamada, K. (2002). Study on removal by soil of pollutants discharged from road surface during storm events. *Proceedings, 6th International Conference on Diffuse Pollution, IWA*, 616–621.

Shiga Prefectural Government (2004). http://www.pref.shiga.jp/index.html.

Trace metal levels in sediments deposited in urban stormwater management facilities

J. Marsalek*, W.E. Watt** and B.C. Anderson**

*National Water Research Institute, Burlington, ON L7R 4A6, Canada (E-mail: *jiri.marsalek@ec.gc.ca*)
**Queen's University, Department of Civil Engineering, Kingston, ON, K7L 3N6, Canada

Abstract Characteristics of solids recovered from stormwater best management practice (BMP) facilities, including stormwater ponds, constructed wetlands, an infiltration basin, a biofilter, a stormwater treatment clarifier, and three-chamber oil and grit separators were described with respect to their metal chemistry. The reported trace metal concentrations in BMP sediments were assessed against the Ontario Sediment Quality Guidelines. Between 80 to 100% of all samples were marginally-to-intermediately polluted by Cd, Cr, Cu, Fe, Pb, Mn, Ni and Zn. Severe pollution of sediments was noted for Cr (122 μg/g), Cu (151 and 196 μg/g), Mn (1,259 and 1,433 μg/g), and Zn (1,116 μg/g), at several facilities studied, and even higher levels of metals were reported in the literature for certain oil and grit separators. With respect to individual BMPs, the severe pollution was found in sediments from oil and grit separators (for Cd, Cr, Cu, Pb and Zn), the stormwater clarifier sludge (Cu, Mn and Zn), a biofilter (Cu and Mn), an industrial area stormwater pond (Cu only), and a commercial/residential pond (Cr only). Finally, the chemical pollution of pond sediment triggered toxicity testing at some of the facilities studied, and sediment toxicity was confirmed at several sites.
Keywords Best management practices; sediment; sediment quality guidelines; stormwater management; trace metals

Introduction

Management and control of solids is one of the primary tasks of stormwater management arising from the need to mitigate impacts of suspended solids and sediment on the operation of drainage systems and water and habitat quality in receiving waters (MOE, 2003). Sources of solids in urban areas are numerous and include soil erosion, attrition/corrosion of pavements and other surfaces, vehicular traffic, atmospheric deposition, vegetation, litter, spills, and street sanding (Sartor and Boyd, 1972). In wet weather, surface runoff scours solids deposited on urban surfaces and transports them to the drainage system, and/or directly to the receiving waters. Solids accumulations are best controlled by source controls, but depending on the success of such measures, significant quantities of solids do enter stormwater and are transported to, or deposit in, various components of the drainage system. The presence of solids and sediment in the drainage system causes concerns about physical effects on system operation (blockage of inlets and sewers; clogging of such stormwater BMPs as infiltration facilities and filters; reduction of active storage of wetlands and ponds, etc.), and the water quality and aesthetic impacts. Among the water quality impacts, the most common problem is the transport of pollutants, including pathogens. Association of pollutants with urban solids was reported by many authors, referring to street surface residue (Stone and Marsalek, 1996), deposits in sewers, and accumulations in various BMPs, including ponds (Marsalek *et al.*, 1997b), wetlands (Bishop *et al.*, 2000), filters (Mothersill *et al.*, 2000), and oil and grit separators (Schueler and Shepp, 1993). The occurrence of solids deposits and accumulations affects stormwater BMPs in two ways: (a) the need to mitigate adverse impacts on BMP operation by implementation of more complex and costly treatment trains, and (b) degradation of

doi: 10.2166/wst.2006.051

habitat functions of BMPs by accumulation of contaminated sediment, with a further risk of release of mobile pollutant fractions (Bishop et al., 2000; vanLoon et al., 2000). The issues of BMP sediment contamination by heavy metals and the need to account for such environmental risks by proper maintenance are discussed for various types of BMP facilities, including ponds, constructed wetlands, an infiltration basin, a biofilter, a stormwater clarifier and three-chamber oil and grit separators.

Sources of data: BMP facilities analysed

The description of the stormwater BMPs studied, or BMPs for which data were adopted from the literature, is limited to basic characteristics, with references to more detailed information presented elsewhere. The selection of the BMPs discussed herein follows from an overview of numerous studies of BMPs, including graduate thesis projects and BMP assessment studies.

Kingston stormwater pond

The Kingston (Ontario, Canada) Pond is an on-line stormwater management pond, which was built in 1982 to reduce peak stormwater flows from a shopping mall. The two-cell pond consists of a permanent wet pond (area $5,200\,m^2$ and 1.2 m average depth) and a dry pond (area $5,000\,m^2$) that floods during larger storm events. Figure 1 shows the pond layout and the location of instrumentation and weirs at the inlets and outlet. Research findings on the Kingston Pond performance and processes were presented elsewhere (Anderson et al., 2002; Van Buren et al., 1997); only the basic findings concerning sediment issues are summarised herein. Marsalek et al. (1997b) reported that pond bottom sediments had accumulated at an average rate of 0.02 m/year and comprised gravel, sand, silt and clay; the gravel and sand accumulated only by the inlet whereas the silt and clay were spread throughout the pond and represented up to 45% and 54% of the total sediment respectively. Marsalek (1997) estimated the volume of the inlet sand spit at $150\,m^3$, and the corresponding sediment mass at 160 t, accumulated over 15 years. The water content of the sediment (by volume) ranged from 48% by the inlet to 75% at the outlet. Anderson et al. (2002) noted a disadvantage of on-stream stormwater ponds built on urbanizing catchments – such ponds accumulate sediment at relatively high rates and will require more frequent sediment removal than off-line facilities.

Stormwater wet pond 1

The pond WP1 consists of two irregularly shaped cells in series with a total storage of $68,000\,m^3$. The upstream cell (with a surface area of $1,750\,m^2$) is fed by three sewers draining two residential communities and a business park. Flow travels from the upstream cell into the downstream cell (surface area of $3,000\,m^2$) through three culverts. Sediment samples were collected in the downstream cell (vanLoon et al., 2000).

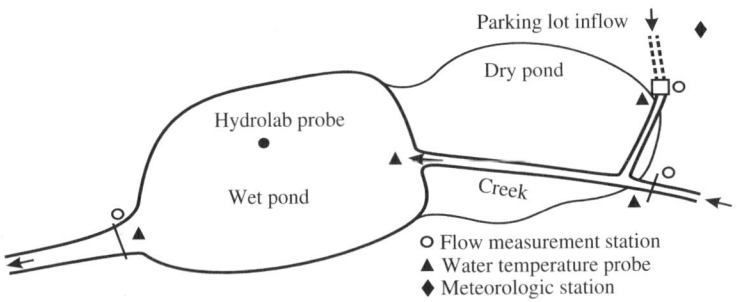

Figure 1 Kingston stormwater management pond

Stormwater wet pond 2

The pond WP2 consists of two elongated rectangular pond cells, in-series, each with dimensions of about 400 m × 25.5 m (length × width), and a total pond storage of 23,000 m^3. The pond serves principally a residential area, with additional input from about half a dozen agricultural drains. The facility is operated in batch mode and retains stormwater for up to 72 hours (vanLoon et al., 2000).

Harding Park pond

A relatively small pond (surface area = 0.7 ha; volume = 3,000 m^3, comprising 1,000 m^3 permanent pool storage and 2,000 m^3 active storage; maximum depth = 1.5 m) receiving stormwater runoff from a residential area of 16.8 ha in north-east Toronto (Ontario, Canada). The pond consists of three cells – a sediment forebay, a quiescent settling permanent pool and a wetland (Bishop et al., 2000).

Col. S. Smith reservoir

The CSS pond serves a catchment of 340 ha, with industrial, commercial, residential and transportation land uses. It receives runoff from a major multilane divided highway with traffic densities over 100,000 vehicles/day. The pond is of a rectangular shape (38 m × 75 m, length x width), with a surface area of about 0.28 ha and average depth of 2.4 m (nominal volume of 7,000 m^3) (Mayer et al., 1996).

Constructed wetlands

The chemistry of sediments from constructed wetlands serving for stormwater management was reported by Bishop et al. (2000). To extend the discussion of BMP sediment quality, the data reported for four constructed wetlands in Guelph (Ontario, Canada) were included here. These wetlands serve to control stormwater from residential areas, and their surface areas range from 2,500 to 7,700 m^2, and depths from 0.2 to 0.7 m. All four facilities yielded similar concentrations for individual metals, which were described by average values for the whole set.

Infiltration basin (IB)

The basin was constructed in an abandoned sand pit, with a capacity of 6,500 m^3. Soils in the basin area are sandy to great depths and underlain with clay. The basin substrate is formed by slightly alkaline crushed stone, forming a 0.3 m layer over 1.25 m of sand. The sand is separated from native soils by a non-woven filter cloth. The facility is equipped with sub-drains. Roadways and other impervious areas form about 25% of the contributing area, another 25% are industrial lands, and the remaining 50% is divided among undeveloped land, a golf course, commercial and other land uses (vanLoon et al., 2000).

Stormwater biofilter

The biofilter was installed at the Kingston Pond outlet and used to polish the pond effluent. It was formed by filling a polyethylene tank (1 m × 1 m × 1 m) with inert expanded schist media, with a nominal diameter of 3–6 mm. When hydraulic losses increased over a threshold, the filter was backwashed. Samples of solids captured by the filter were collected by removing samples of media after three years of operation (Mothersill et al., 2000).

Etobicoke stormwater clarifier

Stormwater treatment by constant high-rate clarification was studied in Etobicoke (Toronto-area, Ontario, Canada) at a site upstream of the Col. S. Smith Reservoir.

The clarifier was fed with a submersible pump from a 2.5 m diameter storm sewer draining an area of about 300 ha, comprising industrial, commercial, and residential land. The rectangular clarifier was 3 m long, 1.4 m wide and 2 m deep. Stormwater at this site was found to be polluted, with typical concentrations falling between the mean and the 90th percentile of the US NURP data (Wood *et al.*, 2004).

Three-chamber oil and grit separators

Oil and grit separators (OGSs) were developed for control of contaminated sediments and free oil spills at inlets to storm sewers. The original designs used in Maryland comprised three chambers, serving for sediment settling and oil retention. These designs were susceptible to sediment washout and were later replaced by manhole-type designs. The chemistry of sediments retained in OGSs was reported by Schueler and Shepp (1993) and included here as average data for 19 facilities.

Results and discussion

Metal burdens in BMP sediments were analyzed for up to 10 metals and inorganics, which are commonly listed in sediment quality guidelines (e.g. MOEE, 1992) and are of interest in environmental studies. Specifically, this list contains the three most ubiquitous anthropogenic metals, Cu, Pb and Zn, which are known to occur in urban areas at high concentrations and may exert toxic effects, and consequently, they are included on the US EPA list of 129 Priority Pollutants. All three were detected in at least 91% of all NURP runoff samples (US EPA, 1983). The next group of five chemicals, As, Cd, Cr, Hg and Ni, are also included in the US EPA Priority list, but their occurrences in urban areas are significantly less frequent (except for Hg, in >40% of samples) (US EPA, 1983). Finally, Fe and Mn are included in the Ontario guidelines for assessing sediment quality (MOEE, 1992).

The Ontario sediment quality guidelines (MOEE, 1992), described by the Lowest Effect Level (LEL) and the Severe Effect Level (SEL), and metal concentrations in BMP sediments are presented in Table 1. For brevity, As, Hg and Fe concentrations were omitted; in the case of As and Hg, no data exceeded the corresponding LELs of 6 and 0.2 μg/g, respectively, and in the case of Fe, limited data ranged from 1.8 to 3.0% (LEL = 2%, SEL = 4%).

Table 1 Sediment quality guidelines and metal concentrations in BMP sediment

Guideline or BMP Type	Constituent concentration [μg/g]						
	Cd	Cr	Cu	Mn	Ni	Pb	Zn
MOE-LEL (sediment quality guideline)	0.6	26	16	460	16	31	120
MOE-SEL (sediment quality guideline)	10	110	110	1,100	75	250	820
Kingston Pond (outlet)(Kingston)	1.4	**122**	80	485	34	149	406
WP1 (pond) (Ottawa area)	0.46	42	28	–	25	20	127
WP2 (pond) (Ottawa area)	0.53	31	22	–	15	22	95
Harding Park pond (Toronto)	< 1	24	37	484	24	24	117
Col. Sam Smith pond (Toronto)	4.2	45	**151**	693	–	202	610
Constructed wetlands (average)(Guelph)[1]	0.07	16	39	–	11	45	397
Infiltration basin (Ottawa area)	1.9	55	86	–	28	107	514
KP biofilter (Kingston)[2]	1.0	49	73	**1,433**	43	82	352
Stormwater clarifier (Toronto)[3]	1.7	61	**196**	1,259	–	200	**1,116**
Three-chamber OGS (average)(USA)[4]	**19.4**	**300**	**346**	–	–	**599**	**2,100**

[1]Constructed wetlands = average of four facilities, data from Bishop *et al.* (2000); [2]KP biofilter = biofilter used to polish effluent from the Kingston Pond; [3]Stormwater clarifier = a clarifier operated upstream of the Col. S. Smith pond; [4]Three-chamber Oil Grit Separators (average) = average of 19 facilities sampled by Schueler and Shepp (1993)

Environmental significance of observed data

The presence of high concentrations of heavy metals creates an environmental hazard, but the actual risk is difficult to assess without assessing the mobility and bioavailability of such burdens (Mikkelsen et al., 2001). There are a number of sediment quality guidelines that help assess the degree of sediment pollution. In this study, the Ontario aquatic sediment quality guidelines (see Table 1) were adopted as the most appropriate for most of the locations studied. These guidelines serve as triggers, which if exceeded, initiate further action including biotesting. Further discussion of potential ecotoxicological risks of sediment from various BMPs follows and focuses on individual BMPs, chemical levels, metal speciation, and toxicological data.

Comparison of individual BMPs

Among the sites studied by the authors, two stand out in terms of metal concentrations – the stormwater clarifier and Col. S. Smith pond, both located on the same sewer trunk and collecting sediment from the same sources. Undoubtedly, this catchment produces fairly polluted stormwater and sediments (Wood et al., 2004) significantly exceeding the level of pollution for the US NURP median site (US EPA, 1983). The runoff from the area drained appears to be heavily polluted by highway runoff and possibly by older *in situ* deposits arising from industrial operations in this catchment. However, the metal concentrations in the clarifier sludge are generally lower than those reported for highway runoff sediment by Marsalek et al. (1997a): Cu = 314, Ni = 56, Pb = 402, and Zn = 997 µg/g, and much lower than the fine fraction of such sediment (< 45 µm), Cu = 737, Ni = 126, Pb = 527, and Zn = 1,634 µg/g. The quality of highway runoff sediments is fairly similar to that of sediments collected by Schueler and Shepp (1993) from oil and grit separators: Cd = 19.4, Cr = 300, Cu = 346, Pb = 599, and Zn = 2,100 µg/g (average of 19 sites, with various land use). Such data exceed the Ontario SELs by about three times, and reflect the nature of the trapped sediments – sediments associated with road runoff, with high content of volatile solids (5–45%) and presence of hydrocarbons. The lowest metal concentrations were observed for three residential ponds, WP1, WP2 (vanLoon et al., 2000) and the Harding Park Pond, and four constructed wetlands (Bishop et al., 2000).

In the overall assessment of Ontario data, only 8.1% (six cases) of site mean concentrations listed in Table 1 exceeded the SEL, with two thirds of those originating in the Etobicoke catchment. Exceedances of Cu, Zn and Mn limits were consistent with the pollution sources in the areas studied, including highway runoff. Among the remaining sites, only the outlet concentrations of Cr in the Kingston Pond exceeded the SEL, but those elevated concentrations were affected by the local geology and represented largely non-bioavailable Cr, as documented by the high residual Cr fraction at this site, up to 60% (Marsalek et al., 1997b). The last exceedance of the SEL was observed for Mn in the biofilter sediment. There is a possibility that the filter medium retained the finest Mn enhanced sediments, with Mn concentrations about twice those found in the pond sediment. The remaining metal concentrations were typically significantly below the SEL, but greater than the LEL.

To synthesise the BMP sediment pollution data, a pollution index was introduced, relating individual metal concentrations to the corresponding SEL, and characterising each BMP by an average index value. Further simplification was possible by grouping all residential pond data together. BMP sediment pollution index values are shown in Figure 2.

The data in Figure 2 indicate that the least polluted sediments are found in residential ponds and constructed wetlands with low input of road runoff. Increased inputs of

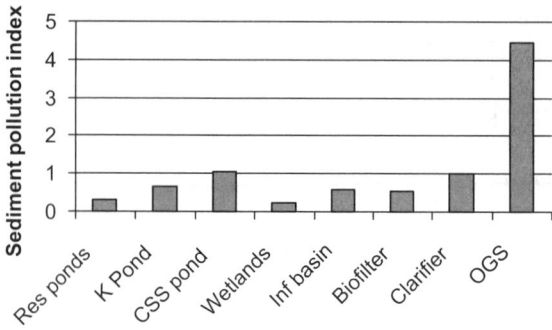

Figure 2 Sediment pollution index (SPI) for various BMPs (SPI = Σ (C_i/SEL_i)/n, i = 1−n)

polluted sediments from road/highway runoff (or similar sources) are reflected in lower quality of sediments in ponds or "treatment" devices processing stormwater (biofilter, clarifier, OGS separators).

While sediment chemistry data may indicate relatively high burdens, such loads represent environmental risks only if bioavailable, or susceptible to transformations which would make them bioavailable. To address the issues of bioavailability, various approaches are used, starting with speciation of metal fractions using sequential analysis, followed by biotesting either *in situ*, or in the laboratory. Some experience with such approaches is reviewed below.

Sequential analysis proposed by Tessier *et al.* (1979) leads to determination of five particular metal fractions operationally defined as exchangeable (Fraction 1), bound to carbonate (Fraction 2), bound to iron and manganese oxides (Fraction 3), bound to organic matter (Fraction 4), and residual (Fraction 5). Using this procedure, sediment samples from seven sites were analysed and the results are summarised in Table 2.

The data in Table 2 indicate large differences in speciation of metals in BMP sediments. The most potentially mobile is Cd, followed by Cu and Pb. The least potentially mobile (and bioavailable) is Cr. It appears from Table 2 that for most metals, the oxidizing and reducing conditions would be most critical with respect to release of metals from sediments, and this indicates the importance of maintaining a stable oxygen regime in stormwater management ponds. Another way of studying bioavailability of accumulated trace metals in pond sediments is by studying metal uptake by caged freshwater mussels. Anderson *et al.* (2004) found no correlation between total metals in pond sediments and the accumulated burden in the mussels. Results also suggested that Pb was a possible concern due to bioavailability, Ni, Cr and Cd did not appear to be in bioavailable forms, and Cu had some limited bioavailability. It was suggested that this type of evaluation tool also be considered by managers in determining operation and maintenance for BMP facilities.

Table 2 Sequential analysis results for sediment samples from seven sites

Source	Sediment fractions [%] (mean of seven sites)				
	Cd	Cr	Cu	Pb	Zn
Fraction 1	12.0	0.7	1.2	1.8	0.3
Fraction 2	38.5	1.2	3.5	14.7	15.8
Fraction 3	25.3	7.7	15.0	43.2	19.5
Fraction 4	12.5	23.3	53.5	22.3	28.8
Fraction 5	11.8	67.5	26.3	26.3	35.8

Sediment particle size affected the sediment chemistry, with fine particles (silt and clay) by the Kingston pond outlet (line 5 in Table 1) containing up to four times higher concentrations of metals than sandy particles by the inlet (Marsalek et al., 1997b). The presence of organic carbon would be particularly important for Cu, which seems to have a significant fraction bound to organic matter. For Kingston pond sediment, organic carbon varied from 3.8% in bottom sediment to 5.8% in suspended particulate (Marsalek et al., 1997b); much higher levels were noted in the stormwater clarifier (Wood et al., 2004), with VSS representing on average 26% of TSS. Also, the OGS data in Table 1 indicate that sediments with higher organic content contain higher metal levels (Schueler and Shepp, 1993).

Besides trace metals, other chemicals in BMP sediment are also of interest in connection with toxicological considerations. Earlier research indicated two major groups of potential toxicants – hydrocarbons (particularly polycyclic aromatic hydrocarbons, PAHs) and pesticides. For the Kingston Pond, Marsalek et al. (2002) reported total PAH concentrations of 16.37 µg/g, well below the SEL level in Ontario guidelines (specified as 10,000 µg/g of organic carbon). Dutka et al. (1994) studied sediment toxicity in four Toronto area ponds, and found incidences of sediment toxicity, which appeared related to pesticides (triazine, metolachlor) in late spring sampling. Without specialized TIE (Toxicity Identification Evaluation) analysis, it appears impossible to clearly identify the sources of toxicity. Most likely, the synergistic effect of various chemicals, and the ambient BMP conditions, lead to sediment toxicity and the development of chemical protocols should focus on known sources of toxicants in the areas studied.

Sediment toxicity testing was reported for the Kingston Pond (Marsalek et al., 1999) and five Toronto ponds (Dutka et al., 1994; Mayer et al., 1997; Rochfort et al., 2000). A variety of bioassays were employed in these tests, including the well known Microtox™ solid phase test, the direct solid-phase SOS-chromotest (for genotoxicants), *Panagrellus redivivus* (nematode), and benthic toxicity tests (*H. azteca, C. riparius, Hexagenia* spp., *T. tubifex*; Rochfort et al., 2002). Such tests indicate the overall toxicity of sediments, without identification of causal constituents. In all studies, some toxic effects were noted. In the Kingston Pond, Microtox™ test results indicated toxicity in pond flow sources – an upstream creek and runoff from a commercial plaza, strong toxicity of sediments in the pond, and no toxicity in the creek downstream of the pond (Marsalek et al., 1999). Even higher toxicity was indicated by Microtox™ for sediments in the Rouge River Pond (Toronto), receiving runoff from a major multi-lane divided highway (Marsalek et al., 1999). In four Toronto ponds (including Col. S. Smith reservoir), sediment solvent extracts showed toxic responses, suggesting presence of organic toxicants/genotoxicants. The highest incidence of responses was found in the Col. S. Smith pond (Mayer et al., 1997), and on a seasonal basis, in spring. The Dutka et al. (1994) study also showed the presence of toxicants and genotoxicants in sediments, indicating the presence of bioavailable toxicants and promutagens, with no seasonal patterns. Finally, benthic toxicity testing indicated no toxicity in the Harding Pond (residential pond), but toxicity indications in the receiving creek, both upstream and downstream of the pond outlet, and therefore attributable to other sources (Rochfort et al., 2000).

Conclusions

Stormwater BMP treatment trains are designed to immobilize suspended solids and sediment in various train elements to protect downstream BMPs and drainage elements against clogging and the downstream waters against pollution. These designs lead to accumulation of solids in BMPs, including ponds, wetlands, biofilters, infiltration basins, and oil and grit separators. Such solids are fairly polluted by metals and inorganics, as indicated by the data reported in this study, and by other chemicals, as reported in the literature. The highest concentrations of heavy metals were found in two facilities serving an area with industrial land and a highway corridor. Identification of chemicals exerting toxic effects remains a challenge and, consequently, direct toxicity testing may be more effective. Circumstantial evidence points to highway runoff sediment as a major contributor to BMP sediment toxicity. Guidelines for operation and maintenance of BMPs neglect the ecotoxicological risks associated with contaminated BMP sediments and should be amended to account for, and to minimise, such risks.

References

Anderson, B.C., Bell, T., Hodson, P., Marsalek, J. and Watt, W.E. (2004). Accumulation of trace metals in freshwater invertebrates in stormwater management facilities. *Wat. Qual. Res. J. Can.*, **39**(4), 362–373.

Anderson, B.C., Watt, W.E. and Marsalek, J. (2002). Critical issues for stormwater ponds: learning from a decade of research. *Wat. Sci. Tech.*, **45**(9), 277–283.

Bishop, C.A., Struger, J., Barton, D.R., Shirose, L.J., Dunn, L. and Campbell, G.D. (2000). Contamination and wildlife communities in stormwater detention ponds in Guelph and the Greater Toronto Area, Ontario, 1997 and 1998. Part II – contamination and biological effects of contamination. *Wat. Qual. Res. J. Can.*, **35**(3), 437–474.

Dutka, B.J., Marsalek, J., Jurkovic, A., McInnis, R. and Kwan, K.K. (1994). A seasonal ecotoxicological study of stormwater ponds. *Zeitschrift fuer Angewandte Zoologie*, **80**(3), 361–381.

Marsalek, P.M. (1997). *Special characteristics of an on-stream stormwater pond: winter regime and accumulation of sediment and associated contaminants*. M.Sc. Thesis, Department of Civil Engineering, Queen's University, Kingston, Ontario, Canada.

Marsalek, J., Brownlee, B., Mayer, T., Lawal, S. and Larkin, G.A. (1997a). Heavy metals and PAHs in stormwater runoff from the Skyway Bridge, Burlington, Ontario. *Wat. Qual. Res. J. Can.*, **32**(4), 815–827.

Marsalek, J., Rochfort, Q., Grapentine, L. and Brownlee, B. (2002). Assessment of stormwater impacts on an urban stream with a detention pond. *Wat. Sci. Tech.*, **45**(3), 255–263.

Marsalek, J., Rochfort, Q., Mayer, T., Servos, M., Dutka, B. and Brownlee, B. (1999). Toxicity testing for controlling urban wet-weather pollution: advantages and limitations. *Urban Water*, **1**, 91–103.

Marsalek, J., Watt, W.E., Anderson, B.C. and Jaskot, C. (1997b). Physical and chemical characteristics of sediments from a stormwater management pond. *Wat. Qual. Res. J. Can.*, **32**(1), 89–100.

Mayer, T., Dutka, B. and Marsalek, J. (1996). Toxicity and contaminant status of suspended and bottom sediments from urban stormwater ponds. In *Proc. 23rd Annual Aquatic Toxicity Workshop*, Goudey, J.S., Swanson, S.M., Treissman, M.D. and Niimi, A.J. (eds), Oct. 7–9, Calgary, Alberta, pp. 159–165.

Mayer, T., Marsalek, J. and Delos Reyes, E. (1996). Nutrients and metal contaminants status of urban stormwater ponds. *J. Lake and Reservoir Mgmt.*, **12**(3), 348–363.

Mikkelsen, P.S., Baun, A. and Ledin, A. (2001). Risk assessment of stormwater contaminants following discharge to soil, groundwater or surface water. In *Advances in stormwater and agricultural runoff source controls*, Marsalek, J., Watt, E., Zeman, E. and Sieker, H. (eds), Kluwer Academic Publishers, NATO Science Series, Dordrecht/Boston/London, pp. 69–80.

Ministry of the Environment (MOE) (2003). *Stormwater management planning and design manual*. Toronto, Ontario.

Ministry of the Environment and Energy (1992). *Guidelines for the protection and management of aquatic sediment quality in Ontario*, MOEE, Toronto.

Mothersill, C.L., Anderson, B.C., Watt, W.E. and Marsalek, J. (2000). Biological filtration of stormwater: field operations and maintenance experiences. *Wat. Qual. Res. J. Can.*, **35**(3), 541–562.

Rochfort, Q., Grapentine, L., Marsalek, J., Brownlee, B., Reynoldson, T., Thompson, S., Milani, D. and Logan, C. (2000). Using benthic assessment techniques to determine combined sewer overflow and stormwater impacts on the aquatic ecosystem. *Wat. Qual. Res. J. Can.*, **35**(3), 365–397.

Sartor, J.D. and Boyd, G.B. (1972). *Water pollution aspects of street surface contaminants*, US EPA report EPA-R2-72-081, Washington, DC.

Schueler, T.R. and Shepp, D. (1993). *The quality of trapped sediments and pool water within oil grit separators in suburban Maryland (interim report)*, Metropolitan Washington Council of Governments, Washington, DC.

Stone, M. and Marsalek, J. (1996). Trace metal compositions and speciation in street sediment: Sault Ste. Marie, Canada. *Wat. Air Soil Poll.*, **87**, 149–169.

Tessier, A., Campbell, P.G.C. and Bisson, M. (1979). Sequential extraction procedure for the speciation of particulate trace metals. *Anal. Chem.*, **51**(7), 844–850.

US EPA (1983). *Results of the Nationwide Urban Runoff Program. Volume I – Final report.* US EPA report PB84-185552, Washington, DC.

Van Buren, M.A., Watt, W.E. and Marsalek, J. (1997). Removal of selected urban stormwater constituents by an on-stream pond. *J. Environmental Planning and Management*, **40**(1), 5–18.

vanLoon, G., Anderson, B.C., Watt, W.E. and Marsalek, J. (2000). Characterizing stormwater sediments for ecotoxic risk. *Wat. Qual. Res. J. Can.*, **35**(3), 341–364.

Wood, J., Dhanvantari, S., Yang, M., Rochfort, Q., Chessie, P., Marsalek, J., Kok, S. and Seto, P. (2004). Feasibility of stormwater treatment by conventional and lamellar settling with and without polymeric flocculant addition. *Wat. Qual. Res. J. Can.*, **39**(4), 406–416.

Lead isotope ratios in urban road runoff

M. Shinya*, K. Funasaka*, K. Katahira*, M. Ishikawa** and S. Matsui***

*Osaka City Institute of Public Health and Environmental Sciences, 8-34 Tojo-cho, Tennoji-ku, Osaka 543-0026, Japan (E-mail: *m.shinya@iphes.city.osaka.jp*)

**Department of Civil Engineering, Osaka Institute of Technology, 5-16-1 Ohmiya, Asahi-ku, Osaka 535-8585, Japan

***Graduate School of Global Environmental Studies, Kyoto University, Yoshida-honmachi, Sakyo-ku, Kyoto 606-8501, Japan

Abstract Lead isotopic analyses of road runoff and airborne particulate matter have been carried out to elucidate sources of lead pollution at urban and suburban sites. While lead is often observed in road runoff in suspended form, suspended particle size had no relation to the lead isotopic distribution, as a result of comparison between runoff samples with total suspended solids and those with minute particles passed through a 75 μm sieve. Lead isotope ratios in airborne particulate matter in urban areas fell within a wider range than those in road runoff. Since there was little difference of the ratios between a heavy traffic-flow site and residential sites, airborne lead derived from vehicle exhaust was found to make little contribution to the contamination of road runoff. On the other hand, the ratios in road runoff at a suburban site showed the same range as those at an urban site. Lead in road runoff was therefore suggested to be produced on site by traffic related substances, such as tire wear, other than vehicle exhaust.

Keywords Road runoff; airborne particulate matter; lead isotope ratio; inductively coupled plasma mass spectrometry (ICP-MS)

Introduction

Lead pollution in urban environment has been widely studied because of its ubiquitous distribution and serious effects on human health (Nriagu, 1988). The amount of anthropogenic lead emitted into the atmosphere has declined as a result of the phasing out of leaded gasoline products (Mukai *et al.*, 1993). However, although unleaded gasoline has been in use for several decades in Japan, runoff from the road surface contains significant loads of hazardous lead, especially in urban areas. Lead in runoff water is mainly particulate-bound and its concentration is highest in the initial phases of runoff water regardless of rainfall conditions (Shinya *et al.*, 2000), often reaching values 10 to 100 times above the Japanese environmental water quality standard ($0.01\,\mathrm{mg\,L^{-1}}$). The source of lead in urban road runoff has however not been identified.

Many studies have sought to determine the source of lead in the urban atmosphere. The potential anthropogenic sources of airborne lead, such as mining, smelting, coal combustion, waste incineration, and the use of lead additives in petrol, exceed the contribution of natural sources on a global scale (Nriagu, 1996). Atmospheric deposition of lead has led to pollution of surface soils (Hansmann and Koppel, 2000; Prohaska *et al.*, 2000) and lake sediments (Renberg *et al.*, 2002).

Measurement of stable lead isotopes in atmospheric aerosols gives valuable information about their sources, since different sources often have a dissimilar lead isotope signature. This variation derives from the use of many different ore bodies in the various industrial uses of lead. For lead isotopic analysis, an alternative to thermal ionization mass spectrometry (TIMS), which is the traditional technique for determining isotope ratio, is provided by inductively coupled plasma mass spectrometry (ICP-MS), which offers ease of operation, high sample throughput, and widespread availability, although with limited precision compared to TIMS. The use of ICP-MS in source determination by isotopic analysis has been increasing year by year in the field of environmental research.

The objective of the present study was to identify the source of lead in urban road runoff by isotopic analysis using ICP-MS and to elucidate the contribution to road runoff of airborne particulate matter by comparing the lead isotope ratio of the two.

Methodology

Sampling sites

The urban sampling site, located in Osaka city, was the same as in the previous study (Shinya et al., 2000). As a suburban control a site in Ritto city was selected, which has several heavily trafficked routes including Route 1, along which are located many factories and warehouses, but which is sparsely populated and has a large area of fields away from the roads. Runoff samples in Ritto were collected at the side of Route 1.

Airborne particulate matter samples were also collected for the purpose of comparison with runoff samples. Collection was performed on filters with a high-volume air sampler at six monitoring sites in Osaka City: one site, Dekijima, was located at the roadside of Route 43, which has heavy traffic flow with a large volume of diesel vehicles, and the others in commercial and residential areas. The sampling sites are shown in Figure 1.

Sampling methods

At the Osaka site, five different runoff events were investigated in 1999 and three in 2000, and at the Ritto site ten events in 2002. Runoff samples were collected in the same manner as described in the previous study (Shinya et al., 2003). All runoff samples were passed through a 2 mm-mesh stainless sieve then stored in a refrigerator at 4°C until analysis. In the 2000 study, a 75 μm-mesh sieve was also used to examine the behavior of the minute particles that are easily flushed out even by lower levels of rainfall and may be transported by air. These particles are rich in lead, and the cumulative runoff load derived from them accounts for half of the total load (Shinya et al., 2002).

Using a high-volume air sampler (Model-120, Kimoto Electric) without impactor, airborne particulates were collected monthly from October 2002 to March 2004 on a quartz fiber filter (QR-100, Advantec) at a flow rate of $1{,}000 \, \text{L} \, \text{min}^{-1}$. The sampling period was

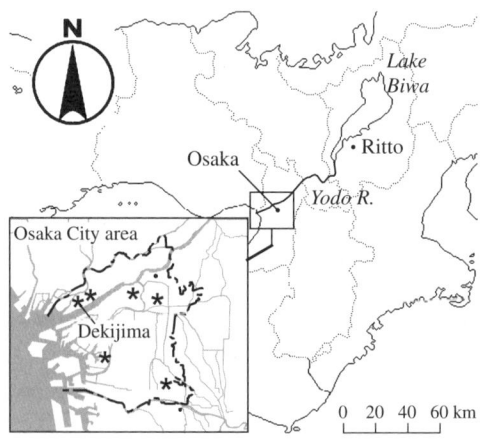

Figure 1 Sampling site for road runoff (●) and airborne particulate matter (★); the Dekijima site is a roadside site with heavy traffic-flow while the others are located in commercial and residential areas

24 h for each sample. After collection, the filters were stored in a freezer at −20°C until analysis.

Analytical procedures

Runoff samples were digested with nitric acid and perchloric acid by heating on a commercially available hot plate and filtered (5C, Advantec). The filtrate was reheated until almost dry and finally dissolved in 0.1 mol l^{-1} HNO$_3$. The filter samples with airborne particulates were prepared using the same procedure as in the previous study (Funasaka et al., 2003).

Lead determination of these samples was performed with an electrothermal atomic absorption spectrometer (4100ZL, Perkin-Elmer) for lower concentration or an inductively coupled plasma emission spectrometer (IRIS 1000, Nippon Jarrel-Ash) for higher concentration. After determination, the lead concentration in the sample solution was adjusted to around 100 μg L^{-1}, as this concentration allowed isotopic analysis with satisfactory precision (Shinya et al., 2005).

Lead isotope analysis of the solution was performed by quadrupole ICP-MS (HP4500, Yokogawa Analytical Systems). Operating conditions have been described in detail elsewhere (Shinya et al., 2005). Four stable isotopes, ^{204}Pb, ^{206}Pb, ^{207}Pb and ^{208}Pb, were monitored; only ^{204}Pb is non-radiogenic, ^{206}Pb, ^{207}Pb, and ^{208}Pb are continuously formed by the radioactive decay of ^{238}U, ^{235}U and ^{232}Th, respectively. Each sample was measured ten times, and the average lead isotope ratio and the relative standard deviation of the sample were calculated. NIST SRM981 was used as a reference to determine the mass bias correction factors of the lead isotope ratios.

Results and discussion

Lead isotope ratios of road runoff samples in Osaka

The lead isotope composition of urban road runoff has been described in detail (Shinya et al., 2005), with typical values for each lead isotope ratio in the road runoff indicated for the duration of all runoff events regardless of sampling season. Obtained by averaging the lead isotope ratio of runoff samples from five events in Osaka in 1999, the typical values for the ratios of ^{207}Pb/^{206}Pb, ^{208}Pb/^{206}Pb, and ^{206}Pb/^{204}Pb were 0.8656 ± 0.0034, 2.108 ± 0.007, and 18.02 ± 0.07, respectively.

Average lead isotope ratios of runoff samples from three events in 2000 are shown in Table 1, in which maximum lead concentration refers to the concentration in the first flush sample. The observed isotope ratios of runoff samples passed through a 2 mm-mesh sieve, which included total suspended solids (TSS), were very close to those observed in 1999. Lead isotope ratios of runoff samples passed through a 75 μm-mesh sieve, which included minute particles only, are also shown in Table 1. In all events, the average value in samples with minute particles was very close to that in a sample with TSS.

Table 1 Average lead isotope ratios of road runoff in Osaka

Event date	Mesh size*	No. of samples	Range of Pb concentration [mg L^{-1}]	^{207}Pb/^{206}Pb	^{208}Pb/^{206}Pb	^{206}Pb/^{204}Pb
9/Jun/00	2 mm	9	0.007–0.211	0.8650 ± 0.0015	2.110 ± 0.002	18.08 ± 0.10
	75 μm	9	0.007–0.149	0.8617 ± 0.0015	2.106 ± 0.004	18.10 ± 0.07
25/Jul/00	2 mm	7	0.023–1.616	0.8678 ± 0.0018	2.115 ± 0.002	18.03 ± 0.07
	75 μm	7	0.021–0.298	0.8669 ± 0.0020	2.114 ± 0.005	18.04 ± 0.09
8/Sep/00	2 mm	5	0.019–1.638	0.8648 ± 0.0016	2.108 ± 0.002	18.07 ± 0.04
	75 μm	5	0.012–0.354	0.8648 ± 0.0018	2.109 ± 0.004	18.03 ± 0.06

*Mesh size of the sieve through which the samples passed

Geochemists prefer to use lead isotope ratios incorporating ^{204}Pb due to the mathematical simplicity of using a non-radiogenic isotope. However, environmental scientists tend to use ^{206}Pb/^{204}Pb vs. ^{206}Pb/^{207}Pb or ^{206}Pb/^{207}Pb vs. ^{208}Pb/^{206}Pb because of their better analytical precision (Monna *et al.*, 1997). Here we use ^{207}Pb/^{206}Pb vs. ^{208}Pb/^{206}Pb for comparability with recent studies conducted using ICP-MS. We also present conventional ^{206}Pb/^{204}Pb vs. ^{207}Pb/^{204}Pb diagrams.

Figure 2 illustrates diagrams of ^{207}Pb/^{206}Pb vs. ^{208}Pb/^{206}Pb and ^{206}Pb/^{204}Pb vs. ^{207}Pb/^{204}Pb in road-runoff samples collected in Osaka in 2000. It is clear that the plotted area of the samples with minute particles is very close to that of the samples with TSS in both diagrams. Larger-sized particles are flushed out when it rains heavily in what is called the first flush phenomenon, so that most particles larger than 75 μm are flushed out within the first 10 min even with a runoff duration of 60 min, as in the event of 25th July, 2000 (Shinya *et al.*, 2002). In other words, very few larger-sized particles are observed after the first flush. As described above, runoff characteristics differed according to particle size, but lead-isotopic characteristics were almost the same. This suggests that runoff water includes particles of various size derived from a specific source of lead.

Comparison of lead isotope ratios of first-flush samples between urban and suburban areas

Lead isotope analysis was carried out with runoff samples from Ritto, a site with suburban location but heavy traffic flow. Since analytical results from the Osaka samples showed that the values for isotope ratio were almost constant for the duration of all events, only the first flush samples with the highest concentration of lead were measured because of their superior analytical precision. Table 2 shows lead isotope ratios of first flush samples from eight events in Osaka in 1999 and 2000 and from ten events in Ritto in 2002. First flush samples contained the richest lead, but, as shown in Table 2, the average lead concentration was one order lower in the Ritto samples than in the Osaka samples as the former location is less polluted. The mean values for lead isotope ratios at the Ritto site were closely similar to those at the Osaka sites.

The lead isotope ratio diagrams shown in Figure 3 indicate that lead isotope compositions for Ritto overlapped fully with those for Osaka. The specific lead isotope compositions observed at both urban and suburban sites, therefore, suggest that traffic is a source of lead in road runoff.

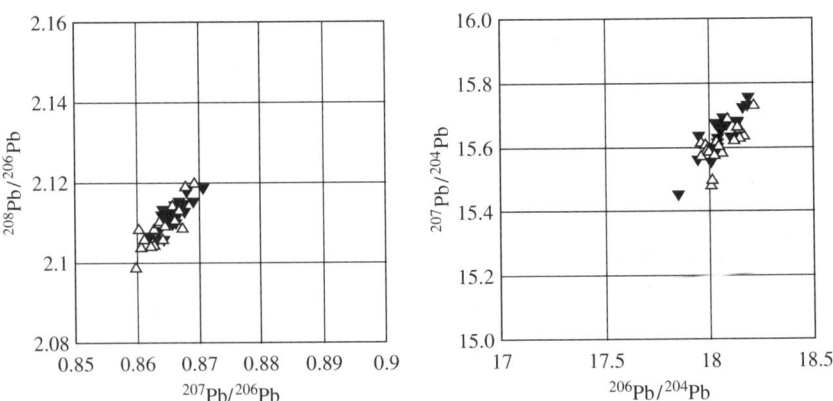

Figure 2 ^{207}Pb/^{206}Pb vs. ^{208}Pb/^{206}Pb and ^{206}Pb/^{204}Pb vs. ^{207}Pb/^{204}Pb of road runoff in Osaka, (▼) runoff samples passed through 2 mm-mesh sieve (including total suspended solids), (△) particles passed through 75 μm-mesh sieve (including minute particles)

Table 2 Average lead concentration and isotope ratios of first flush samples of road runoff

Site	Category of site	Year	No. of samples	Mean Pb concentration [mg L^{-1}]	^{207}Pb/^{206}Pb	^{208}Pb/^{206}Pb	^{206}Pb/^{204}Pb
Osaka	Urban	1999–2000	8	0.745	0.8647 ± 0.0027	2.108 ± 0.007	17.99 ± 0.14
Ritto	Suburban	2002	10	0.084	0.8644 ± 0.0033	2.109 ± 0.006	18.04 ± 0.07

Lead isotope ratios in airborne particles

Lead isotope ratios were also measured in airborne particulate matter collected in Osaka. The results are shown in Table 3. The ratio of both ^{207}Pb/^{206}Pb and ^{208}Pb/^{206}Pb in airborne particulates showed somewhat larger values than in road runoff samples at the Osaka site. The isotopic characteristics of airborne lead from the heavy-traffic roadside site, Dekijima, were not distinct from those at commercial and residential sites, with no evident difference in lead isotope ratio, although higher lead concentration was observed at Dekijima (Funasaka *et al.*, 2003). Although the airborne samples at each site were taken during different seasons, no large variation in isotope ratio was observed.

Road runoff samples were not collected from January to April, as the region is affected during this season by yellow sand dust from China (Funasaka *et al.*, 2003) and continental lead could therefore be present in airborne particulate samples. However, even when samples collected during this season were excluded, lead isotope ratio did not vary between the Dekijima and the others. Mukai *et al.* state that there is no large seasonal variation in isotope ratios of airborne lead in Japan because industrial emissions also do not show seasonal variation (Mukai *et al.*, 1993).

Diagrams of isotope ratios of airborne lead are shown in Figure 4. Compared with the road runoff in Figures 2 and 3, lead isotope ratios of airborne particulate matter fell within a wider range which included the isotope ratio range of road runoff. This finding suggests that the source of airborne lead included not only the sources of lead in road runoff but other various sources. In other words, lead in airborne particulate matter does not contribute greatly to lead in road runoff. It is concluded that lead in road runoff is produced on site as a consequence of traffic.

Characterization of the lead source in urban road runoff

Figure 5 was obtained by comparing the lead isotope ratios in the present study to those in the literature, focusing on the radiogenic lead isotope ratio of ^{207}Pb/^{206}Pb, which is generally measured in various studies. Examination of the figure leads to the inference

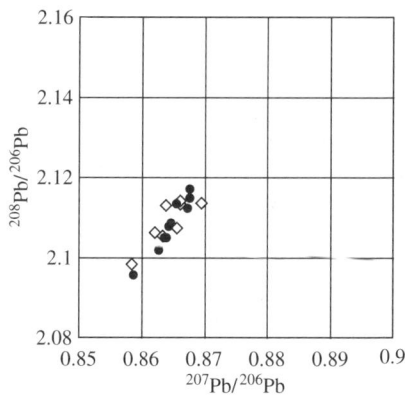
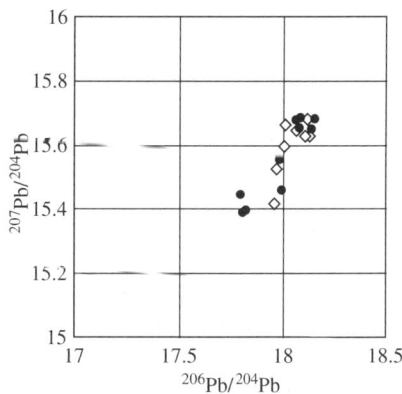

Figure 3 ^{207}Pb/^{206}Pb vs. ^{208}Pb/^{206}Pb and ^{206}Pb/^{204}Pb vs. ^{207}Pb/^{204}Pb in first-flush samples of road runoff in Osaka (◇) and Ritto (●)

Table 3 Average lead isotope ratios of airborne particulate matter collected in Osaka

Category of site	Season	No. of samples	$^{207}Pb/^{206}Pb$	$^{208}Pb/^{206}Pb$	$^{206}Pb/^{204}Pb$
Heavy traffic roadside	All	18	0.8710 ± 0.0060	2.118 ± 0.011	17.91 ± 0.16
	except for winter–spring*	11	0.8700 ± 0.0052	2.118 ± 0.012	17.98 ± 0.08
Residential and commercial	All	90	0.8706 ± 0.0072	2.117 ± 0.014	17.91 ± 0.17
	except for winter–spring*	55	0.8689 ± 0.0063	2.116 ± 0.015	17.97 ± 0.13

*January, February, March and April

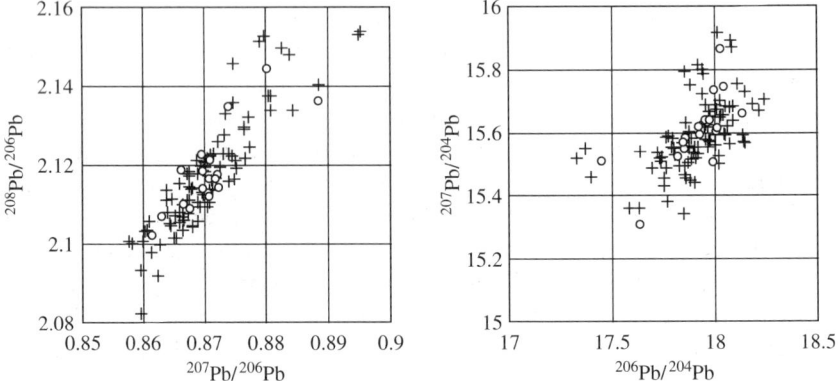

Figure 4 Lead isotope ratios ($^{207}Pb/^{206}Pb$ vs. $^{208}Pb/^{206}Pb$ and $^{206}Pb/^{204}Pb$ vs. $^{207}Pb/^{204}Pb$) in airborne particulate matter collected in Osaka; (○) heavy traffic-flow roadside site, (+) residential and commercial sites

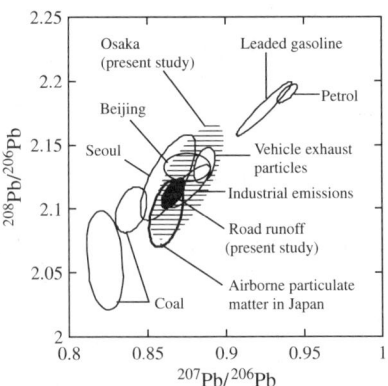

Figure 5 Comparison of lead isotope ratios from other environmental samples and sites; airborne particulate matter in Japan (Mukai et al., 1993), in Beijing (Mukai et al.,1993), in Seoul (Mukai et al.,1993), coal (Bacon, 2002; Mukai et al.,1993), industrial emissions (Bacon, 2002; Monna et al., 1997), vehicle exhaust particles (Mukai et al., 1993; Zheng et al., 2004), leaded gasoline (Monna et al., 1997; Mukai et al.,1993) and petrol (Bacon, 2002).

that the source of lead in road runoff could be airborne particulate matter or industrial emissions. While it is likely that industrial emissions have a widely varying lead isotopic composition due to the multiple sources of lead (as shown in several studies), lead in incinerator ashes is generally considered to be an acceptable surrogate as incinerated wastes represent all the sources of industrial lead (Bacon, 2002). Even though incinerator fly ash may well be deposited on the road surface by way of the atmosphere, it is difficult to determine whether it is the source of the lead in road runoff because of the small contribution of airborne particulate matter to road runoff, as described above.

Traffic as a potential lead source was mentioned above, but the lead isotope ratio of vehicle exhaust particles is quite far removed from that of road runoff, as shown in Figure 5. These particles are therefore not proposed as the source of the lead. Leaded gasoline had not been in use for several decades in Japan. Lead in road runoff is therefore suggested to be produced by wear of other traffic-related substances, for example tires and asphalt. Further study is necessary to clearly explain the lead source of road runoff.

Conclusions

Lead isotopic measurement of road runoff and airborne particulate matter in the Osaka urban area was applied to investigate the contribution of airborne lead to lead in urban road runoff and to estimate the source of lead in urban road runoff. The following conclusions could be drawn:

The lead isotope composition of road runoff was specific regardless of the particle sizes included in the samples and of the sampling sites, urban or suburban. They fell within a narrower range than airborne lead isotopes, suggesting little contribution of airborne particulate matter to road runoff. While no definite lead source for urban road runoff was determined, it appeared to be produced on site by traffic-related substances other than vehicle exhaust.

References

Bacon, J.R. (2002). Isotopic characterization of lead deposited 1989–2001 at two upland Scottish locations. *J. Environ. Monit.*, **4**, 291–299.

Funasaka, K., Sakai, M., Shinya, M., Miyazaki, T., Kamiura, T., Kaneko, S., Ohta, K. and Fujita, T. (2003). Size distributions and characteristics of atmospheric inorganic particles by regional comparative study in urban Osaka, Japan. *Atmos. Environ.*, **37**, 4597–4605.

Hansmann, W. and Koppel, V. (2000). Lead isotopes as tracers of pollutants in soils. *Chem. Geol.*, **171**, 123–144.

Monna, F., Lancelot, J., Croudace, I.W., Cundy, A.B. and Lewis, J.T. (1997). Pb isotopic composition of airborne particulate material from France and the southern United kingdom; implication for Pb pollution sources in urban areas. *Environ. Sci. Technol.*, **31**, 2277–2286.

Mukai, H., Furuta, N., Fujii, T., Ambe, Y., Sakamoto, K. and Hashimoto, Y. (1993). Characterization of sources of lead in the urban air of Asia using ratios of stable lead isotopes. *Environ. Sci. Technol.*, **27**, 1347–1356.

Nriagu, J.O. (1988). A silent epidemic of environmental metal poisoning? *Environ. Pollut.*, **50**, 139–161.

Nriagu, J.O. (1996). A history of global metal pollution. *Science*, **272**, 223–224.

Prohaska, T., Watkins, M., Latkoczy, C., Wenzel, W.W. and Stingeder, G. (2000). Lead isotope ratio analysis by inductively coupled plasma sector field mass spectrometry (ICP-SMS) in soil digests of a depth profile. *J. Anal. At. Spectrom.*, **15**, 365–369.

Renberg, I., Brannvall, M.-L., Bindler, R. and Emteryd, O. (2002). Stable lead isotopes and lake sediments – a useful combination for the study of atmospheric lead pollution history. *Sci. Total Environ.*, **292**, 45–54.

Shinya, M., Tsuchinaga, T., Kitano, M., Yamada, Y. and Ishikawa, M. (2000). Characterization of heavy metals and polycyclic aromatic hydrocarbons in urban highway runoff. *Wat. Sci. Technol.*, **42**(7–8), 201–208.

Shinya, M., Konishi, T., Miyanishi, H. and Ishikawa, M. (2002). Characterization of urban highway runoff. *Jpn J. Water and Waste*, **44**, 207–213 (in Japanese).

Shinya, M., Tsuruho, K., Konishi, T. and Ishikawa, M. (2003). Evaluation of factors influencing diffusion of pollutant load in urban highway runoff. *Wat. Sci. Technol.*, **47**(7–8), 227–232.

Shinya, M., Katahira, K., Ishikawa, M. and Matsui, S. (2005). Determination of stable lead isotope ratios in urban highway runoff by inductively coupled plasma mass spectrometry, submitted.

Zheng, J., Tan, M., Shibata, Y., Tanaka, A., Li, Y., Zhang, G., Zhang, Y. and Shan, Z. (2004). Characteristics of lead isotope ratios and elemental concentrations in PM10 fraction of airborne particulate matter in Shanghai after the phase-out of leaded gasoline. *Atmos. Environ.*, **38**, 1191–1200.

The characteristics and measuring technique of refractory dissolved organic substances in urban runoff

K. Wada*, S. Yamanaka**, M. Yamamoto** and K. Toyooka**

*Lake Biwa – Yodo River Water Quality Preservation Organization, 1-1-30, Kitahama, Chuo-ku, Osaka, 541-0041, Japan (E-mail: wada@byq.or.jp)
**Shiga Prefectural Government, 4-1-1, Kyomachi, Otsu, Shiga, 520-8577, Japan
(E-mail: yamanaka-sunao@pref.shiga.lg.jp; yamamoto-masanori@pref.shiga.lg.jp; toyooka-koji@pref.shiga.lg.jp)

Abstract It is considered that refractory dissolved organic substances have caused an increase in the COD concentration in Lake Biwa in recent years. We investigated the organic matter in the first flush of stormwater runoff from a road in the watershed area of the lake, and studied the possibility of improvement in the water environment from that aspect. After percolating the stormwater through soil, we analyzed organic substances fractionated by using GPC-TC. And we examined the effect of removal of organic substances by comparing the peak height before and after percolation. In the result of the experiments, we found that soil infiltration reduced the refractory dissolved organic substance and we successfully designed a system for a simple and easy experimental facility to treat urban runoff.
Keywords First flush runoff; GPC-TC analysis; organic substance; refractory; soil; stormwater runoff

Introduction

In the Lake Biwa Basin, they have implemented a system for total effluent control and water quality management for point source pollution. The water quality of Lake Biwa has had a decreasing BOD concentration since 1984, but it has an increasing COD concentration (Shiga pref., 2001). We call it the divergent phenomenon of BOD and COD (Figure 1). In regard to the divergence of the COD and the BOD concentration, it is considered that refractory dissolved organic substances cause the increase in the COD not the BOD. This substance may increase the amount of organic substance which cannot be decomposed by microorganisms (Imai *et al.*, 1998; Yonebayashi, 2004). Additionally, the water quality that flows into Lake Biwa from the rivers doesn't progress in terms of improvement, because the COD concentration remains unchanged or increases under the aging phenomena. As for the cause of the increase of COD, it is implied from the study that there is a possibility of significant influence of the catchment area.

However, the catchment basin of Lake Biwa has undergone rapid urban development in recent years, and impervious land surface has increased. In the result, we can't disregard the non-point source pollution from urban runoff as a significant influence on the water environment of Lake Biwa. So it's imperative that we have to take effective action against the non-point source pollution from the impervious surfaces such as the buildings and the roads, because the sediment deposited in those areas is discharged into the river and channels along with the urban runoff at a time (Barrett *et al.*, 1998; Great Lakes Science Advisory Board, 2000).

The aim of this study is to reduce non-point pollution sources from the urban area. In this paper we discuss refractory dissolved organic substances that we investigated and their characteristics within the urban runoff. Finally we put forward a proposal on effective means to reduce the pollutant load.

doi: 10.2166/wst.2006.053

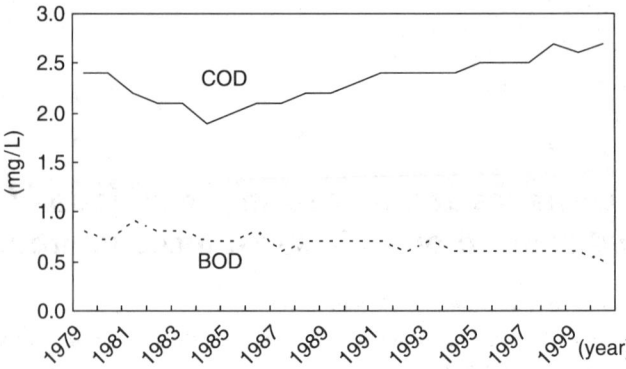

Figure 1 Divergent phenomenon of BOD and COD concentration of Northern Lake Biwa

Methodology

The characteristics of urban runoff

The site used is situated on the southeast area of Lake Biwa which is the lakeshore road in Kusatsu City. The road usually carries 12,000 vehicles per 12 hours (Shiga Pref., 1999). This area from which runoff is collected is assumed to be a parallelogram of approximately $750 m^2$. We investigated the characteristics and dynamics of stormwater runoff from this road surface. The runoff was collected over time intervals during one rainfall. We analyzed the samples collected using the method outlined below.

Gel Permeation Chromatography - Total Carbon Analysis (GPC-TC Analysis). The sample is filtrated using a 0.45 μm membrane-filter, and the filtrate is evaporated at 10–60 times using freeze concentration. The measured amount of this sample and the solution were injected into the GPC-TC analysis. The operating conditions for GPC-TC are presented in Table 1, and these conditions were established in accordance with those by Yonebayashi (1995), Higashi *et al.* (1997) and Shiga Pref. Report (1997). GPC-TC chromatogram is shown as the organic volume and peaks of higher molecular weight substances appear at an earlier stage of the chromatogram. The most delayed peak, (3) shown in Figure 2, is the inorganic dissolved carbon, and according to the report, the water of Lake Biwa has a peak at retention time of around 50 min (1) and three peaks at retention times from 70 to 80 min (2); (1) including the products by photosynthesis is biodegradable. However, (2) is said to include refractory dissolved organic substances because it remains after the biodegradable test (Figure 2).

Biodegradable test (Fukushima et al., 1995, 1996, 1997). A road runoff sample batch was placed into a 5,000 ml brown glass container, and the bottle was sealed using sterile silicon. Next the bottle was placed into a centrifuge at a moderate speed. This machinery was covered with sheets to shut out the light completely. The sample was then kept for

Table 1 Analytical conditions of GPC-TC

	Measuring condition
Column	Asahipak GS-320 (50 cm)
Mobile phase	30 mmol L^{-1}-Na_2HPO_4 (pH 6.0)
Flow rate	1.5 mL/min
Injection volume	1 mL
Detector	Total carbon analyzer

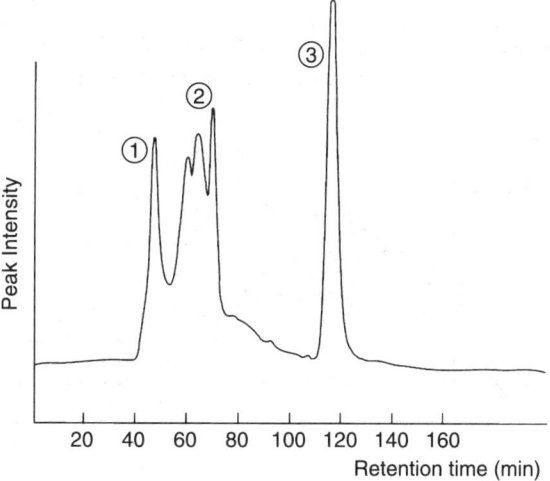

Figure 2 GPC-TC chromatogram; numerals in the circle indicate main three peaks

one hundred days with a water temperature set at 20°C. The sample then underwent a biodegradable test. Each sample was tested before the biodegradable test and one hundred days after the biodegradable test, these samples were analyzed for TOC and organic substances and fractionated using GPC-TC.

Field experiment for water purification

As the technique reducing pollutants in the first flush runoff from the road surface, we have taken the soil into consideration. In our previous research and papers, we stated that the soil has mechanisms that are physical, chemical and biological. And these mechanisms have proven to be effective in reducing various pollutants, especially phosphorus and D-COD (Tomioka et al., 1988; Wada et al., 2001). According to their reports, the soil had capacity to absorb the D-COD for three years in purifying the lake water.

So, we have conducted a field experiment with the intention of looking at the first flush pollutants in the stormwater runoff from the road surface and then studied effective techniques in reducing pollutants by using the capacity of the soil. The experiment site is the Kunobe overbridge of the Kibe-Yasu Line in Yasu Shiga prefecture.

Stormwater runoff from the road is transferred from the drainage area to the experimental facility (Figure 3) under the bridge. This facility is designed to selectively collect the first flush runoff and to percolate it through soil, based on the foregoing and other study results (Shiga pref. report, 2000). We used a decomposed granite soil that is low cost and a good capability of water purification.

Stormwater runoff was sampled four times during the experimental period (September 14, 2001–January 17, 2002) on Kunobe overbridge. The samples collected are the first flush runoff and treated effluent through the system filled with the soil, and these samples were then analyzed.

Results and discussion

The characteristics of urban runoff

Organic substances in stormwater runoff from the road. At the lakeshore road in Kusatsu City, we collected samples for 0–15 minutes where the starting flush (drainage volume of runoff was 185 l, catchment area was 750 m^2, equivalent to about 2 mm rainfall), and used the biodegradable test. In the result, the TOC concentration remained

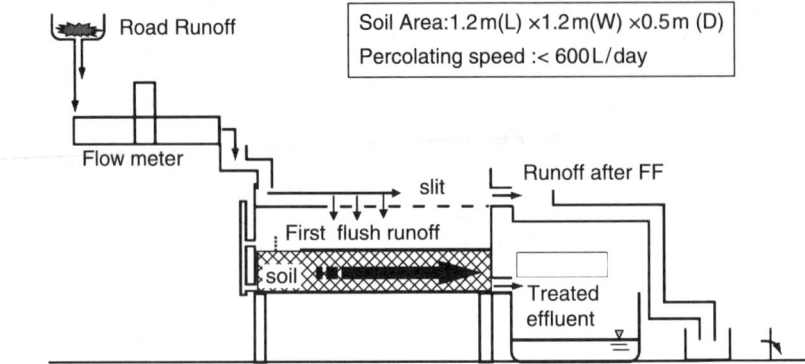

Figure 3 Schematic diagram of the experimental facility (Arrows indicate the stream of water)

at approximately 80% in the first flush runoff. And the DOC concentration remained at 50%; the POC concentration remained at 90%.

The GPC-TC chromatograms of before the biodegradable test and one hundred days after the biodegradable test are presented in Figure 4. The peak at approximately 50 minutes of the retention time indicates clearly that decomposable organic substances disappear after one hundred days, but the peak of 70 to 80 minutes shows that the refractory organic substances have a much smaller rate of decrease. Consequently the first flush runoff includes refractory organic substances and this suggests that it is the main pollutant load found in the water environment.

Characteristics of runoff from road surface. We conducted a survey seven times to measure the characteristics of the stormwater runoff during January 1999–September 2000. We collected the samples from the barrow pit at certain times and analyzed organic substance concentration.

The relationship between the COD and the TOC concentration of the stormwater runoff are presented in Figure 5. They show a strong correlation between the COD and the TOC concentration ($y = 1.0286x - 1.5673$; $R^2 = 0.9876$). Therefore, in this study we see no problem regarding the COD concentration in relation to the organic substances shown in the GPC-TC chromatogram.

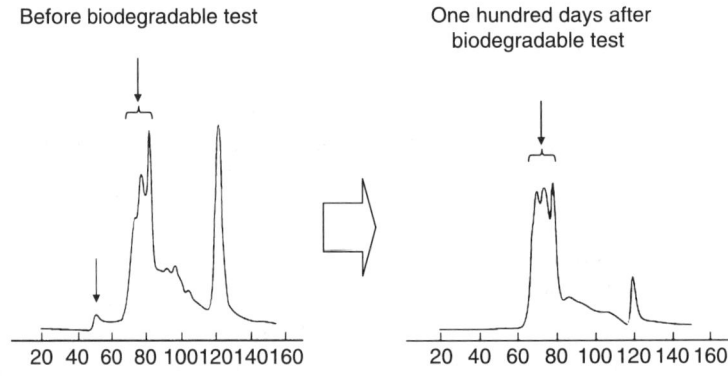

Figure 4 Biodegradable test of road runoff (24 Jan. 1999: Sinhayama Bridge Road in Kusatsu). Arrows indicate significant peaks

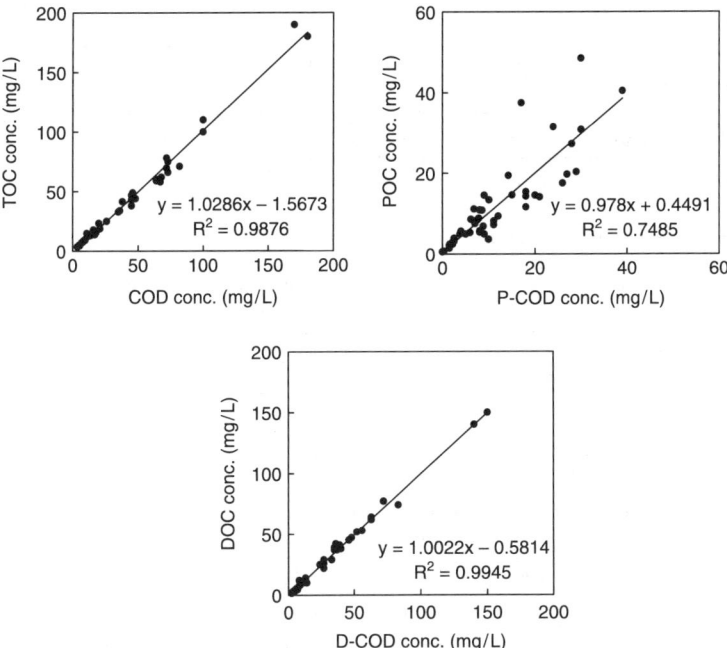

Figure 5 Relationship between COD and TOC concentration

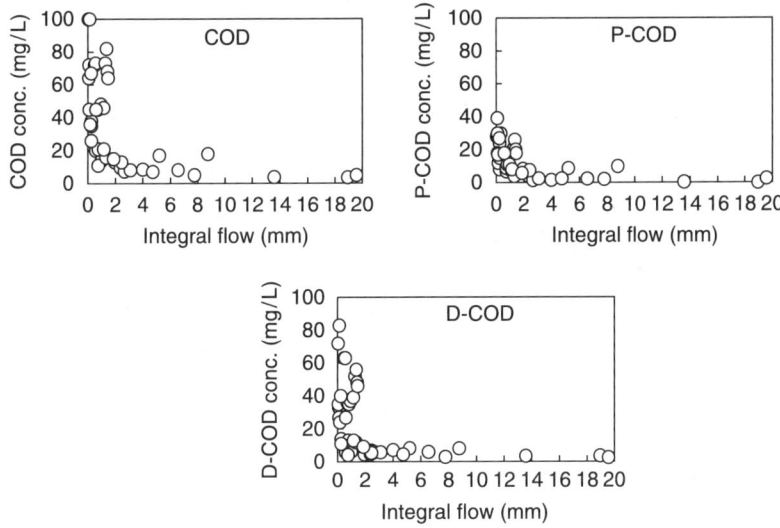

Figure 6 Relationship between COD and integral flow

Next, it is important to look at the relationship between the COD concentration of the stormwater runoff and integral flow shown in Figure 6. It is commonly known as a major pollutant of stormwater runoff during the time of the immediate flush. These surveys show the same tendency. Moreover there is a very strong relationship between dissolved organics i.e. D-COD and DOC. There is a possibility that the COD concentration of the first flush runoff exceeds five times that of the outflow drainage after the first flush when the concentration values are compared. As characteristics of the runoff from the road surface, we confirmed that a high concentration of pollutant continued until approximately 2.0 mm of integral flow.

Table 2 Characteristics of site in experiments

Location	Type	ADT* Vehicles/12 h	Surface type
Kunobe overbridge (Kibe-Yasu Line in Yasu)	Urban	11,966 (Compact vchicles) 1,036 (Heavy vehicles)	Asphalt

*Average daily traffic

Table 3 Summary of the observed rainfall

Data	Catchment area (m²)	Days to rain (day)	Rainfall (mm)	Period of raining (h)	Drainage volume of runoff (mm)	Volume of treated runoff (mm)
30 Sep. 2001	285	2.1	11.0	9.7	9.5	3.4
16 Oct. 2001	285	3.2	5.0	7.0	3.8	2.7
03 Nov. 2001	285	5.6	13.0	9.8	9.8	3.8
29 Nov. 2001	285	17.5	11.0	10.4	5.7	3.2

Judging from the results of this survey, the stormwater runoff from the road surface includes refractory dissolved organic substances, and the organic substance contained in the first flush runoff is of a high concentration. It is, therefore, necessary to reduce the runoff pollutant loadings from the road surface in order to reduce the COD concentration. Considering the measures to reduce non-point source pollution, we can carry out these measures effectively and efficiently by implementing solutions to cover outflow drainage from the starting flush to 2.0 mm of integral flow.

Field experiment for water purification

Results of the water quality survey. Characteristics of experiment sites conducted under the Kunobe overbridge in Yasu and the summary of observed rainfall in the form of a survey are presented in Tables 2 and 3. As the volume of treated runoff is close to the first flush of 2.0 mm, we confirmed that this system is a selective intake of the first flush runoff.

COD concentration of the monitoring in four rainfalls is shown in Figure 7, and concentrations of COD components accumulated in the samples are shown in Figure 8.

The COD concentration of the first flush runoff is 2–4 times higher than that found in the runoff after the first flush, and is 4–6 times higher than that found in the treated effluent. Runoff after the first flush is the same or about two times higher than that of the treated effluent. So we have been able to effectively reduce the pollution in the first flush runoff. Additionally, the first flush runoff has a high proportion of particle species and accounts for approximately 50% of all particles found.

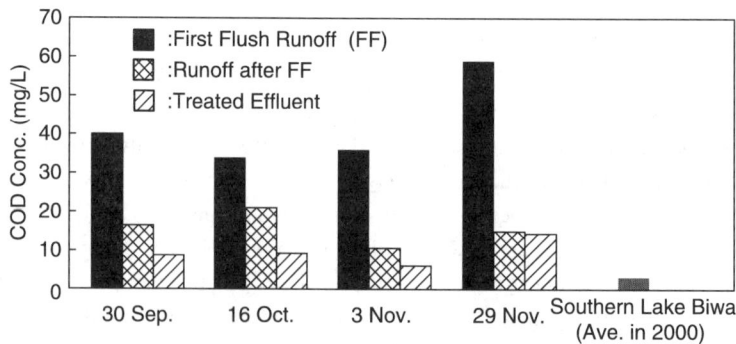

Figure 7 COD concentration of the monitoring

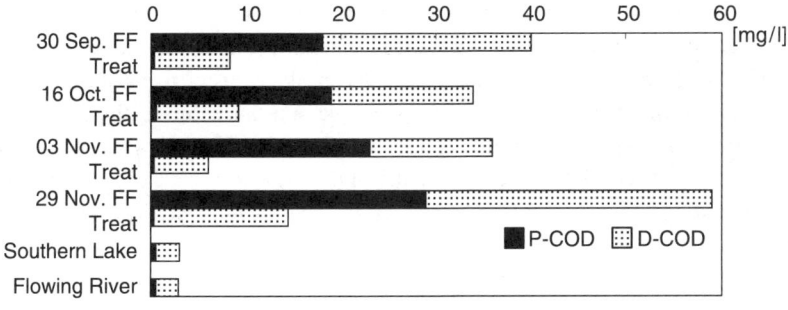

Figure 8 Concentration of COD components accumulated in the samples (FF: First Flush Runoff)

As the removal ratio of P-COD is more than 95%, the P-COD concentration can be reduced to several percent by percolating the runoff through the soil. Moreover, the P-COD concentration of the treated effluent is similar to the water quality of Lake Biwa and rivers flowing into Lake Biwa. The removal ratio of the monitoring is COD 77.5%, P-COD 97.7%, and D-COD 53.7%.

According to the results, we can reduce the first flush runoff loading by about 4–6 times using the methods in this experiment, though the COD concentration of the first flush is high compared with that of Lake Biwa and the rivers flowing into Lake Biwa.

GPC-TC chromatograms. The GPC-TC chromatograms of the first flush runoff and the treated effluent of each experiment are shown in Figure 9. For all of the survey, the peak intensity of GPC-TC chromatogram of the treated effluent in total retention times (for 50–120 minutes) was low. Both the easily decomposable dissolved organic substances and the refractory dissolved organic substances were reduced.

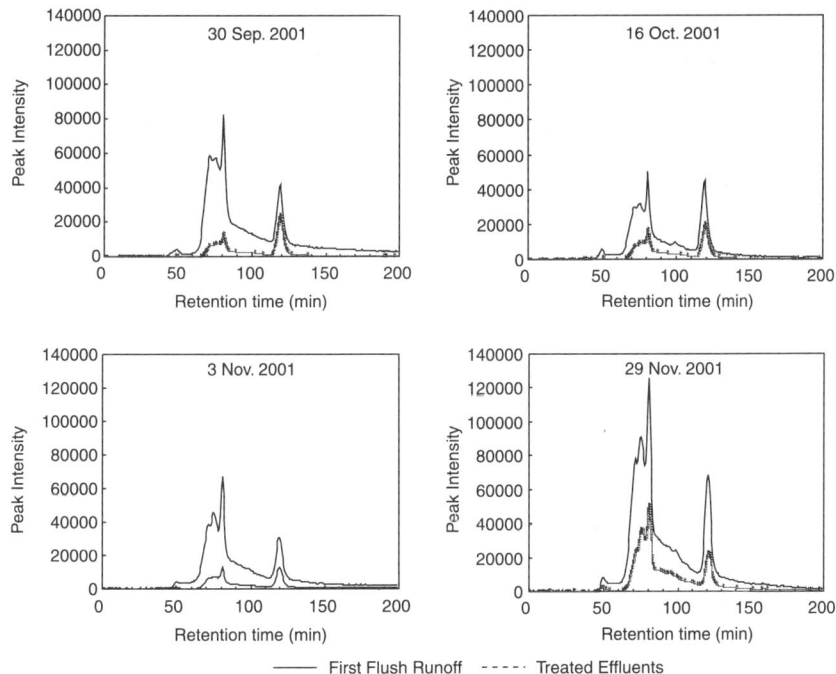

Figure 9 GPC-TC chromatograms of the experiments

If the peak intensity of the retention time of 70–80 minutes corresponds to the amount of the refractory dissolved organic substances, we can estimate the removal ratio of the refractory dissolved organic substances from the peak intensity using the TC analyzer. Therefore, we estimated that the refractory organic substances using this system reduce by approximately 70%; however, this is just the qualitative data. So we found that percolating through soil reduced the refractory dissolved organic substances and we successfully designed a system as a simple and easy experimental facility for urban runoff treatment.

Conclusions

Refractory dissolved organic substances are considered to be significant components causing an increase in the COD concentration in Lake Biwa. From our study, we found that refractory dissolved organic substances were high in road runoff and their concentrations were very high especially in the first stage of runoff. And taking advantage of the adsorption capability of organic matters by soil infiltration, we have suggested an effective countermeasure and instrument for reducing the supply of organic substances from road runoff to the lake. Detailed results obtained in our study were as follows.

- From GPC-TC analysis, easily decomposable and refractory dissolved organic substances were found to be contained in stormwater of road runoff. And refractory dissolved organic substances were considered to be less biodegradable.
- From field surveys of road runoff in rainfall events, COD concentration was found to be high especially in the first flush runoff, i.e. from the runoff start to 2.0 mm of integral flow.
- In experiments using the facility that selectively collects the first flush runoff and percolates it through soil, removal ratios of COD were found to be 77% for T-COD, 98% for P-COD and 54% for D-COD. And GPC-TC analysis confirmed that refractory dissolved organic substances were reduced by percolating the runoff through soil.

We consider that percolating the first flush runoff through soil is an effective and successful countermeasure for urban runoff treatment from the viewpoint of reducing COD components due to refractory dissolved organics. In future work toward its practical application, we consider it necessary to solve some maintenance issues, e.g. clogging of the soil surface, which is of critical importance for controlling its service life.

Acknowledgements

This work was supported by funding from Ministry of Environment of Japan. The authors wish to thank TORAY-TECHNO CO., Ltd for collaboration on practicing the experiments.

References

Barrett, M.E., Irish, L., Jr, Malina, J.F., Jr and Charbeneau, R.J. (1998). Characterization of highway runoff in Austin, Texas, Area. *Journal of Environmental Engineering*, February, pp. 131–137.

Fukushima, T., Imai, A., Matsushige, K., Aizaki, M. and Park, J. (1995). Fractionation of aquatic dissolved organic matter by mini-cartridge columns and its relation to biodegradability. *Journal of Japan Society on Water Environment*, **18**(4), 332–337.

Fukushima, T., Park, J., Imai, A. and Matsushige, K. (1996). Dissolved organic carbon in a eutrophic lake; dynamics, biodegradability and origin. *Aquatic Sciences*, **58**, 138–157.

Fukushima, T., Aizaki, M., Matsushige, K. and Imai, A. (1997). Index concerning on organic matter in lakes. *Journal of Japan Society on Water Environment*, **20**(4), 238–245.

Great Lakes Science Advisory Board (2000). Nonpoint sources of pollution to the great lakes basin; *Report to the International Joint Commission*.

Higashi, K., Wakida, S., Yamane, M., Takeda, S., Siroma, Z. and Taira, N. (1997). Dynamic behavior of organic substances in lake water and their origins: The study of effective separation method for the organic substances. *Technical Report of Ministry of International Trade and Industry, Agency of Industrial Science and Technology*, 97-II-1-15.

Imai, A., Fukushima, T., Matsushige, K., Inoue, T. and Ishibashi, T. (1998). Fractionation of dissolved organic carbon from the waters of Lake Biwa and its inflowing rivers. *Jpn. J. Limnol.*, **59**, 53–68.

Shiga prefecture (1999). Road traffic census. Shiga pref., JP.

Shiga prefecture (2001). White paper on environment. Shiga pref., JP.

Shiga prefecture report (1997). Investigation of organic pollutant component for 1997. *Department of Lake Biwa and the Environment* (in Japanese).

Shiga prefecture report (2000). Study of effluent treatment facility for road runoff for 1999. *Department of Public Works and Transportation Road Management Division* (in Japanese).

Tomioka, N., Matsushige, K., Yagi, O. and Sudo, R. (1988). Improvement of water quality by land application (IV): change of water quality by land application for a long term. *Res. Rep. Natl. Inst. Environ. Stud., Jpn.*, No 118, 67–88.

Wada, K., Horino, Y., Tainaka, Y., Itasaka, H. and Haruki, F. (2001). Demonstration experiment on water purification by utilizing adsorption capacity of soils. *The 9th International Conference on the Conservation and Management of Lakes Conference Proceedings Session 3–2*, pp. 189–192.

Yonebayashi, K. (1995). Isolation, fractionation and characterization of humic substances. *Journal of Japan Society on Water Environment*, **18**(4), 257–260.

Yonebayashi, K. (2004). Humic substances in pedosphere and aquasphere -ROSE in Biwa Lake-. *Journal of Japan Society on Water Environment*, **27**(2), 75.

Runoff and loads of nutrients and heavy metals from an urbanized area

H. Shirasuna*, T. Fukushima*, K. Matsushige**, A. Imai** and N. Ozaki***

*Graduate School of Life and Environmental Sciences, University of Tsukuba, Tsukuba 305-8572, Japan (E-mail: s0430320@ipe.tsukuba.ac.jp; fukusima@arsia.geo.tsukuba.ac.jp)
**Water and Soil Environment Division, National Institute for Environmental Studies, Tsukuba 305-8506, Japan (E-mail: matusige@nies.go.jp; aimai@nies.go.jp)
***Graduate School of Engineering, Hiroshima University, Higashihiroshima, 739-8527 Japan (E-mail: ojaki@hiroshima-u.ac.jp)

Abstract To investigate the run-off characteristics of dissolved and particulate substances from a heavily urbanized area (basin area: 95 ha, percentage of impervious surfaces: 60%), sensors for measuring water level, water temperature, DO, pH, electric conductivity (EC), turbidity and ammonium ion were placed in the channel connecting storm sewers and natural river, together with water sampling for analyzing SS, nutrients and metals. While both turbidity and EC showed apparent "first flush", the peaks of EC were always earlier than those of turbidity. In a similar manner, dissolved nutrients and metals exhibited earlier "first flush" compared with particulate nutrients and acid-extractable metals. Significantly positive correlations between EC and dissolved substances as well as those between turbidity and particulate (acid-extractable minus dissolved) substances were usually observed, and two distinct different regressions were found between the two datasets separated before and after the concentration peaks. Using these relationships, the total loads during the respective rainfall events were calculated on the basis of EC and turbidity changes. The total loads of nitrogen, zinc, etc. were nearly proportional to the lengths of non-rainfall periods before the events, indicating that these loads derived from the atmospheric deposition.

Keywords Electric conductivity; metal; nutrient; storm sewer; turbidity; urban runoff

Introduction

Urban stormwater has now been recognized as a substantial source of pollutants to receiving waters (Novotony and Olem, 1994). A number of investigators have found significant levels of metals (Sansalone and Buchberger, 1997; Gromaire-Mertz et al., 1999), nutrients (Wahl et al., 1997; Lee et al., 2002), and toxic organic substances (Fukushima et al., 2003) in runoff from urbanized areas. Prediction of their runoff patterns and total loads is, however, rather difficult due partly to differences in precipitation characteristics, their sources in the urban environment, their dynamics during runoff processes etc. and/or due partly to the great difficulty of their continuous measurement.

From the latter viewpoint, turbidity and electrical conductivity could be candidates for tracing the respective runoff of particulate and dissolved substances. For example, the modeling of dissolved pollutants yielded equations connecting runoff parameters, which were able to approximate chemographs of conductivity and dissolved heavy metal (Robien et al., 1997).

Hence, the goals of the present study are to investigate the characteristics of dissolved and particulate substances run off from a heavily urbanized area using sensor-monitored data and to construct a model for estimating their total loads during each run-off event.

Method

Research site

The research basin with a drainage area of 95 ha and a maximum altitude difference of 5 m was located in the central commercial and business district of Tsukuba, Japan (Figure 1). Impervious surfaces, e.g., streets, rooftops, parking lots, etc. occupied 60% ± 10% of the basin. The average traffic volume in Street A (also similar level in Streets B and C) was about 30,000 vehicles per day. The storm sewers mainly from the impervious parts of the basin were connected to a natural river named the Hanamuro through a straight channel with a length of about 200 m.

Measurement and sampling

The sensors for measuring water level (WT-HR 500, Tru Track Co. Ltd.), water temperature, electric conductivity (EC), turbidity, pH, DO and ammonium ion (NH_4^+) (Environmental Monitoring System 6600; YSI Co. Ltd.) were placed at about 2 cm above the bottom of the channel during two nearly 1-month periods. The discharge rate was calculated using the water level with the regression curve between them, which was obtained for several periods. On the occasions of a number of rainfall events (Table 1), water samples were also taken into 1 L acid washed polycarbonate bottles at intervals of 10 or 15 min for several hours after the initial rise of water discharge. They were cooled in an icebox and then taken to the laboratory. During each rainfall event, rainwater was also collected with a stainless steel bucket and analyzed in the same way as the above samples.

Analysis of samples

Dry weights (SS) were measured gravimetrically. The sub-samples for nutrients were obtained by passing through pre-combusted (450°C for 4 hours) Whatman GF/C glass-fiber filters (pore size 1.2 μm) and those for metals were prepared with Milex-FH membrane filters (pore size 0.45 μm, Millipore Co. Ltd.), respectively. The sub-samples for total nitrogen (TN), dissolved total nitrogen (DTN), total phosphorus (TP), and dissolved total phosphorus (DTP) were digested with potassium peroxodisulfate and then determined with TRACCS800 (Bran + Luebbe Co. Ltd.), together with the sub-samples for PO_4^{3-}-P, NO_3^--N, NO_2^--N, NH_4^+-N (Otsuki, 1982). Dissolved and acid-extractable (filtrated by the membrane filter after digestion) metals (Zn, Cu, Fe, Al, Mn etc.) were determined with ICAP-757 (Jarrel-Ash Co. Ltd.). The digestion for the acid-extractable

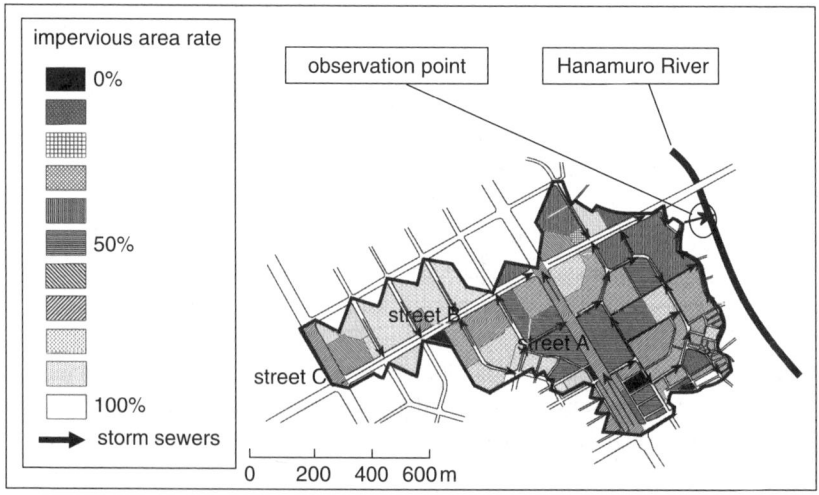

Figure 1 Research basin in the central commercial and business district of Tsukuba, Ibaraki, Japan

Table 1 Characteristics of the observed rainfall events

Date Y/M/D	duration of precipitation (h)	ADWP (day)	Total precipitation (mm)	Maximum precipitation intensity (mm/h)	Effluent rate of water (%)	Comment
2003/5/15	31	6.1	22	3	46.8	WWS
2003/5/31	32	10.1	51	11	52.9	
2003/9/20	41	24.0	91	6	55.3	WWS
2003/9/24	11	2.6	16	6	28.6	
2003/10/28	12	4.6	18	4	45.2	WWS
2003/11/5	10	7.9	22	5	55.8	

ADWP: antecedent dry weather period
WWS: without water sampling (only monitoring with sensors)

component was done in a bath with water of >95°C for 15 min after addition of 5 ml 1 + 1 high-purity HCl into 100 ml sub-sample (Standard Methods for the Examination of Water and Wastewater, 1995). Their particulate concentrations were calculated as acid-extractable ones minus dissolved ones.

Results and discussion

Hydrology

The effluent rate of water (amount of water discharge divided by precipitation amount) increased with precipitation towards 60%, indicating that almost all the water precipitated on the impervious surfaces discharged into the channel (Table 1). The time of peak discharge after the start of rainfall ranged from one to two hours (Figure 2), which was rather reasonable based on the scale of the basin.

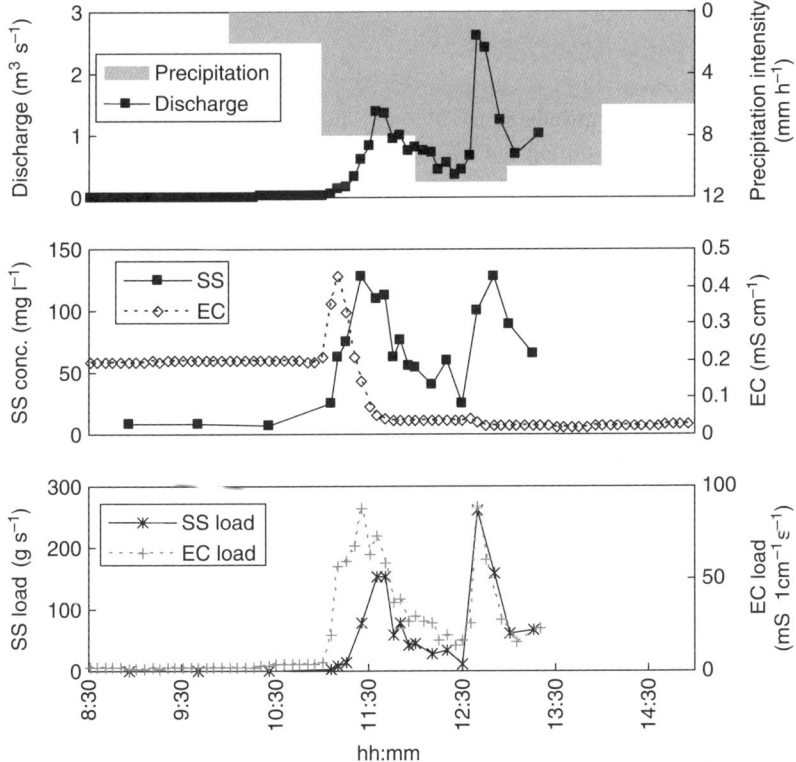

Figure 2 Temporal changes in precipitation, water discharge concentrations and loads of SS and EC (May 31, 2003)

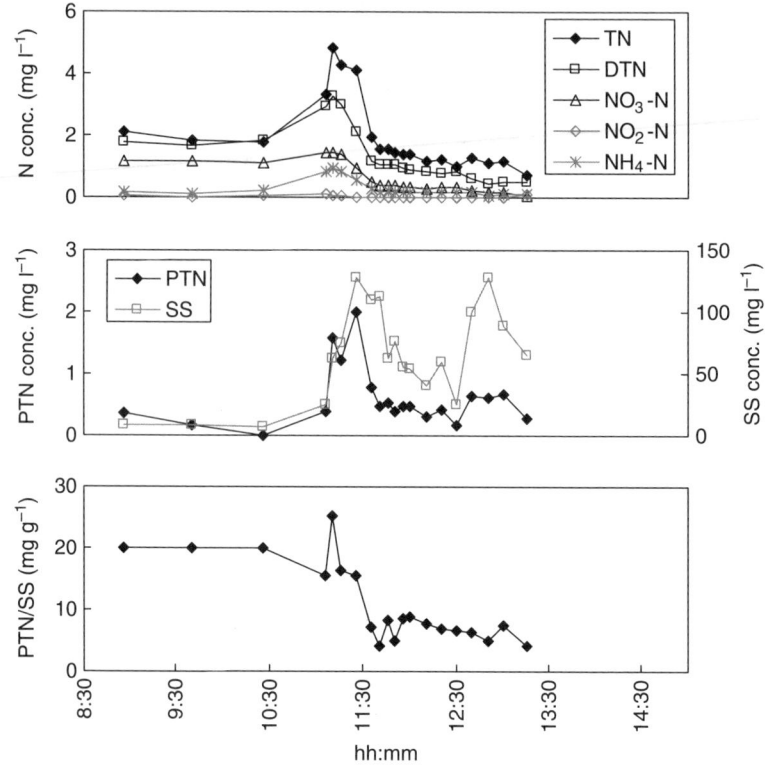

Figure 3 Temporal changes in various components of nitrogen (May 31, 2003)

Water quality runoff curves

The initial period of stormwater runoff during which the concentration of pollutants is substantially higher than during the later periods is called the first flush phenomenon. Although both SS (also turbidity shown later) and EC showed apparent first flushes, the peaks of EC were always earlier than those of SS (Figure 2). In a similar manner, dissolved nutrients and metals exhibited earlier first flush compared with particulate nutrients (Figures 3 and 4) and metals (Figure 5). Lee *et al.* (2002) evaluated the first flush by analyzing the relationship between dimensionless cumulative pollutant mass vs. dimensionless cumulative runoff volume. They showed that PO_4-P and total Kjeldahl nitrogen, i.e. dissolved component, usually showed the maximum first flush from urbanized areas, coinciding with our results. The peaks of the runoff loads were observed between the peaks of their concentrations and the discharge (Figure 2).

The dissolved components of nitrogen dominated by NO_3-N and NH_4-N always exceeded the particulate ones (D/P = 1.93 ± 0.30; D: dissolved, P: particulate); on the other hand the particulate ones were predominant in the case of phosphorus (P/D = 1.61 ± 0.37) (Figures 3 and 4). For Fe, Al, and Mn, the particulate components predominated (P/D = 26.2 ± 3.4, 16.6 ± 3.0, 5.1 ± 0.4, respectively), while the dissolved ones were comparable to the particulate ones for Zn and Cu (P/D = 1.1 ± 0.1, 0.5 ± 0.1, respectively) (Figure 5). This tendency agrees with the results observed for the collected runoff water from urban highway pavement (Sansalone *et al.*, 1996). In addition, the concentrations of N, P and Zn usually exceeded the water quality standards (N: 0.4 mg l^{-1}, P: 0.03 mg l^{-1} for Class III and Zn: 0.03 mg l^{-1}), suggesting adverse influences on the water environment.

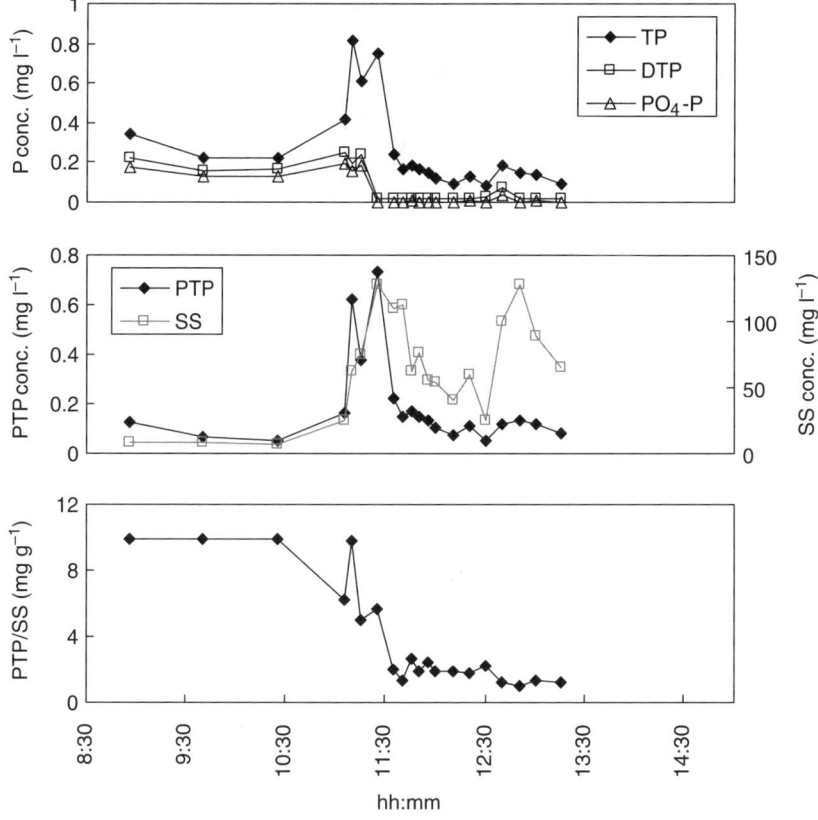

Figure 4 Temporal changes in various components of phosphorus (May 31, 2003)

Comparison with SS and EC

Significantly positive correlations between EC and dissolved nutrients (Figure 6 and Table 2; $r^2 = 0.43$ for P, 0.72 for N) were observed as well as those between SS and particulate nutrients in which we found two distinctly different regressions (Figure 7) for the two datasets separated before ($r^2 = 0.58$ for P and 0.852 for N) and after the concentration peaks ($r^2 = 0.64$ for P, 0.47 for N). This is probably due to the content change in the runoff particulate matter (Figures 3–5), which resulted from the gradual increase in transporting capacity of suspended matter before the discharge peak and the dependency of nutrient content on particle size, since most of the nitrogen and phosphorus was attached to finer particles (Vase and Chiew, 2002).

We found higher metal correlations between particulate concentrations and SS ($r^2 = 0.91$ for Fe, 0.86 for Al, 0.67 for Mn, 0.33 for Cu, 0.74 for Zn) (Figure 8 and Table 2) than those between dissolved concentrations and EC ($r^2 = 0.41$ for Fe, 0.18 for Al, 0.47 for Mn, 0.08 for Cu, 0.67 for Zn) (Figure 9). The data separation before and after the peaks substantially raised the correlation for particulate and dissolved Zn. The increase in particulate Zn would be due to nearly the same causes as mentioned above for nutrients (Sansalone and Buchberger, 1997). The increase in dissolved Zn (Table 2) indicates the differences in sources and dissolving processes between EC and dissolved Zn; the main sources for Zn were considered to be rooftop (Gromaire-Mertz et al., 1999; Garnaud et al., 1999; Van Metre and Mahler, 2003) and building siding (Davis et al., 2001), runoff, tire wear (Davis et al., 2001), car washes (Sorme and Langerkvist, 2002), and atmospheric deposition (Garnaud et al., 1999) etc.

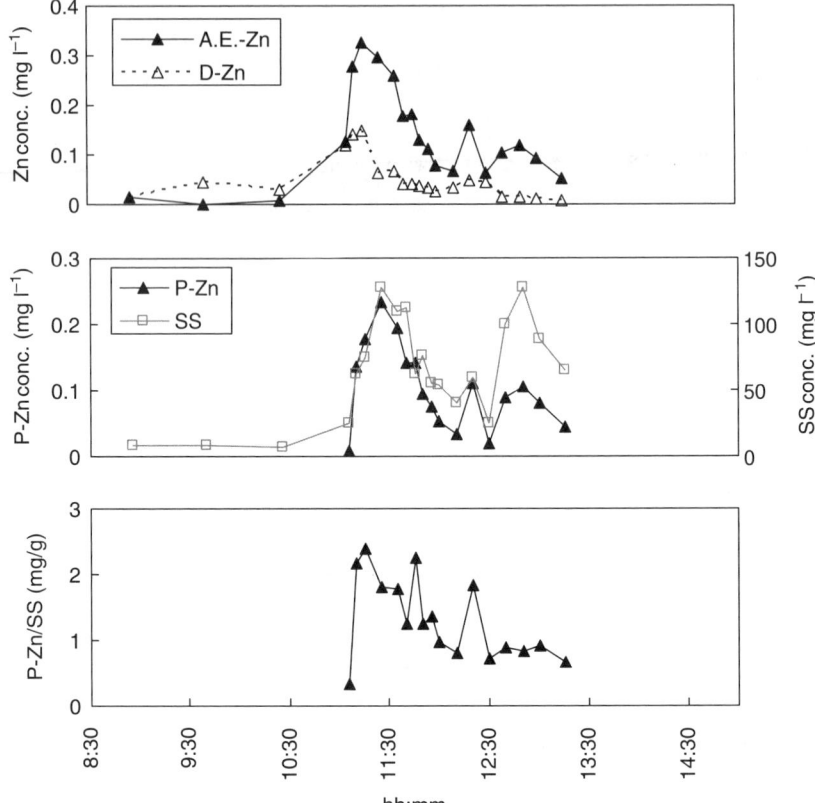

Figure 5 Temporal changes in various components of zinc. A.E.: acid-extractable (May 31, 2003)

Because turbidity had a fairly good correlation with SS ($r^2 = 0.81$) and because turbidity and EC could be monitored with sensors continuously, they would be appropriate candidates for tracing the runoff of particulate and dissolved substances, respectively. To make this monitoring system more accurate and robust, we, however, need more information on pollution sources and dynamics in the basin, element by element.

Factors determining runoff curve and load

The peak SS and EC concentrations increased with the length of antecedent dry weather periods (ADWP) (Figure 10), but approached upper limits, indicating the accumulation on the impermeable surfaces and the limitation of transporting and/or dissolving capacity. In addition, the time from rainfall start to SS peak and that from rainfall start to EC peak

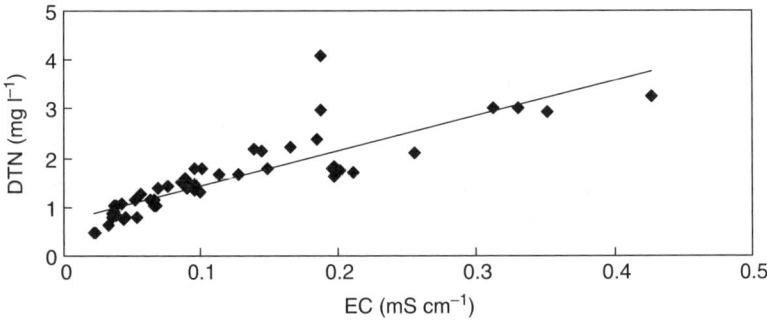

Figure 6 DTN conc. vs. EC

Table 2 Relationships between dissolved substances and EC, and those between particulate substances and SS concentrations

	Relationships with EC (mS·cm^{-1})			Relationships with SS conc. (mg·l^{-1})	
Substances	Regression equation	r^2	Substances	Regression equation	r^2
constant equations			constant equations		
DTN	7.2 × EC + 0.72	0.71	P-Fe	0.030 × SS + 0.15	0.91
DTP	0.94 × EC − 0.011	0.43	P-Al	0.030 × SS − 0.059	0.86
D-Cu	0.012 × EC + 0.0078	0.079	Before first SS peak		
D-Mn	0.080 × EC + 0.00070	0.47	PTN	0.015 × SS + 0.037	0.85
D-Fe	0.19 × EC + 0.022	0.41	PTP	0.0047 × SS + 0.023	0.58
D-Al	0.11 × EC + 0.044	0.18	P-Zn	0.0016 × SS + 0.0015	0.86
Before first SS peak			P-Cu	0.00012 × SS + 0.0038	0.40
D-Zn	0.45 × EC − 0.046	0.77	P-Mn	0.0015 × SS + 0.0098	0.72
After first SS peak			After first SS peak		
D-Zn	0.33 × EC + 0.021	0.55	PTN	0.0056 × SS + 0.13	0.47
			PTP	0.0012 × SS + 0.032	0.64
			P-Zn	0.0011 × SS + 0.0086	0.67
			P-Cu	0.00008 × SS + 0.0018	0.47
			P-Mn	0.0011 × SS − 0.00021	0.92

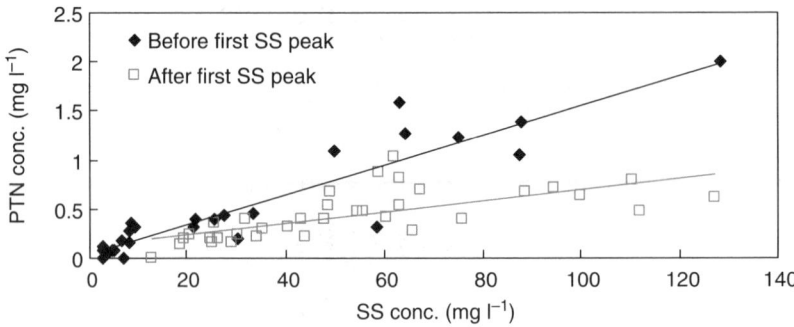

Figure 7 PTN conc. vs. SS conc

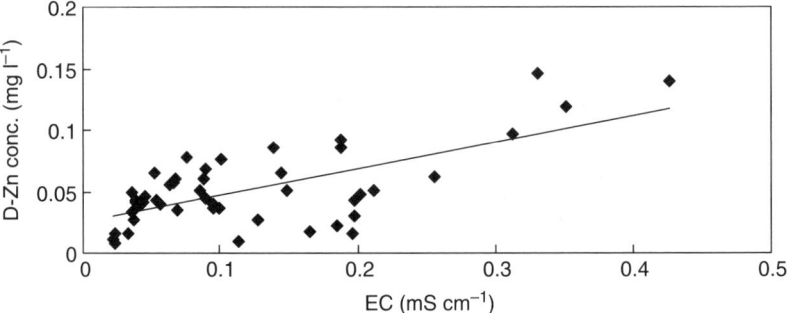

Figure 8 D-Zn conc. vs. EC

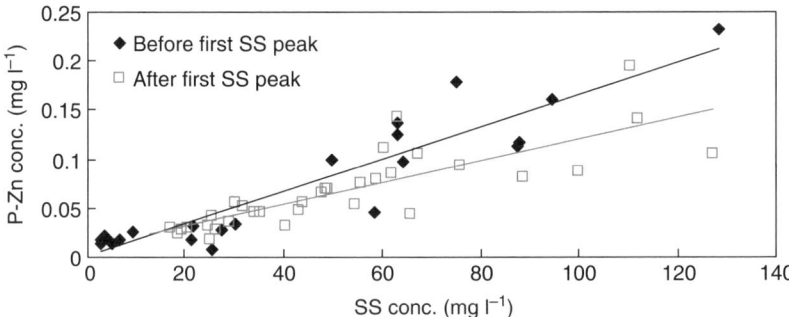

Figure 9 P-Zn conc. vs. SS conc

also increased with initial precipitation intensity (IPI for initial 1 h), suggesting a change in the water discharge rate from the basin (Figure 11).

Using the relationships between dissolved components and EC and those between particulate ones and turbidity, the total loads during each rainfall event were calculated on the basis of EC and turbidity changes. The total loads of N, Zn, etc. were nearly proportional to the lengths of ADWP before the events, indicating that these loads accumulated during the periods, e.g. the atmospheric deposition (Figure 12). The averaged accumulation rates for SS and N were calculated to be 200 and 6.1 kg ha^{-1} y^{-1}, respectively, which were close to the reported ones measured in other basins, 105–2,390 kg ha^{-1} y^{-1} for SS and 4.5–39.6 kg ha^{-1} y^{-1} for N, respectively (Ministry of Land, Infrastructure and Transport, 1999). In addition, the SS accumulation rates slightly exceeded the atmospheric deposition rates 73 kg ha^{-1} y^{-1} observed at the University of Tsukuba which is located in a less urbanized area about 5 km apart from the basin (Shanningrahi et al., 2003).

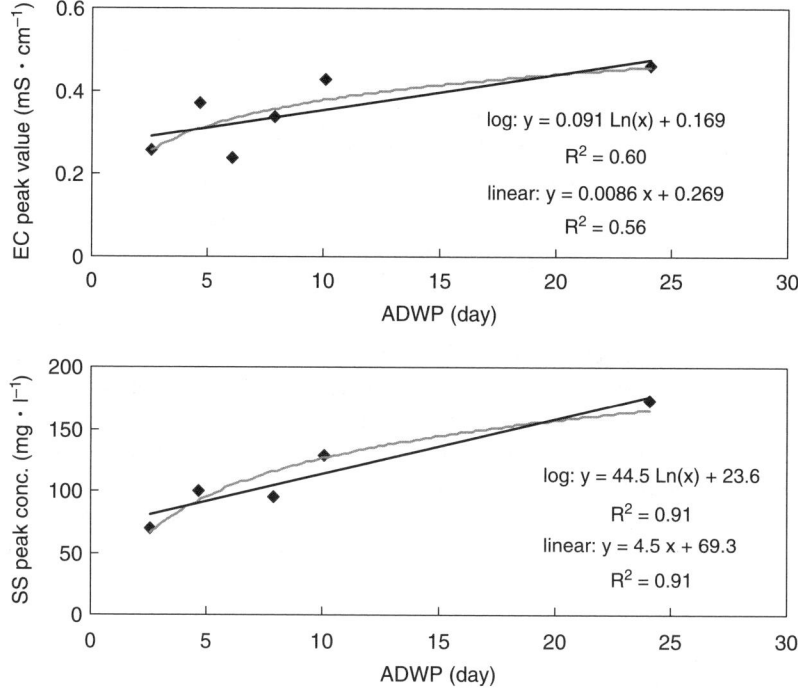

Figure 10 SS and EC peak conc. vs. ADWP

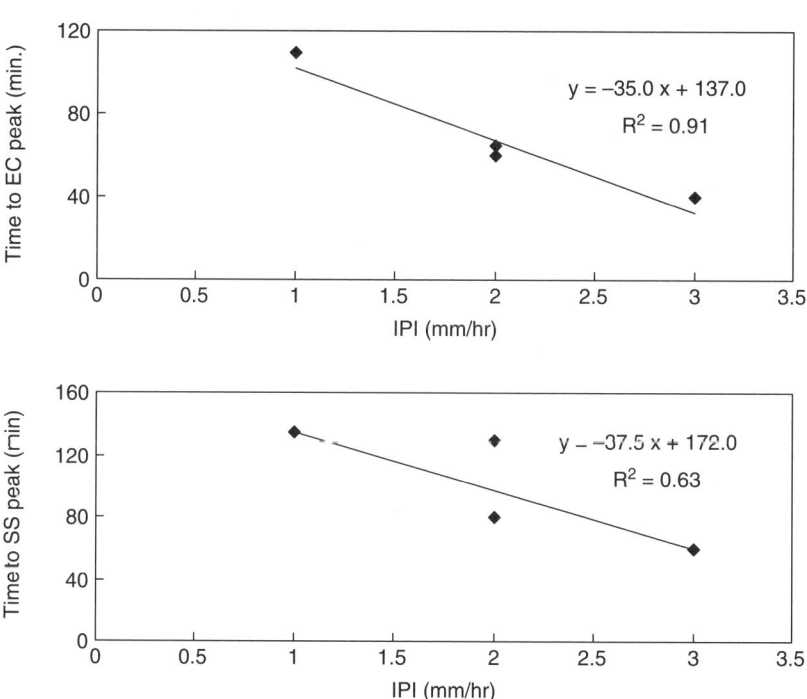

Figure 11 Time from rainfall start to SS and EC peak conc. vs. IPI

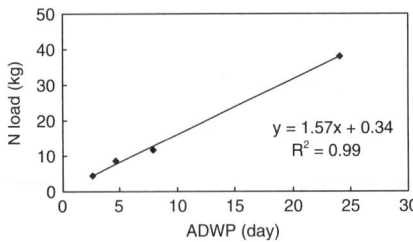 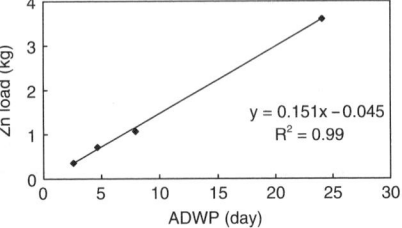

Figure 12 Nitrogen and zinc loads vs. ADWP

Conclusions

(1) "First flushes" were clearly observed for dissolved substances as well as particulate substances. The former peaks were always earlier than the latter peaks.
(2) The dissolved and particulate concentrations of nutrients, metals, etc. could be predicted using the relationships between EC and dissolved substances and those between turbidity and particulate substances, respectively.
(3) The total runoff loads of nutrients, metals, etc. were usually related to the length of antecedent dry weather periods, implying load accumulation, e.g. their atmospheric deposition.

Acknowledgement

Financial support provided by the Sewage Works Promoting Fund is gratefully acknowledged.

References

Davis, A.P., Shokouhian, M. and Ni, S. (2001). Loading estimates of lead, copper, cadmium, and zinc in urban runoff from specific sources. *Chemosphere*, **44**, 997–1009.

Standard Methods for the Examination of Water and Wastewater (1995). 19th edn, American Public Health Association/American Water Works Association/Water Environment Federation, Washington, DC, USA.

Fukushima, T., Ozaki, N. and Hamada, T. (2003). PAHs dynamics in Hiroshima Bay and its watershed. *Asian Waterqual-2003* (CD-ROM).

Garnaud, S., Mouchel, J., Chebbo, G. and Thevenot, D.R. (1999). Heavy metal concentrations in dry and wet atmospheric deposits in Paris district: comparison with urban runoff. *Sci. Total Environ.*, **35**, 235–245.

Gromaire-Mertz, M.C., Garnaud, S., Gonzalez, A. and Chebbo, G. (1999). Characterization of urban runoff pollution in Paris. *Water Sci. Tech.*, **39**(2), 1–8.

Lee, J.H., Bang, K.W., Ketchum, L.H., Choe, J.S. and Yu, M.J. (2002). First flush analysis of urban storm runoff. *Sci. Total Environ.*, **293**, 163–175.

Ministry of Land, Infrastructure and Transport (ed.) (1999). *Guideline of Sewage Planning*, JSWA, Tokyo, pp. 61–63 (in Japanese).

Novotony, V. and Olem, H. (1994). *Water Quality: Prevention, Identification, and Management of Diffuse Pollution*, John Wiley & Sons, New York.

Otsuki, A. (1982). Determination of nutrients. JAWQP (ed.), In *Guideline for lake survey*, Kogai-Taisaku-Gijyutsu-Doyukai, Tokyo, pp. 121–145 (in Japanese).

Robien, A., Striebel, T. and Herrmann, R. (1997). Modeling of dissolved and particle-bound pollutants in urban street runoff. *Water Sci. Tech.*, **36**(8–9), 77–82.

Sansalone, J.J., Buchberger, S.G. and Al-Abed, S.R. (1996). Fractionation of heavy metals in pavement runoff. *Sci. Total Environ.*, **189/190**, 371–378.

Sansalone, J.J. and Buchberger, S.G. (1997). Characterization of solid and metal element distributions in urban highway stormwater. *Water Sci. Tech.*, **36**(8–9), 155–160.

Shannigrahi, A.S., Fukushima, T., Ozaki, N. and Hasegawa, S. (2003). Settling flux of PAHs on the ground surface. *Res. Rep. of TERC, University of Tsukuba*, **4**, 41–50.

Sorme, L. and Langerkvist, R. (2002). Sources of heavy metals in urban wastewater in Stockholm. *Sci. Total Environ.*, **298**, 131–145.

Van Metre, P.C. and Mahler, B.J. (2003). The contribution of particles washed from rooftops to contaminant loading to urban streams. *Chemosphere*, **52**, 1727–1741.

Vase, J. and Chiew, F.H.S. (2002). Experimental study of pollutant accumulation on an urban road surface. *Urban Water*, **4**, 379–389.

Wahl, M.H., McKellar, H.N. and Williams, T.M. (1997). Patterns of nutrient loading in forested and urbanized coastal streams. *J. Exp. Mar. Biol. Ecol.*, **213**, 111–131.

Dispersion and dry and wet deposition of PAHs in an atmospheric environment

N. Ozaki*, K. Nitta** and T. Fukushima***

*Dept. of Civil and Environ. Engineering, Faculty of Engineering, Hiroshima University, 1-4-1 Kagamiyama, Higashi-Hiroshima, Japan. 739-8527 (E-mail: *ojaki@hiroshima-u.ac.jp*)
**Hiroshima City-bureau, Japan
***Institute of Geoscience, University of Tsukuba, Japan

Abstract The atmospheric concentration and dry and wet deposition were measured for particulate matter (PM) and polycyclic aromatic hydrocarbons (PAHs) from August to December in Higashi-Hiroshima City, Japan. PM concentration of fine particles (0.6–7 μm) was 5.7–75.1 $\mu g\,m^{-3}$, and coarse particles (>7 μm) was 2.2–22.3 $\mu g\,m^{-3}$. Total PAHs concentration of fine particles was 0.14–16.3 $ng\,m^{-3}$, and coarse particles was 0.01–0.77 $ng\,m^{-3}$. Their concentration increased on non-rainy days and decreased rapidly on rainy days. For seasonal fluctuations of PAHs, their concentrations decreased from summer to winter, and the rate of decrease was more distinct for fine particles. For total (dry + wet) depositions, the PM flux was 1.9–11.2 $mg\,m^{-2}\,d^{-1}$, and the total PAHs flux was 1.9–97.2 $ng\,m^{-3}\,d^{-1}$. From these measurements, the yearly total loading of PAHs was estimated for the particle phase. Total loading was 28 $\mu g\,m^{-2}\,y^{-1}$ for the dry deposition and 52 $mg\,m^{-2}\,y^{-1}$ for the wet deposition. The loading of the wet deposition was comparable to those of the dry deposition for all ring numbers.

Keywords Atmospheric concentration; PAHs; particulate matter; polycyclic aromatic hydrocarbons; total deposition

Introduction

Polycyclic aromatic hydrocarbons (PAHs) are a class of organic compounds composed of two or more fused aromatic rings. Several compounds of this class are believed to be human carcinogens. PAHs are emitted into atmospheric environments mainly due to incomplete combustion processes, deposited on ground surfaces, and discharged into aquatic environments. Many researchers have estimated PAH emissions on a nationwide scale, and the atmospheric concentrations and depositions have been measured (Baek *et al.*, 1991; Fukushima *et al.*, 2003). According to the results, vehicle emissions, especially diesel exhausts, are one of the major sources of PAH emissions in urban areas. Higher molecular weight PAHs are adsorbed into particulate matter (PM), and move with PM. In discussing the discharges from atmospheric environments into aquatic environments, the relation of the atmospheric condition and the deposition onto ground surface would be the major concern. These relations, however, have not been scientifically determined yet. Moreover, the estimation of temporal and spatial variations in the atmospheric concentration has made for basic uncertainties in discussing the total loading on a larger scale. In the present study, the atmospheric concentration and deposition of particulate matter and PAHs were measured in a suburban area. The measurements of atmospheric concentrations and depositions of particulate matter and PAHs formed the basis of a discussion on the temporal variations in the atmospheric concentration and ground surface loading.

doi: 10.2166/wst.2006.055

Experimental methods
Sampling campaigns
The sampling site was located at the Saijo campus of Hiroshima University in Higashi-Hiroshima, Japan (34°23′11″N, 132°43′00″E). Sampling apparatuses were put on the roof of an eight storey building in the campus (height: 30 m). The population of Higashi-Hiroshima is 1.2×10^5. The city center district is located 3 km northeast from the sampling site. Atmospheric particulate matter was collected using a high-volume air sampler (HVS-500-5; Shibata Kagaku Co., Ltd.) with an impactor system followed by a glass fiber filter. Each sampling was conducted for 24 hours, and the sampling rate was 500 L min^{-1}. Using the impactor system, particles of >7 μm (coarse particulate matter; CPM) were trapped on a stainless plate. After sieving, particles of 0.6–7 μm (fine particulate matter; FPM) were collected on a glass fiber filter (GB-100R, Advantec Co., Ltd.). Prior to the experiment, the stainless plate and glass fiber filter were precombusted at 110 °C for 24 hours. The deposited particulate matter for total and wet deposition was collected in a bucket (24 cm in diameter and 35 cm height). For total deposition, the sampling tray was put in for 72 hours, and for the wet deposition, the sampling tray was put in during the rainy periods. At the end of the sampling for non-rainy periods, the bucket and stainless plate for holding coarse particle matters were rinsed with Milli-Q water (Millipore Co., Ltd.). Collected water was then filtered through a glass fiber filter (GF/C, Whatman Co., Ltd.), which was precombusted at 450 °C for 4 hours. For both atmospheric and deposited particulate matter, each sampling began at 14:00 and lasted 24 hours. For rainy periods, dissolved PAHs were also measured. (The PAHs concentration dissolved from collected particles in the water used for rinsing the non-rainy period sampling was negligible and so this concentration was omitted.) Overall, the sampling session was performed from 10th August to 28th December, 2003. The periods of sampling of each item are shown below.

Meteorological measurements
The meteorological data were collected from the meteorological data acquisition system (http://home.hiroshima-u.ac.jp/hirodas) in Hiroshima University. The observation point was at a 380 m distance from the PAHs sampling site.

PAHs extraction and analysis
For particulate matter, the filters collecting particles were dried for 2 days under dark conditions at room temperature and subsequently extracted with dichloromethane in a sonication water bath. For the dissolved phase, a silica column (Sep-Pak Plus C18, Waters Co., Ltd.) was used for extraction. The collected water was passed through the silica column for PAHs entrapment at the rate of 3 mL min^{-1}. After the entrapment, 10 mL DCM (dichloromethane) was passed for extraction at the rate of 1 mL min^{-1}. After the extraction, these DCM extracts were concentrated into 2 mL by N_2. PAHs were then analyzed on a gas chromatograph (GC17A, Shimadzu Co., Ltd.) equipped with a mass selective detector (QP5050, Shimadzu Co., Ltd.) operated in the single-ion monitoring mode. Injection was split with detector and inlet temperatures at 230 °C. The initial temperature was 80 °C held 2 min, ramped at 30 °C min^{-1} to 210 °C, ramped at 5 °C min^{-1} to 295 °C, and ramped at 2 °C min^{-1} at 315 °C. Sixteen unsubstituted PAHs were measured: acenaphthalene (Ace), acenaphthene (Act), fluorene (Flu), phenanthrene (Phe), anthracene (Ant), fluoranthene (Flt), pyrene (Pyr), benzo(a)anthracene (B(a)A), chrysene (Chr), benzo(b)fluoranthene (B(b)F), benzo(k)fluoranthene (B(k)F), benzo(e)pyrene (B(e)P), benzo(a)pyrene (B(a)P), dibenzo(ah)anthracene (D(ah)A), benzo(ghi)perylene (B(ghi)P), and indeno(123-cd)pyrene (Ind(123)P). The detection limit was set at the

level of 3 in the SN ratio. Detection limits ranged from 1 to 5 ng for individual PAHs. Within this level, the CV ratio of each of the compounds was less than 20%. Quality of extraction was checked using dried marine sediments (HS-3B, National Research Council of Canada Institute for Marine Biosciences). The recovery averaged 50–80% for all PAHs, and the repetition error was 5–10%.

Results
Atmospheric concentrations

The climatic conditions and averaged atmospheric concentrations and deposition fluxes of PM and total PAHs in particulate matter are listed in Table 1. In the sampling periods, temperature and precipitation were moderate for this sampling site. For particulate matter, the atmospheric concentration of FPM (fine particulate matter) was 22.0–29.2 μg m^{-3}, and that of CPM (coarse particulate matter) was 7.0–11.1 μg m^{-3}. For total PAHs, the atmospheric concentration of PAHs in FPM was 0.80–5.16 ng mm^{-3}, and that in CPM was 0.04–0.17 ng mm^{-3}. For FPM and CPM concentrations, they were stable from summer to winter. For PAHs, on the other hand, the concentration in FPM increased from summer to winter. This seasonal fluctuation is commonly observed in many atmospheric measurements (Fukushima et al., 2003; Ozaki, 2002; Wu and Fang, 2001; Panther et al., 1999). For the concentration in CPM, on the other hand, no seasonal variation was observed. Effects of seasonal changes on PAHs in atmospheric environments would be different between FPM and CPM.

An example of the results of PM and PAH fluctuation is shown in Figure 1 (from 6th Nov. to 6th Dec.). By comparing these results with the precipitation profile, the dependencies of PM concentration on precipitation can be observed (Figure 1a). The FPM and CPM concentrations increased during non-precipitation periods and decreased rapidly during precipitation. This tendency is apparent for the period from 6th to 13th of November. When we compare the FPM and CPM, we see that the pattern of fluctuation is fairly similar (Figure 2; R = 0.67 for all the samplings). The concentration profile of total PAHs is shown in Figure 1b. Compared to the PM, the extent of fluctuation is larger in PAHs (the CV values of all the samplings were 43% and 49% for FPM and CPM, and 92% and 130% for PAHs in FPM and CPM) and a sharp increase was occasionally observed. In comparing the total PAHs in FPM and CPM, the fluctuation pattern was basically similar (the R of total PAH concentration between FPM and CPM was 0.72).

Total and wet deposition of PM and PAHs

Deposition samplings were conducted from 6th November to 5th December in 2003. During this period, precipitation occurred four times (9th to 10th Nov., 19th to 21st Nov., 26th to 28th Nov., and 5th Dec.). Figures 1c and 1d show the results of the PM and total PAH flux. (For wet deposition, the dissolved phase PAHs were not included in these results.) The PM total flux was 1.9–11.2 mg m^{-2} d^{-1}. The level of the total flux did not change significantly in the rainy periods. The deposition in the rainy periods was similar to those of the non-rainy periods.

In the third precipitation (26th to 28th Nov.), the bucket for collection was changed twice during the sampling. From this sampling, the first flush phenomenon was clearly observed. Also, the PM flux decreased after the precipitation, and increased in the non-rainy period. The fluctuation is, however, not so clear. This may be due to the sampling time duration (three days). The flux values were averaged over three days and the effects may be masked in these longer sampling periods. For atmospheric concentration, fairly clear oscillation was observed from the end of one precipitation to another (Figure 1a).

Table 1. Climatic condition and atmospheric concentration of PM and total PAHs

	August	September	October	November	December
Air temperature (°C)	25.4 (18.8–28.3)	22.6 (15.9–26.7)	14.5 (9.3–21.5)	11.9 (5.2–17.6)	4.6 (10.9–0.1)
Precipitation (mm)	114.5	59.5	12.5	130.0	44.0
Wind speed (m s^{-1})	0.6 (0.0–1.7)	0.3 (0.0–2.3)	n.d	n.d	n.d
PM concentration	n = 28	n = 20	n = 12	n = 28	n = 28
FPM (μg m^{-3})	22.0 (5.7–62.7)	22.9 (8.6–64.9)	29.2 (15.5–48.2)	22.7 (5.8–52.7)	25.7 (12.0–75.1)
CPM (μg m^{-3})	7.0 (2.2–14.3)	7.7 (3.6–14.3)	11.1 (6.1–10.6)	8.4 (1.3–17.3)	8.8 (2.2–22.3)
Total PAHs	n = 20	n = 18	n = 0	n = 28	n = 8
in FPM (ng m^{-3})	1.0 (0.14–5.0)	0.80 (0.17–2.88)	n.d	4.1 (0.4–14.9)	5.16 (0.94–16.3)
in CPM (ng m^{-3})	0.05 (0.01–0.02)	0.04 (0.02–0.05)	n.d	0.17 (0.01–0.77)	0.06 (0.02–0.09)

Air temperature, wind speed: daily average; in the parenthesis is the min. to max.
FPM: fine particulate matter, DPM: coarse particulate matter
n: number of sampling; n.d: no data

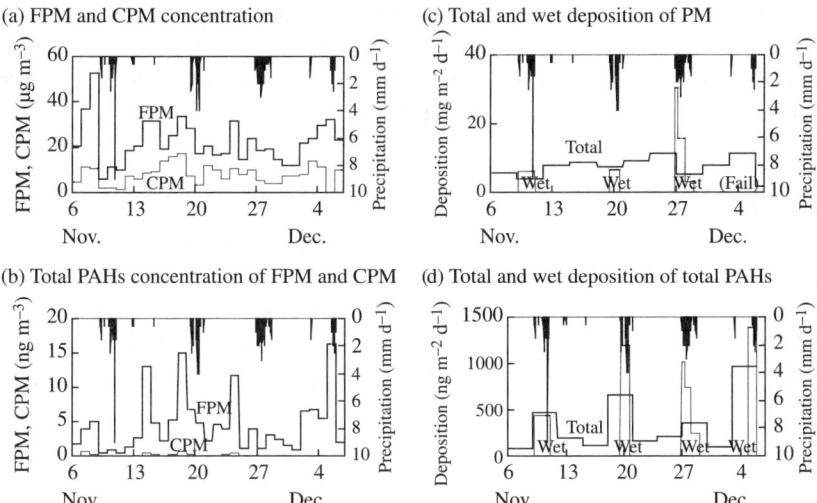

Figure 1 Profile of atmospheric and deposition PMs and PAHs (6 Nov. to 6 Dec. in 2003)

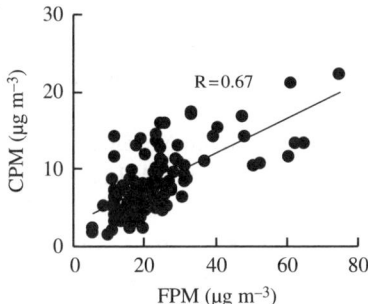

Figure 2 FPM vs. CPM concentration

It is reasonable to suppose that this oscillation affects the deposition. To elucidate this, the relation between the atmospheric PM (FPM + CPM) concentration and the dry deposition flux was shown (Figure 3; the three days average was taken for atmospheric PM in accordance with the period of deposition sampling). The dependence of the flux on the atmospheric concentration can be seen (R = 0.51; Figure 3), and further, more distinct dependence was observed for CPM concentration (R = 0.79).

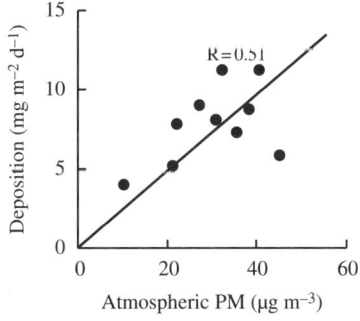

Figure 3 Atmospheric concentration and deposition

For the deposition of PAHs, the tendencies in the rainy periods are different (Figure 1d) from those of PM (Figure 4). The deposition increased sharply in the rainy periods. From this observation, it can be conjectured that the PM will have a different result of clean-up by precipitation than will PAHs.

Clean-up ratio of wet deposition

In order to estimate the effect of clean-up by precipitation, the mass balance was taken for before, during, and after precipitation. For the estimation of the total mass in the atmosphere, using the mixing height, the concentration profile in height was hypothesized as follows: from the ground level to the mixing height, the concentration is constant, and over the mixing height, the concentration is zero (Figure 4). From this hypothesis, the total mass per unit area in atmosphere can be calculated as follows:

Total Mass in air $[g\,m^{-2}] = H_m \cdot C$

where H_m: mixing height [m], C: atmospheric concentration $[g\,m^{-3}]$. The mixing height was calculated from the irradiation intensity for daytime (Kim et al., 2001), and the monthly average value was taken for this study. The estimated averaged mixing height was 805–1,186 m and the value had seasonal fluctuation. (The value was higher in summer.) From this estimation, the mass balance of PM and PAHs was taken for ten major precipitations from 11th August to 6th November (Figure 5; the dissolved phase is excluded for the calculation). For estimation of PAHs, they were divided into four groups by ring number (3-rings: Ace, Act, Flu, Phe, Ant, 4-rings: Flt, Pyr, B(a)A, Chr, 5-rings: B(b)F, B(k)F, B(e)P, B(a)P, D(ah)A, and 6-rings: B(ghi)P, Ind(123)P) and they were summed for each group. Results of 3 and 6-rings were shown in Figure 5.

For the precipitations of 8th Nov. and 18th Nov., the total mass before precipitation is considerably higher for both PM and PAHs. This is probably due to the error of the calculation of H_m. Except for these two events, the balance seems to have been successfully achieved. From the results, in order to discuss the relation of the atmospheric mass and precipitation loading, the clean-up ratio (CR) for the particulate phase was defined as follows:

$$CR = \frac{\text{Wet deposition}}{\text{Atmospheric Mass}(= \text{Atmospheric conc.} \times \text{Mixing height})}$$

The averaged CR value was 0.38, 0.86, 0.30, 0.07, and 0.11 for PM, 3-, 4-, 5-, and 6-ring PAHs respectively. The CR value can be supposed to change with the precipitation condition (e.g. the total precipitation mass, intensity). No clear relation, however, was found in this study.

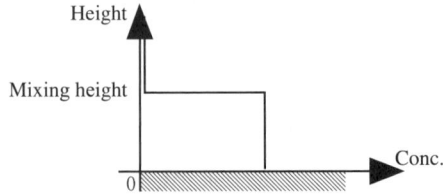

Figure 4 Hypothesis of concentration profile of PM and PAHs

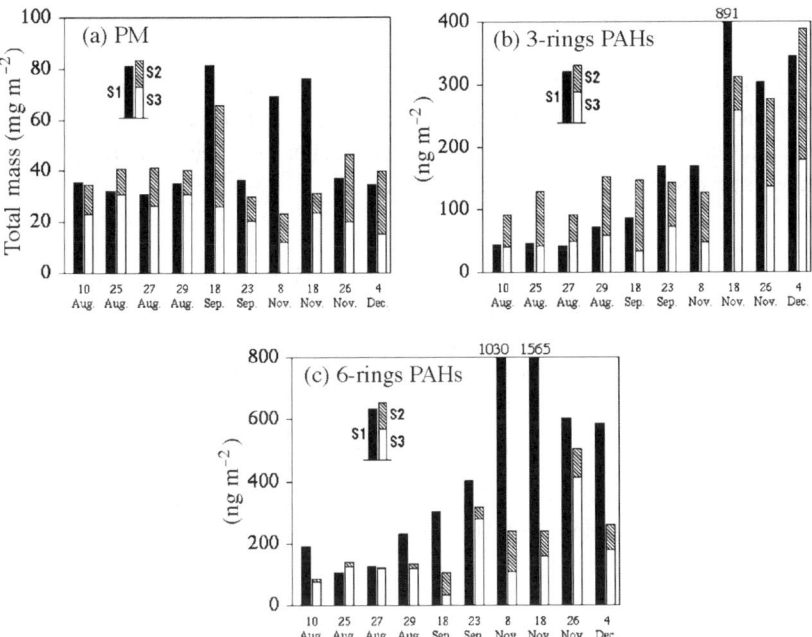

Figure 5 Mass balance of PM and PAHs in the precipitations. S1: total mass in the air before precipitation, S2: Precipitation load, and S3: total mass in the air after precipitation

Estimation of yearly total loadings of dry and wet deposition (particulate phase)

In order to estimate the yearly total loadings of the dry and wet deposition, the mechanisms of atmospheric concentration and deposition changes in time are hypothesized as follows (the scheme is described in Figure 6):

(a) Atmospheric PM mass. The atmospheric PM mass increases with time in the non-rainy period and decreases in the precipitation period. For the following calculations, the PM is defined as the sum of FPM and CPM (PM = FPM + CPM).

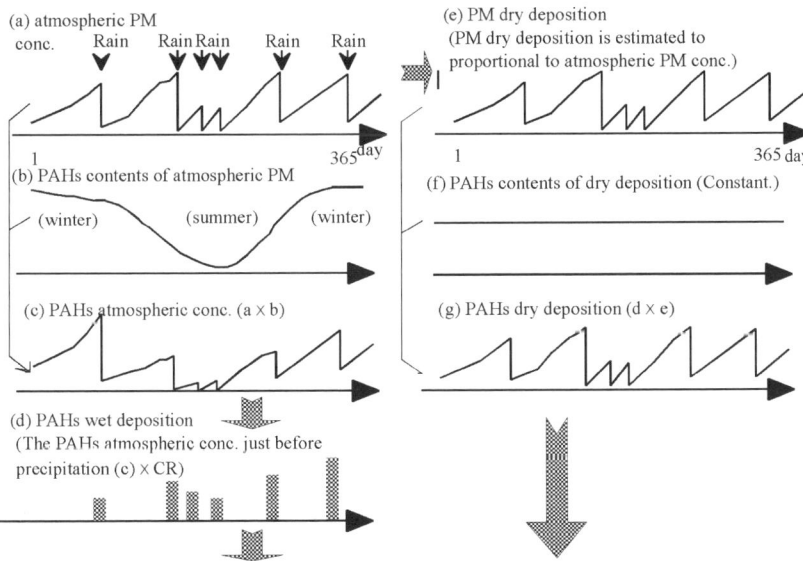

- Total deposition loading of PAHs was estimated as the summation of dry and wet deposition.

Figure 6 Scheme of estimation of yearly total deposition loading of PAHs (a-1) The rate of increase of PM concentration is dependent on the wind speed (Figure 7).

(a-2) For the calculation of the rate of decrease of PM mass in precipitation, the averaged (1-CR) for PM applied. The precipitation event is defined on a daily basis, and when it exceeds 5 mm, it is recognized as "precipitation". Consecutive precipitation over two days is recognized as one precipitation event.

(b) The PAH contents of atmospheric PM. The PAH contents of atmospheric PM are hypothesized to have seasonal fluctuation. The atmospheric concentration was measured one year from June 2000 to June 2001 at the identical sampling site (Ozaki, 2002). These obtained PAH values were summed for each ring number, log-normalized, and fitted with sine curves (Figure 8).

(c) The PAH atmospheric mass. The PAH atmospheric mass is the product of the atmospheric PM mass (a) and the PAH contents of atmospheric PM (b).

(d) The PAH wet deposition. The PAH wet deposition is calculated as the product of the PAH atmospheric mass (c) and the averaged CR values of each PAH ring group.

(e) The PM dry deposition. The PM dry deposition is hypothesized to be proportional to the PM concentration (Figure 3).

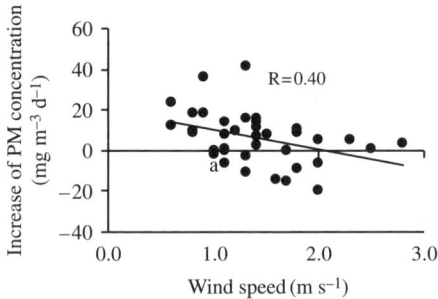

Figure 7 Wind speed and the increase of PM concentration

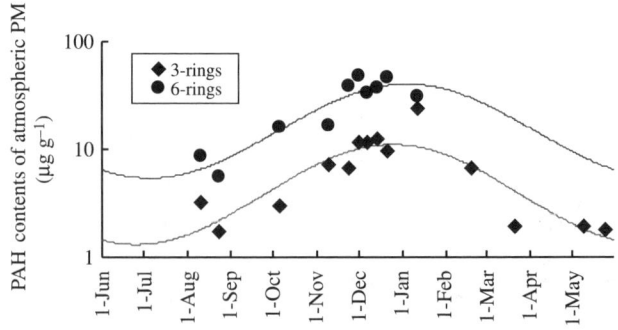

Figure 8 Yearly changes of PAH contents of atmospheric PM. The 3- and 6-rings PAH results are shown for example

(f) The PAH content of dry deposition. The PAH content of dry deposition is hypothesized to be constant. The value was estimated from the June 2000 to June 2001 measurements (Ozaki, 2002).

(g) The PAH dry deposition. The PAH dry deposition is the product of the PM dry deposition (e) and the PAH contents of dry deposition (f).

From these hypotheses, the dry and wet precipitation can be estimated. From 1st January to 31st December, the yearly profile was calculated. The yearly profile of estimation dry and wet deposition for PM is depicted in Figure 9. Along with the estimated values, the measured values are also shown in Figure 9a. The PM concentration does not have seasonal variations (Figure 9a). The PAH concentration is, on the other hand, higher in winter (Figure 9d). In accordance with this, PAH dry deposition is stable in seasons (Figure 9e), and wet deposition has seasonal fluctuation (Figure 9f). The total mass of PAHs is summarized and shown in Figure 10. The total PAH loading was 28 $\mu g\,m^{-2}\,yr^{-1}$ for dry deposition, and 52 $\mu g\,m^{-2}\,yr^{-1}$ for wet deposition. From the

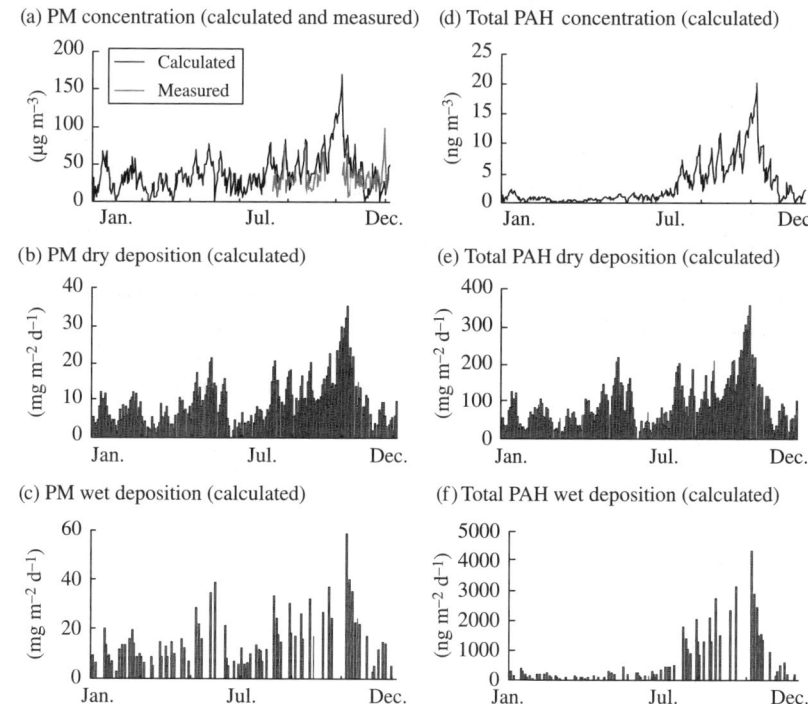

Figure 9 The calculated yearly profile of PM and PAH atmospheric concentration and deposition

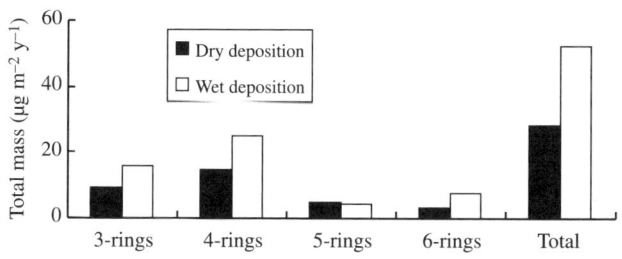

Figure 10 The yearly total loading of PAHs

results, the loading for each PAH ring group was in the same order, and the wet deposition was higher than the dry deposition.

Conclusions

From the measurement of the atmospheric PAH concentration at the sampling site of a suburban area, the dispersion and deposition profile were clarified. Using the data, a model describing the atmospheric concentration and deposition profile was established and the yearly total atmospheric loading was estimated. The total PAH loading was 28 $\mu g\,m^{-2}\,yr^{-1}$ for dry deposition, and 52 $\mu g\,m^{-2}\,yr^{-1}$ for wet deposition. From the results, the loading for each PAH ring group was in the same order, and the wet deposition was higher than the dry deposition.

This study focused on particle PAHs, and the load from the vapor phase and the dissolved phase were not estimated due to the lack of sampling of the vapor phase in the atmosphere. The contribution of the dissolved phase may be higher especially for lower molecular-weight PAHs. In our measurements, the load of wet deposition in the dissolved phase was in the same order as those of the particulate phase for 3- and 4-rings PAHs (not shown). For further study, in order to establish a more accurate estimation, the contribution of the dissolved phase should be considered.

References

Baek, S.O., Field, R.A., Goldstone, M.E., Kirk, P.W., Lester, J.N. and Perry, R. (1991). A review of atmospheric polycyclic aromatic hydrocarbons: Sources, fate and behavior. *Water, Air and Soil Pollut.*, **60**, 279–300.

Fukushima, T., Ozaki, N. and Hamada, T. (2003). PAHs dynamics in Hiroshima Bay and its watershed. *Asian Waterqual*-2003 (CD-ROM).

Kamens, R.M., Guo, Z., Fulcher, J.N. and Bell, D.A. (1988). Influence of humidity, sunlight, and temperature on the daytime decay of polyaromatic hydrocarbons on atmospheric soot particles. *Environ. Sci. Technol.*, **22**(1), 103–108.

Kim, D.Y., Yamaguchi, K., Kondo, A. and Soda, S. (2001). Study on relationship between photochemical oxidant concentration and primary pollutants emission amounts in Osaka and Hyogo Regions. *Journal of Japan Society for Atmospheric Environment*, **36**(3), 156–165 (in Japanese).

Ozaki, N. (2002). Rate of photolysis of polycyclic aromatic hydrocarbons (PAHs) in suspended particulate matter and dry deposited particulate matter. In *Proceedings of International Conference on Civil and Environmental Engineering (ICCEE-2003)*, 147–155.

Panther, B.C., Hooper, M.A. and Tapper, N.J. (1999). A comparison of air particulate matter and associated polycyclic aromatic hydrocarbons in some tropical and temperate urban environments. *Atmos. Environ.*, **33**, 4087–4099.

Wu, Y.S. and Fang, G.C. (2001). Mutagenicity and PAH-analysis of airborne particulate matter in rural site of central Taiwan, Sha-Lu. *Toxicol. Environ. Chem.*, **80**, 217–225.

Characteristics of litter waste in highway storm runoff

L.-H. Kim*, J. Kang**, M. Kayhanian***, K.-I. Gil****, M.K. Stenstrom** and K.-D. Zoh*****

*Dept. of Civil and Environmental Engineering, Disaster Prevention Research Center, Kongju National University, Kongju-si, Chungnam-do 314-701, Korea (E-mail: *leehyung@kongju.ac.kr*)

**Dept. of Civil and Environmental Engineering, University of California, Los Angeles, California 90095-1593, USA (E-mail: *joohyon@ucla.edu, stenstro@seas.ucla.edu*)

***Center for Environmental and Water Resources Engineering, Dept. of Civil and Environmental Engineering, University of California, Davis, California 95616, USA (E-mail: *mdkayhanian@ucdavis.edu*)

****Dept. of Civil Engineering, Seoul National University of Technology, Seoul 139-743, Korea (E-mail: *kgil@snut.ac.kr*)

*****Dept. of Environmental & Health, Graduate School of Public Health, Seoul National University, Jongro-Gu, Seoul, 110-799, Korea (E-mail: *zohkd@snu.ac.kr*)

Abstract Litter characterization is an integrated part of the Caltrans First Flush Characterization Study. These data will provide a basis to develop potential treatment technologies and best management practices to control pollutants in runoff from freeways. During monitoring periods in Southern California areas, the first flush phenomenon was evaluated and the impacts of various parameters such as rain intensity, drainage area, peak flow rate, and antecedent dry period on litter volume and loading rates were evaluated. First flush phenomenon was generally observed for litter concentrations, but was not apparent with litter mass loading rates. Total captured gross pollutants, defined as larger than 0.5 cm, was 90% vegetation with only 10% being litter. The normalized cumulative litter loadings were determined from 1.25 to 13.39 kg/ha for dry litter weight and 0.40 to 8.99 kg/ha for dry biodegradable litter weight. The portions of biodegradable litter to non-biodegradable litter were roughly the same across the entire event. Event mean concentrations were ranged 0.0021 to 0.259 g/L for wet gross pollutants, 0.0001 to 0.027 g/L for wet litters and 0.00007 to 0.018 g/L for dry litters. The mass emission rates should be useful to estimate total litter production for developing total maximum daily loads.

Keywords Best management practice; Caltrans; first flush; highways; litter; stormwater

Introduction

Street litter, such as plastic bags, cups, cigarette butts, and candy wrappers, often is accumulated during dry seasons. It gets swept away with stormwater into storm drains and ends up floating in the ocean or washing up on our beaches. A great deal of street litter is made up of plastic, which takes hundreds of years to break down and become harmless to the environment (US EPA, 1994). Therefore, litter is considered one of the major pollutants when protecting receiving waters for beneficial use. The California Water Resources Control Board has identified in their 303(d) list at least 36 water bodies where trash is considered a pollutant of concern (California State Water Resources Control Board, 1999). Recently Los Angeles Region of the California Regional Water Quality Control Board developed a total maximum daily load (TMDL) standard for trash in the Los Angeles River (California Department of Transportation, 2001). Faced with expected future trash regulation, the California Department of Transportation (Caltrans) is actively assessing the characteristics and potential impacts of litter generated from their surface transportation (California Department of Transportation, 2000a). Currently, litter characterization is an integrated part of the Caltrans First Flush Characterization Study (FFCS) where both water quality and litter characteristics during the first flush and the entire storm event are evaluated. As part of this study, litter weight and volume were evaluated from six monitoring sites in the Los Angeles area for up to 17 storm events during the 2000–2002 rainy seasons.

Methods

The locations of the monitoring sites in Southern California area are shown in Figure 1. Rainfall, runoff flow rate and runoff quality were monitored at six freeway sites in Southern California over two rainy seasons. The stations were equipped with a rainfall gage, flow meter and flow-weighted composite sampler. Rainfall and flow data were recorded at one-minute intervals. The monitoring sites were designed to capture litter for off-site evaluation. The circular storm drain outfalls were modified by a metal collar extension to mount and secure litter collection bags with 0.5 cm openings.

Gross pollutant samples were collected during storms. Gross pollutants are the combination of litter and vegetation collected initially in the bags. During the storm event, up to four bags were used at each monitoring site. To the extent possible, bags were collected after the first 30 minutes of stormwater flow, after the end of the first hour, and after the end of the second hour of stormwater flow. The fourth and final bag was collected after the storm event. At the completion of each sample interval, the filled collection bag was removed from the outfall and placed inside a plastic trash bag. The trash bag was secured with a large, plastic tie-wrap and labeled with a Tyvek sample tag with the appropriate sample information. Following the storm event, the collected bags were delivered to the laboratory for analysis. Litter analyses were conducted for weight and volume for the following constituents: gross pollutants, vegetation, wet litter, dry litter, biodegradable dry litter, and non-biodegradable dry litter according to the procedures specified in Caltrans Litter Monitoring Guidance Manual (California Department of Transportation, 2000b). Litter was defined as material larger than 0.5 cm that is not vegetation. Non-biodegradable litter was defined as litter that does not naturally degrade in the environment, such as metals and plastics. Biodegradable litter consists primarily of paper products. Mass balances were used for quality control (Kim et al., 2004).

The mean concentration for each event was used to characterize litter loading, which was calculated from the captured litter mass by dividing by discharged runoff volume. EMCs are frequently used to characterize stormwater loadings and can be multiplied by the runoff volume to estimate the mass discharge (Irish Jr. et al., 1998). The mass emission rate is generally greater at the beginning of rainfall, which is often called a first flush effect. The criteria of a first flush can influence the selection of best management practices (BMPs).

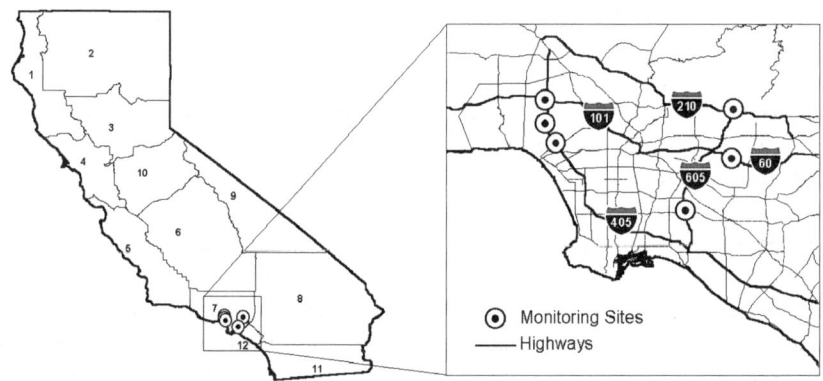

Figure 1 Monitoring sites in Southern California

Results

Continuous flow and rainfall were measured as a minute time interval. Table 1 shows storm event summaries for each monitored event and site. It includes event rainfall, maximum rainfall intensity, total runoff volume and antecedent dry days (ADD). The hydrologic data were used to prepare hydrographs and to calculate event mean litter concentrations. ADD were observed from 1 day to 190 days and event rainfalls were monitored from 0.28 cm to 15.6 cm during the monitoring periods. The total runoff volume varied from 8 m^3 to 1,420 m^3 among the sites. The gross pollutant and litter data for each site were taken and analyzed for all of the storm events. The results of a statistical analysis for normalized weight and volume by area are summarized in Table 2 for the 2000–2002 monitoring seasons.

In evaluating the raw gross pollutant and litter data for each monitoring site, UCLA 2 and URS23 had the highest relative total weight and volume of gross pollutants. The net volume of litter collected from URS23 was more than 10 times greater than other sites because of the watershed area. However, in the site, URS6-20F and URS8-23C have larger pollutant loading rates compared to other sites. The mean mass loadings for wet gross pollutant are 18.63 kg/ha in URS6-20F and 13.97 kg/ha in URS8-23C. The mean mass for wet vegetation was determined to be 16.59 kg/ha in URS6-20F and 11.51 kg/ha in URS8-23C. This means that most of the wastes of the highway runoff are originated from plants near the highways. Of the monitored sites, a high fraction of biodegradable litter was observed at URS6-20F and a high fraction of non-biodegradable litter at UCLA2.

The hydrograph shows the flow rate, rainfall intensity and the time when a litter bag was collected during a storm event. Since the bags were exchanged at preset time intervals, the hydrograph provides a visual representation of what transpired during the event. A litter bag may be collected prior to, during, or after a peak in storm water flow. Figure 2 shows dry litter concentrations and loading rates for the seasonal first storm event. The concentrations are determined by the dry litter mass by dividing by the total flow volume during the time of the litter sample collection. The litter loading rates were calculated as the dry litter mass divided by the elapsed time of litter collection and catchment area.

The gross pollutants are composed of vegetation and litter. Figure 3 shows fraction of wet vegetation to wet gross pollutant weight for each event and site. The fractions of vegetation for all sites and events are ranged from 70% to 95% of the total gross pollutants weight.

The vegetation in highways comes from plants of the road side and hill side. It is usually pulled out by strong rainfall intensity and sometimes by weathering effect. When it is washed off by storm flow to the nearby stream, it may be floated or deposited at the bottom of the water body. For a long time it can be decomposed into nutrients and other pollutants. The first flush phenomenon was evaluated during monitoring periods and the impacts of various parameters such as rain intensity, drainage area, peak flow rate, and antecedent dry period on litter volume and loading rates were evaluated. First flush phenomenon was generally observed for litter concentrations, but was not apparent with litter mass loading rates.

Wet gross pollutant loading rates for each event and site are shown in Figure 4. It was calculated with wet gross pollutant mass dividing by storm duration and catchment area. It can be useful to determine the amount of washed-off mass during storms to receiving water bodies. Usually the rates depend on antecedent dry days, storm duration, rainfall intensity, total rainfall volume, etc. On Jan. 10, 2001, the loading rates show the highest values of all events although antecedent dry days is very short, around 1.9–2 days except site URS6-20F. However, rainfall intensities on the date are ranged from 18.3 to 32.26 mm/h, which is the highest range of all the events. Also the event rainfalls are

Table 1 Hydrological summary for all monitored events

Event date (m/d/y)	UCLA monitoring sites											
	UCLA1				UCLA2				UCLA3			
	Event rainfall (cm)	Max rainfall intensity (mm/h)	Total flow (m³)	ADD (days)	Event rainfall (cm)	Max rainfall intensity (mm/h)	Total flow (m³)	ADD (days)	Event rainfall (cm)	Max rainfall intensity (mm/h)	Total flow (m³)	ADD (days)
10/26/00	2.39	6.10	260.7	33.6	2.39	5.84	200.8	33.6	2.59	4.06	94.7	33.6
01/08/01	0.38	1.78	43.7	69.4	0.51	1.78	52.2	69.4	0.53	2.03	17.9	69.4
01/10/01	12.70	30.23	1,327.4	1.9	15.60	32.26	1,416.2	1.9	12.85	22.35	481.1	2.0
02/10/01	1.32	7.87	155.2	14.2	–	–	–	–	1.55	4.57	58.6	14.2
02/19/01	0.71	2.79	80.9	5.4	2.39	7.11	261.6	4.8	3.02	12.19	112.4	5.3
02/24/01	1.45	1.78	165.6	1.0	1.91	2.29	241.6	1.0	1.14	2.29	38.1	1.0
03/04/01	1.19	3.05	139.1	4.0	0.89	4.83	140.2	4.0	0.51	2.29	11.4	4.0
04/07/01	–	–	–	–	3.02	5.33	501.9	31.5	2.54	7.62	65.2	31.6
04/20/01	0.81	2.79	79.0	13.2	–	–	–	–	–	–	–	–
10/30/01	–	–	–	–	0.33	2.03	47.5	192.20	0.28	2.29	8.1	192.30
11/12/01	–	–	–	–	1.19	9.91	172.3	12.98	0.74	5.33	24.8	12.99
11/24/01	–	–	–	–	5.03	26.67	737.8	11.69	2.97	14.48	108.7	11.60
12/14/01	–	–	–	–	0.36	2.03	52.0	19.73	–	–	–	–
01/27/02	–	–	–	–	3.18	8.13	445.6	27.13	2.46	5.08	92.2	27.14
02/17/02	–	–	–	–	–	–	–	–	0.74	4.32	25.6	20.31
03/07/02	–	–	–	–	–	–	–	–	0.46	3.30	14.4	17.74
03/17/02	–	–	–	–	0.23	2.29	23.53	10.69	1.04	9.40	37.0	10.70

Table 1 (continued)

Event date (m/d/y)	UCLA monitoring sites										
	UCLA1				UCLA2				UCLA3		
	Event rainfall (cm)	Max rainfall intensity (mm/h)	Total flow (m^3)	ADD (days)	Event rainfall (cm)	Max rainfall intensity (mm/h)	Total flow (m^3)	ADD (days)	Event rainfall (cm)	Max rainfall intensity (mm/h)	Total flow (m^3)	ADD (days)
	URS6-20F				URS8-23C				URS23		
10/26/00	0.89	15.20	33.9	33.0	–	7.60	11.2	–	3.20	–	–	33.0
01/08/01	0.23	3.00	2.0	70.6	0.33	–	–	108.0	0.43	–	–	70.4
01/10/01	7.11	18.30	130.6	72.2	10.26	24.40	168.0	2.0	8.74	27.40	1,673.5	2.0
01/26/01	0.71	6.10	10.7	1.7	1.45	6.10	14.6	1.7	1.19	21.30	156.1	1.7
02/10/01	0.71	6.10	4.3	14.5	1.25	12.20	7.3	2.9	0.79	9.10	152.7	14.6
02/19/01	0.46	6.10	7.9	5.6	0.94	1.50	12.0	5.5	1.04	6.10	117.6	5.7
02/24/01	9.04	9.10	80.8	1.1	6.43	9.10	118.3	0.0	9.55	12.20	1,047.4	5.1
04/09/01	1.55	15.20	21.7	31.8	–	24.40	31.4	28.1	2.29	15.20	564.6	31.8

Table 2 Statistical summary of monitored litter wastes

Parameter	Monitoring sites	Gross pollutants		Litter				Biodegradable				Non-biodegradable			
		Wet		Wet		Dry				Dry				Dry	
		Weight (g)	Volume (ml)	Weight (g)	Volume (ml)	Weight (g)	Volume (ml)	Weight (g)	Volume (ml)	Weight (g)	Volume (ml)	Weight (g)	Volume (ml)
Mean	UCLA2	15,777	18,296	775	6,272	517	4,668	154	1,460	220	1,525
	UCLA3	2,134	5,122	166	911	86	1,040	57	668	38	536
	URS6-20F	4,301	18,833	331	2,156	184	1,956	118	1,073	58	763
	URS8-23C	3,047	4,480	479	1,738	335	1,603	151	711	166	797
	URS23	15,512	44,803	1,581	7,153	871	7,226	389	3,014	367	3,481
Median	UCLA2	6,420	7,550	461	1,822	277	1,458	82	663	71	447
	UCLA3	1,911	4,050	160	650	96	897	50	620	31	495
	URS6-20F	1,770	9,500	198	800	126	1,070	92	580	29	420
	URS8-23C	840	1,200	243	840	151	850	90	425	49	375
	URS23	6,537	25,480	1,050	4,500	549	4,850	171	985	110	1,600
Standard deviation	UCLA2	23,800	22,299	751	11,157	504	6,006	205	2,472	276	2,672
	UCLA3	1,208	3,526	106	751	51	886	33	536	23	504
	URS6-20F	4,755	19,156	391	2,926	235	2,354	147	1,243	71	887
	URS8-23C	4,312	7,714	710	2,601	554	2,398	213	1,016	305	1,151
	URS23	19,069	45,910	1,939	7,811	1,131	7,928	564	4,222	616	4,580

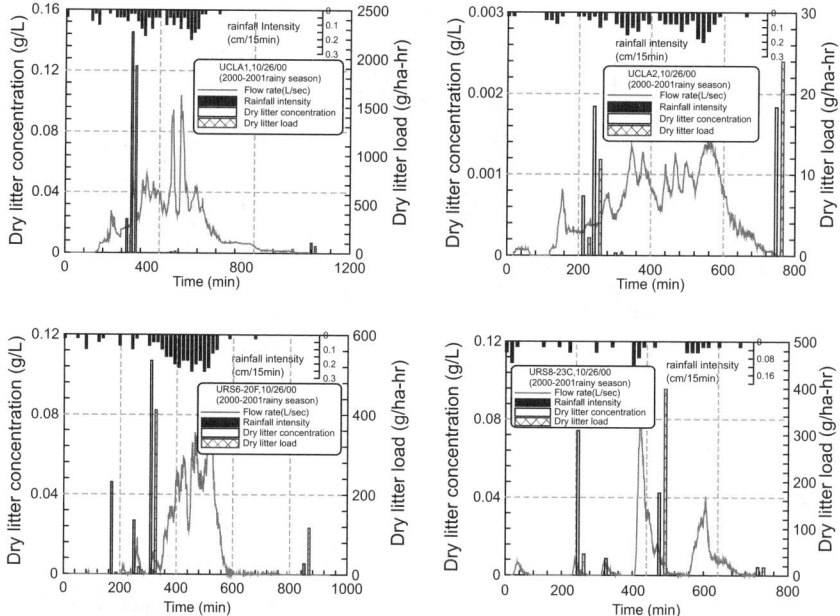

Figure 2 Litter polluto- and load-graphs for first storm event

ranged from 7.11 to 15.6 cm. As a result, it is clear that loading rates are affected by maximum rainfall intensity, event rainfall and total flow.

Litters are finally deposited into receiving water bodies and degraded by microorganism activities for a long time. Therefore it can act on inner pollution sources in the future when the environment such as pH, DO, temperature, etc. between water body and sedimentation layer changes. Figure 5 shows loading rates for wet biodegradable and dry non-biodegradable litters. The loading rates of biodegradable litters are ranged from 1 to 200 g/hr-ha. The ratio of biodegradable and non-biodegradable litters is very similar around 0.5 for all events and sites. The ratio is not affected by maximum rainfall intensity, event rainfall and total flow.

Each pollutant parameter normalized by area was compared with potential affecting factors such as total rainfall, maximum rainfall intensity and antecedent dry days to determine whether there are any potential relationships. The matrix of small figures represents

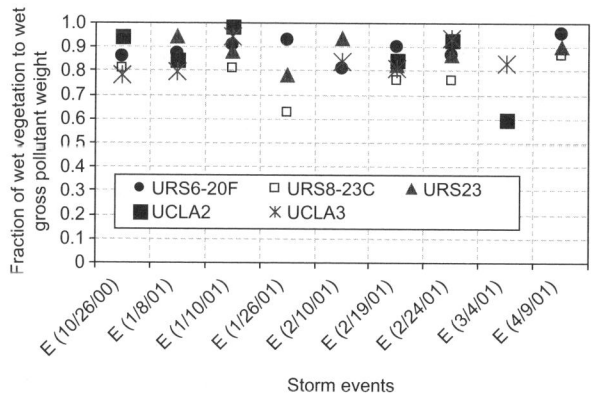

Figure 3 Fraction of wet vegetation to wet gross pollutant weight for each event and site

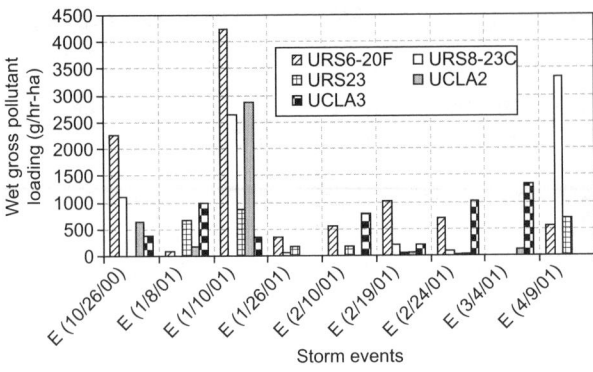

Figure 4 Average wet gross pollutant loading for each event

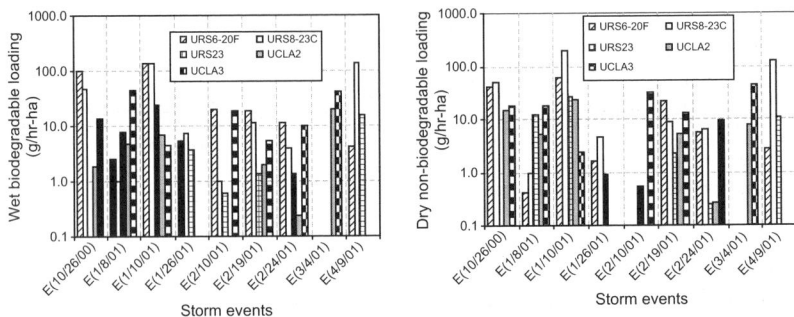

Figure 5 Loading rates for wet biodegradable and dry non-biodegradable litters

mass and volume loadings relationship with affecting parameters. Figure 6(a) shows the mass-based parameters and Figure 6(b) shows the volume-based parameters.

The two lines represent 90% confidence intervals of data. There are no obvious correlations with storm characteristics, such as ADD and total rainfall (TR). The relationship

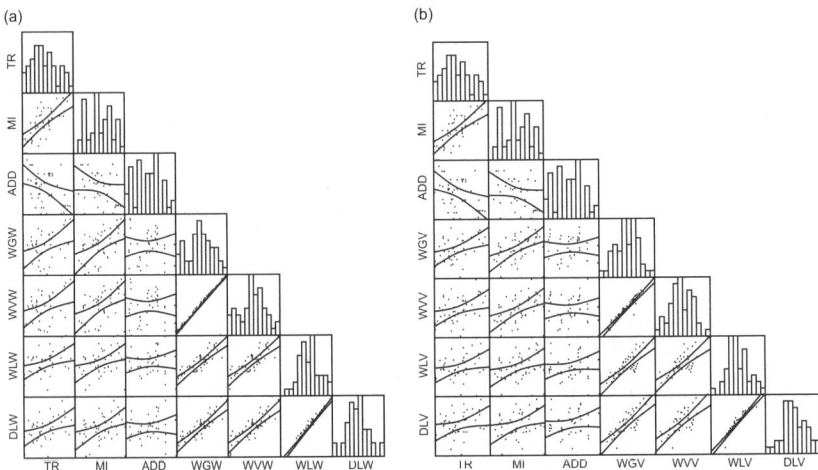

Figure 6 Correlation matrix for mass (a) and volume (b) loading with impacting parameters such as total rainfall, maximum rainfall intensity and antecedent dry days (TR: total rainfall, MI: maximum rainfall intensity, ADD: antecedent dry days, WGW: wet gross pollutant weight, WVW: wet vegetation weight, WLW: wet litter weight, DLW: dry litter weight, WGV: wet gross pollutant volume, WVV: wet vegetation volume, WLV: wet litter volume, DLV: dry litter volume)

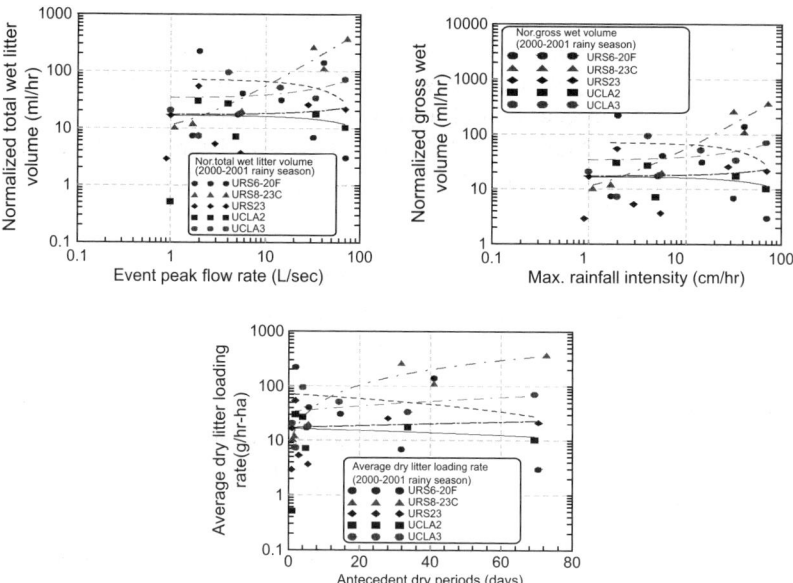

Figure 7 Impact of the hydrological data on loading volume and mass loading

between wet gross pollutant mass or volume and wet vegetation mass or volume is striking. The wet gross pollutant mass is primarily vegetation. There are also significant relationships between wet and dry volumes, which are expected.

Figure 7 shows impact of the hydrological data on loading volume and mass loading. The total volume of litter and gross pollutant collected during each storm event were evaluated to determine if there was any potential impact by event peak flow rate, maximum rainfall intensity, and the antecedent dry period. URS8-23C illustrates a possible positive trend that may be present; however, the data of the other monitoring sites are more widely scattered. In previous research (Kayhanian *et al.*, 2002), however, there appears to be a stronger correlation between the normalized total litter volume and the rain intensity for each site. Also the volume of litter collected at a site was closely related to the rain intensity or relative strength of the storm water flow.

Conclusions

As a part of a large non-point source pollution study, the litter study was performed during 2000–2002 rainy seasons at six different freeway sites located in Southern California. Litter pollutants washed-off from highways have harmful effects on drinking water supplies, recreation, fisheries, and wildlife. Therefore the problems of litter wastes have recently attracted very considerable attention due to Total Maximum Daily Load regulations.

The observation of first flush is important for best management practices. According to litter data analysis, a first flush phenomenon was generally observed for litter concentrations, but was not apparent with litter mass loading rates. The size of a monitoring site drainage area did not impact the total litter mass loading rate. Litter volume and loading rates appear to be directly related to peak storm intensity. The ratio of biodegradable litter to non-biodegradable litter was quite variable. However, a slightly greater percentage of biodegradable litter was usually collected in the first flush. The normalized cumulative litter loadings vary from 1.25 to 13.39 kg/ha for dry litter weight, 0.40 to 8.99 kg/ha for dry biodegradable litter weight, and 0.85 to 6.60 kg/ha for non-biodegradable litter weight. Event mean litter concentrations are determined and compared with antecedent dry days, event rainfall and total flow to find a stronger relationship between litter and impact parameters. The EMC distribution does support that higher litter accumulation is associated with longer antecedent dry days. Generally, the large event rainfall and runoff volume decreases the concentrations of litter because of dilution effect. Event mean concentrations are ranged 0.0021 to 0.259 g/L for wet gross pollutants, 0.0001 to 0.027 g/L for wet litters and 0.00007 to 0.018 g/L for dry litters.

Acknowledgements

This study was supported in part by the California Department of Transportation (Caltrans). The authors are grateful for their continuous support.

References

California Department of Transportation (2000a). *Sampling and analysis plan, Caltrans 2000–2001, first flush characterization study*. Caltrans Document No. CTSW-RT-00-044.

California Department of Transportation (2000b). *Litter monitoring guidance manual*. Caltrans Document No. CTSW-RT-00-025.

California Department of Transportation (2001). *Gross solids removal devices (GSRD) pilot study 2000–2001 interim report*. Caltrans Document No. CTSW-RT-01-047.

California State Water Resources Control Board (CSWRCB, 1999). *1998 California 303(d) list and TMDL priority schedule*. http://www.swrcb.ca.gov/tmdl/docs/303d98.pdf.

Irish, L.B., Jr, Barrett, M.E., Malina, J.F., Jr and Charbeneau, R.J. (1998). Use of regression models for analyzing highway storm-water loads. *J. of Environ. Engineering*, **124**(10), 987–993.

Kayhanian. M., Kummerfeldt. S., Lee-Hyung Kim, Gardiner. N. and Kuen Tsay (2002). Litter Pollutograph and Loadograph, *Proceedings of 9th International Conference on Urban Drainage*, September 8–13, Portland, Oregon.

Kim, L.-H., Kayhanian, M. and Stenstrom, M.K. (2004). Event mean concentration and loading of litter from highways during storms. *Science of the Total Environment*, **330**, 101–113.

US EPA (1994). *Nonpoint sources pollution control program*, US EPA Report 841-F-94-005, USA.

Correlation analysis among highway stormwater pollutants and characteristics

Y.H. Han*, S.L. Lau*, M. Kayhanian** and M.K. Stenstrom*

*Civil and Environmental Engineering Department, University of California, Los Angeles, CA 90095-1593, USA (E-mail: *younghan@ucla.edu; simlin@seas.ucla.edu; stenstro@seas.ucla.edu*)
**Center for Environmental and Water Resources Engineering, Department of Civil and Environmental Engineering, University of California, Davis, CA 95616, USA (E-mail: mdkayhanian@ucdavis.edu)

Abstract Stormwater runoff from highway land use is a common non-point source of pollutants. A large quantity of highway stormwater runoff characteristics were collected in California during the past three years. Correlations among various water quality parameters and constituents were performed using data sets collected over the 2000–2001, 2001–2002, and 2002–2003 wet seasons for 18, 21 and 23 storm events at three highway sites in west Los Angeles, California. In addition, statistical and graphical correlation analysis of the mass first flush ratio (MFF) with storm characteristics was made to determine if the first flush is related to site or storm characteristics. The results and analyses performed indicate that (1) TSS correlates well with most particulate-bound metals. However, TSS was poorly correlated with most other pollutants. (2) Strong correlations were also observed among dissolved and total metals; DOC, COD, TKN and oil and grease; conductivity and Cl. (3) Total metals, COD and DOC were generally well correlated with mass first flush, suggesting that BMPs that treat the early portion of runoff have an opportunity to remove high concentrations of these pollutants.
Keywords Correlation analysis; highway stormwater runoff; mass first flush; storm characteristics

Introduction

The million miles of highway throughout the United States represent a known but unquantified non-point source (NPS) of pollution (Wu *et al.*, 1998), and have been identified as one of the leading causes of the degradation of the quality of receiving waters. Highway runoff pollution recently has been identified as an important source of micropollutants such as heavy metals, hydrocarbons and fuel additives (Barrett *et al.*, 1998; Furumai *et al.*, 2002). These constituents are generated mainly from traffic activities, component wear, fluid leakage, pavement degradation, roadway maintenance and atmospheric deposition (Sansalone and Buchberger, 1997a,b; Shinya *et al.*, 2000).

In an effort to manage a watershed, the impact of highway runoff must be treated as one of the most important components among pollutant sources (Wu *et al.*, 1998). Accurate knowledge of the quantity and quality of runoff is required to assess the impacts of runoff on the environment and to develop appropriate mitigation technologies (Barrett *et al.*, 1998). In addition, knowing the correlation among stormwater pollutants may reduce the huge costs associated with monitoring and hence will assist in evaluating best management practices (BMPs) by using pollutants that may be accompanied by similar correlations in removal rates by BMPs. This paper reports a large quantity of highway stormwater runoff data and characteristics collected in west Los Angeles over the past three years (2000–2003). Correlations among pollutants and storm and site characteristics were examined. The information gained from this correlation analysis may assist transportation agencies to reduce the number of constituents that they are required to monitor, and may be used to evaluate the performance of BMPs.

doi: 10.2166/wst.2006.057

Methods

Site description

Three monitoring sites in west Los Angeles, California, were selected with catchment areas ranging from 0.39 to 1.69 hectares and annual average daily traffic (AADT) of over 260,000 cars per day. These three sites were chosen as typical small catchment sites with heavy traffic load. All sites were equipped with American Sigma (Loveland, Colorado) Ultrasonic 950 Area-Velocity flow meters, tipping bucket rain gauges and composite auto samplers. Table 1 summarizes the three site descriptions and Figure 1 shows their locations.

Sample collection and analysis procedure

In general, five grab samples were collected every 15 minutes in the first hour after the start of detectable runoff. After the first hour, samples were taken at one-hour intervals for the following 7 hours, providing a total of 12 grab samples. For events lasting longer than 8 hours, one or two additional grab samples were collected. The number of the samples collected depended on the storm duration. Table 2 lists basic statistics of all the sampling events conducted between October 2000 and April 2003. A flow weighted composite sample was collected by the American Sigma equipment, which covered the duration of each storm.

All the samples were transported back to the laboratory at UCLA after collection and refrigerated at 4°C until analyzed. This was done to facilitate certain analysis, such as filtration for soluble/total metals and particle size distribution, which must be performed very soon after sample collection. Generally, the samples were transported to UCLA after the first hour and two more times during the following 7 hours. A composite sample was collected after the event.

Correlation analysis

The correlation analysis was conducted using data sets, collected over the 2000–2001, 2001–2002, and 2002–2003 wet seasons for 18, 21 and 23 storm events at three highway sites. Correlations were performed on data from individual sites as well as for the combined data.

In addition, statistical and graphical correlation analysis of the mass first flush ratios (MFF = normalized pollutant mass divided/normalized runoff volume) developed by Ma *et al.* (2002) with storm characteristics was made to determine if the first flush is related to site or storm characteristics. Seven parameters were selected as storm characteristics and MFF was analyzed at five points (10 to 50% by 10% increments). The MFF ratios were also ranked by parameter, and divided into four groups with similar correlation coefficients.

Results

Water quality data collected

The average number of observations for each parameter or constituent was approximately 160 per site and 500 for the combined sites. A total of 44 water quality parameters/

Table 1 Site description summary

Site ID	Location	AADT (Vehicles/day)	Catchment area (hectares)	Number of lanes (each direction)	Approximate impervious (%)
7-201	Hwy 101, Van Nuys	328,000	1.28	6	100
7-202	Hwy 405, Getty Center Exit	260,000	1.69	5	95
7-203	Hwy 405, Santa Monica Blvd.	322,000	0.39	5	100

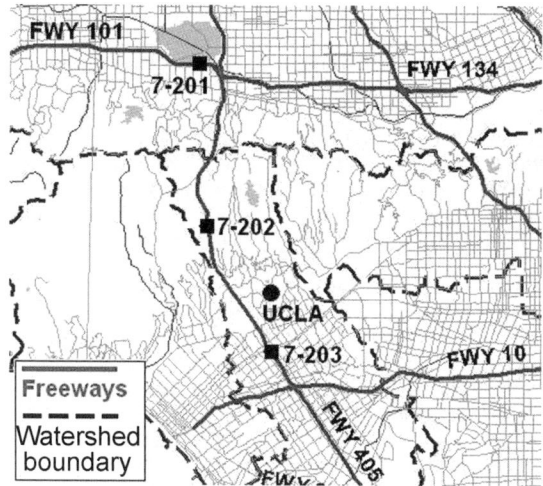

Figure 1 Highway runoff sampling sites

constituents comprising conventional, metals (particulate and dissolved), nutrients, fecal and total coliform, and organics including oil and grease were measured. Table 3 shows the basic statistics of principal constituents among data collected for all sites.

Correlation analysis among constituents

There is a general belief that many stormwater pollutants are sorbed to the surfaces of suspended solids, and previous researchers have correlated metal elements to total suspended solids in highway stormwater runoff (Sansalone and Buchberger, 1997a; 1997b; Shinya et al., 2000). Heavy metals except for nickel showed strong correlation to suspended solids (Shinya et al., 2000).

Table 4a shows correlation among dissolved organic carbon (DOC), total Kjeldahl nitrogen (TKN), chemical oxygen demand (COD) and TSS; Table 4B shows the correlation among metals and TSS. The numbers above the diagonal are Pearson's coefficients, r, and the numbers below the diagonal are probability or P values. TSS was poorly correlated to DOC, COD and TKN, but the DOC, COD and TKN were all well correlated.

Table 2 Basic statistics of storm events

		Total rainfall (mm)	Max. intensity (mm/h)	Antecedent dry day	Storm duration (h)	Ave. rainfall intensity (mm/h)
7-201	No. of storm events	18	18	18	18	18
	Min./Max.	2.0/127.0	3.0/51.8	2.0/69.4	2.0/47.5	0.6/10.7
	Median/Mean	18.3/28.3	15.2/20.3	13.7/19.6	8.6/10.7	1.9/3.0
	Standard Dev.	30.5	17.8	17.5	10.3	2.7
7-202	No. of storm events	21	21	21	21	21
	Min./Max.	1.8/156.0	3.0/61.0	1.0/192.2	0.9/46.5	0.2/11.3
	Median/Mean	19.1/26.1	12.2/17.3	19.8/29.0	6.0/9.3	2.0/3.2
	Standard Dev.	34.9	16.3	41.0	10.0	3.3
7-203	No. of storm events	23	23	23	23	23
	Min./Max.	1.5/128.5	6.1/39.6	0.3/192.3	1.4/47.1	0.2/8.9
	Median/Mean	19.8/28.5	21.3/21.1	14.2/26.4	6.9/9.9	2.5/3.3
	Standard Dev.	34.5	11.2	39.9	9.7	2.6
Combined sites	No. of storm events	62	62	62	62	62
	Min./Max.	1.5/156.0	3.0/61.0	0.3/192.3	0.9/47.5	0.2/11.3
	Median/Mean	19.4/27.6	15.2/19.6	16.3/25.3	7.2/9.9	2.2/3.2
	Standard Dev.	33.0	15.0	35.0	9.8	2.8

Table 3 Basic statistics of principal EMCs calculated for combined sites

Parameters	No. of cases[1]	Mean	EMC/Grab[2] (Mean, %)	Median	EMC/Grab[3] (Median, %)	Minimum	Maximum	Standard dev.
TSS (mg/L)	62/569	67.7	94.9	57.6	125.4	8.8	466.4	63.0
Turbidity (NTU)	62/569	46.8	89.9	33.1	103.8	10.9	170.5	39.2
COD (mg/L)	62/569	252.5	78.6	119.8	86.5	19.1	2,282.8	372.8
DOC (mg/L)	62/569	66.9	82.8	28.9	97.5	2.9	848.8	127.0
Oil & Grease (mg/L)	62/569	14.0	77.4	8.8	83.2	1.5	80.2	14.6
TKN (mg/L)	62/569	9.6	82.8	4.1	87.8	0.8	111.3	16.4
NH_3-N (mg/L)	62/569	4.6	84.1	1.4	105.6	0.0	65.0	9.7
NO_3-N (mg/L)	62/569	2.7	83.3	1.3	90.9	0.3	34.7	5.3
Total P (mg/L)	62/566	0.9	98.0	0.4	85.6	0.1	8.2	1.3
Total Cd (μg/L)	46/362	1.8	86.0	1.1	100.2	0.4	20.2	3.0
Total Cr (μg/L)	62/564	9.7	92.8	8.7	105.5	2.3	40.1	6.3
Total Cu (μg/L)	62/564	92.9	83.5	55.7	88.1	15.9	920.8	125.2
Total Ni (μg/L)	62/563	20.0	86.0	11.2	87.7	2.4	253.7	33.9
Total Pb (μg/L)	62/562	25.8	105.8	22.9	121.1	4.6	151.6	20.5
Total Zn (μg/L)	62/564	506.3	90.2	267.9	100.4	83.3	8,881.3	1,137.0
Diss. Cd (μg/L)	43/363	1.4	90.4	0.6	93.6	0.5	17.8	2.7
Diss. Cr (μg/L)	58/566	2.8	89.9	2.0	96.6	0.5	19.3	2.8
Diss. Cu (μg/L)	62/566	65.9	79.0	35.4	91.9	5.2	735.3	103.9
Diss. Ni (μg/L)	62/565	15.7	83.4	7.9	91.0	0.8	229.2	31.3
Diss. Pb (μg/L)	47/564	4.9	110.7	3.6	145.6	0.5	43.4	6.5
Diss. Zn (μg/L)	62/566	415.3	89.5	178.2	97.2	42.3	8,150.0	1,055.7

[1] Number of events or EMCs over total number of grab samples
[2] Arithmetic mean of the EMCs divided by mean of grab samples
[3] Arithmetic median of the EMCs divided by median of grab samples

TSS showed a relatively good correlation to particulate metals except for Cd. Among particulate metals, Cu-Ni and Cu-Zn were strongly correlated. The correlation results for all three sites are summarized as follows:

- Very strong correlations ($r \geq 0.90$): TSS-VSS, COD-DOC, total P-diss. P, total Cd-total Zn, total Cd-diss. Cd, total Cd-diss. Zn, total Cu-total Ni, total Cu-diss. Cu, total Cu-diss. Ni, total Ni-diss. Cu, total Ni-diss. Ni, total Pb-particulate Pb, total Zn-diss. Zn, diss. Cd-total Zn, diss. Cd-diss. Zn, and diss. Cu-diss. Ni.

Table 4 Correlation analysis results among non-metals and TSS and particulate metals

(a) Non-metals

Para.	TSS	COD	DOC	TKN
TSS	1	0.40	0.34	0.40
COD	0	1	0.92	0.84
DOC	0	0	1	0.81
TKN	0	0	0	1

(b) TSS and particulate metals

Para.	TSS	Part. Cd	Part. Cr	Part. Cu	Part. Ni	Part. Pb	Part. Zn
TSS	1	0.52	0.59	0.60	0.58	0.60	0.61
Part. Cd	0	1	0.72	0.78	0.70	0.68	0.76
Part. Cr	0	0	1	0.78	0.72	0.65	0.70
Part. Cu	0	0	0	1	0.85	0.70	0.83
Part. Ni	0	0	0	0	1	0.70	0.75
Part. Pb	0	0	0	0	0	1	0.74
Part. Zn	0	0	0	0	0	0	1

- above the diagonal: Pearson's coefficient "r"
- below the diagonal: Probability values (P-value)

- Strong correlations ($r \geq 0.80$): Conductivity-hardness, COD-oil & grease (O&G), O&G-TKN, DOC-NH_3-N, cond.-Cl, TKN-Cl, cond.-SO_4, hard.-SO_4, total Cd-total Cu, total Cd-total Ni, total Cd-diss. Cu, total Cd-diss. Ni, total Cr-particulate Cr, total Cu-diss. Cd, total Cu-diss. Zn, total Ni-diss. Cd, total Ni-diss. Zn, total Zn-diss. Cu, total Zn-diss. Ni, total Cu-total Zn, total Ni-total Zn, diss. Cd-diss. Cu, diss. Cd-diss. Ni, diss. Cu-diss. Zn, and diss. Ni-diss. Zn.

Table 5 shows the summary statistics for particulate-bound metals. In general, Cr and Pb are in particulate-bound phase and Cu, Ni and Zn are more associated with the dissolved phase.

Mass first flush and correlation analysis

The concept of the first flush phenomenon was first advanced in the early 1970s. Bertrand et al. (1998) assumed that there is a significant first flush if at least 80% of the total pollutant mass is transported in the first 30% of the volume discharged during rainfall events. However, such instances are extremely rare and found in only 1% of the events (Lee et al., 2003). Ma et al. (2002) previously defined the mass first flush ratio, which describes the fractional mass of pollutants emitted as a function of storm duration. It is defined mathematically as follows:

$$MFF_n = \frac{\frac{\int_0^{t_1} C(t)Q(t)dt}{M}}{\frac{\int_0^{t_1} Q(t)dt}{V}} \quad (1)$$

For a mass first flush to exist (e.g. $MFF_n > 1$), the concentration must not only be greater during the early part of the storm, but the mass emissions must also be greater. The ratio allows convenient characterization of the first flush, and allows the first flush to be analyzed statistically. For example, $MFF_{20} = 2.5$ means that 50% of mass load is contained in the first 20% of the runoff. Bertrand et al. (1998)'s definition can be described as $MFF_{30} > 2.66$.

Figure 2 shows notched bar plots of the MFF ratios for COD, TSS and six total metals for the combined sites. The ranges of MFF_{10} and MFF_{20} by median are 1.250 to 2.511 and 1.232 to 1.897. This suggests that there is a first flush, although the magnitudes are less than suggested by Bertrand et al. (1998). Pb and Cr had the lowest MFF ratios, both of which are shown to be particulate-bound. The value of knowing the first flush is quantifying the additional removal affected by BMPs that can completely capture the early runoff.

In order to assess the runoff characteristics and to compare MFFs for each constituent, the MFFs were ranked by median values. The results are shown in Table 6. In general, TSS, COD, DOC, TKN and NH_3-N had high MFF values, whereas NO_2-N, dissolved Cd, total Cr, dissolved Cr and total Pb had low values. Site 7-203 had higher MFF values than the other sites. Site 7-203 is smaller than the areas of other sites, with shorter rainfall retention time.

Correlation analysis among storm characteristics and MFF_n

Previous researchers have noted interactions between storm characteristics and first flush. Furumai et al. (2002) found a stepwise washoff phenomenon, and the wash-off process did not appear to be linear with the runoff rate and to the mass of pollutant. Gupta and Saul (1996) noted that the maximum rainfall intensity, maximum inflow, rainfall duration, and the antecedent dry weather period were the most important parameters influencing first flush.

Table 5 Summary statistics for particulate-bound metals

		7-201	7-202	7-203	Combined			7-201	7-202	7-203	Combined
No. of cases	Cd	85	147	130	362	Standard dev.	Cd	19.1	31.6	33.7	30.0
	Cr	172	192	200	564		Cr	17.8	15.3	15.5	16.3
	Cu	172	192	200	564		Cu	15.8	20.6	22.8	20.8
	Ni	171	192	200	563		Ni	17.0	22.4	24.6	22.2
	Pb	170	192	200	562		Pb	18.2	15.1	18.7	17.8
	Zn	172	192	200	564		Zn	20.6	26.1	19.1	22.6
Minimum	Cd	6.9	0.0	5.0	0.0	Maximum	Cd	100.0	100.0	100.0	100.0
	Cr	0.2	16.0	30.2	0.2		Cr	100.0	100.0	100.0	100.0
	Cu	0.2	2.3	2.1	0.2		Cu	91.5	84.0	85.3	91.5
	Ni	0.3	0.9	1.4	0.3		Ni	97.5	87.5	90.6	97.5
	Pb	23.2	25.3	7.5	7.5		Pb	100.0	100.0	99.5	100.0
	Zn	0.8	1.2	0.3	0.3		Zn	92.1	93.4	78.7	93.4
Median	Cd	50.0	50.0	50.0	50.0	Mean	Cd	53.4	48.3	52.3	50.9
	Cr	72.5	74.8	75.0	74.3		Cr	68.8	73.8	74.5	72.5
	Cu	26.4	42.4	25.7	31.3		Cu	29.6	42.0	32.7	34.9
	Ni	22.0	31.9	28.8	27.6		Ni	25.7	35.4	35.3	32.4
	Pb	91.7	92.7	81.7	88.1		Pb	84.4	86.6	77.1	82.6
	Zn	22.0	33.6	19.6	24.1		Zn	27.7	36.5	25.3	29.8

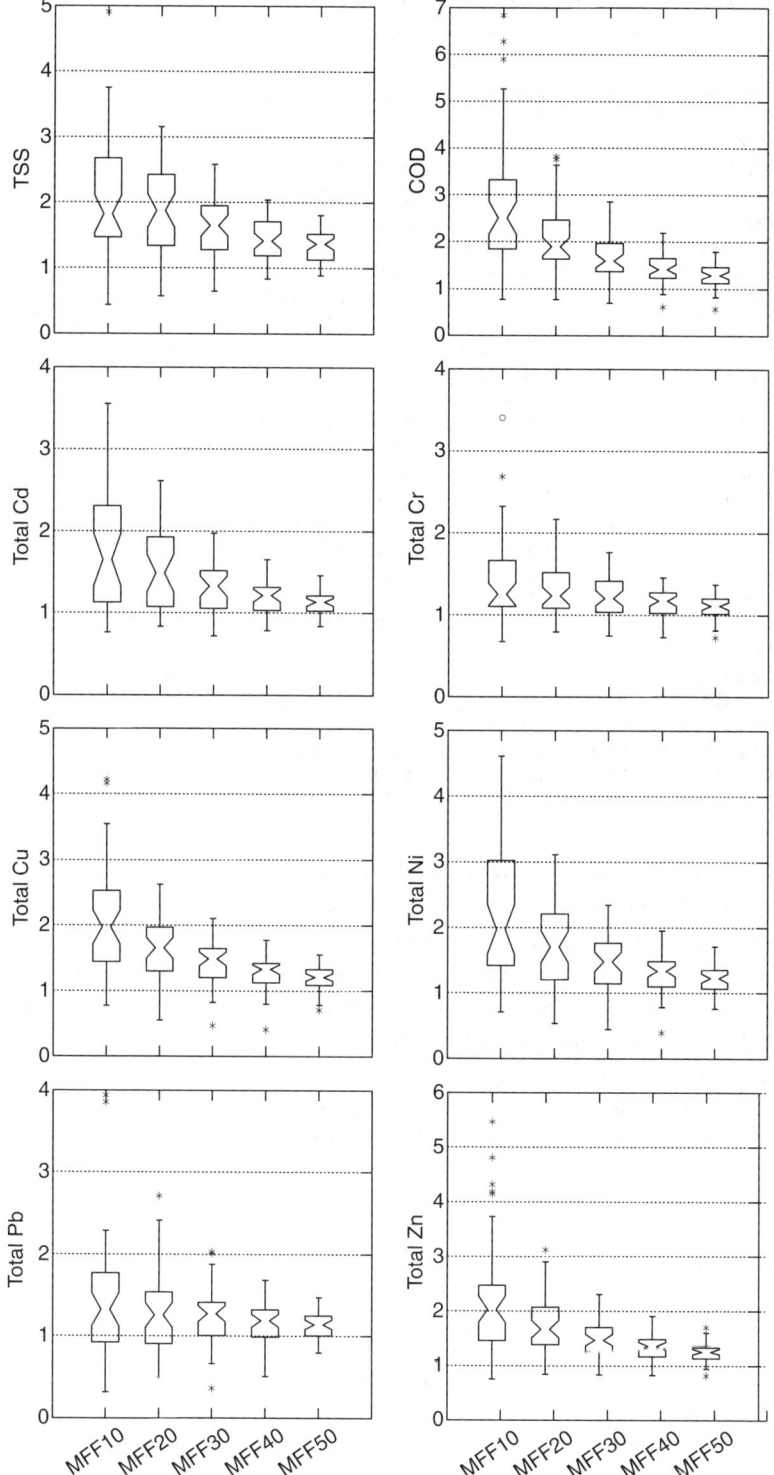

Figure 2 MFF plots for combined sites

Table 6 Ranking of MFF$_{20}$ by median value for all sites

Rank	7-201 Parameters	Median	7-202 Parameters	Median	7-203 Parameters	Median	Combined sites Parameters	Median
1	COD	1.84	PO$_4$-P	2.28	DOC	2.74	DOC	1.95
2	PO$_4$-P	1.77	Total Cd	2.04	Dissolved Ni	2.48	TKN	1.94
3	Total P	1.77	DOC	2.01	COD	2.41	COD	1.90
4	TSS	1.71	TKN	2.01	TKN	2.39	NH$_3$-N	1.88
5	Oil & Grease	1.60	Dissolved Zn	1.99	NH$_3$-N	2.26	TSS	1.87
6	Dissolved Ni	1.57	Dissolved P	1.98	Total Ni	2.17	Dissolved Ni	1.79
7	NH$_3$-N	1.55	Total Zn	1.97	Dissolved Cu	2.12	PO$_4$-P	1.77
8	Dissolved P	1.55	NH$_3$-N	1.96	TSS	1.98	Dissolved P	1.76
9	TKN	1.52	Dissolved Ni	1.94	Total Cu	1.96	Total P	1.75
10	Hardness	1.51	Dissolved Cu	1.91	Oil & Grease	1.91	Total Ni	1.70
11	Conductivity	1.48	Dissolved Pb	1.90	Total Zn	1.80	Oil & Grease	1.69
12	Total Zn	1.46	Dissolved Cd	1.90	Conductivity	1.79	Dissolved Cu	1.69
13	Dissolved Pb	1.46	COD	1.86	Dissolved P	1.79	Conductivity	1.68
14	DOC	1.43	Total Ni	1.84	Total P	1.76	Total Zn	1.67
15	Total Ni	1.41	TSS	1.69	Hardness	1.71	Total Cu	1.65
16	Total Pb	1.40	Oil & Grease	1.69	Dissolved Zn	1.66	Dissolved Zn	1.65
17	Total Cu	1.40	Total Cu	1.68	NO$_3$-N	1.63	Hardness	1.54
18	NO$_2$-N	1.36	Total P	1.62	PO$_4$-P	1.53	Dissolved Pb	1.50
19	Dissolved Cu	1.34	Turbidity	1.43	Total Cd	1.49	Total Cd	1.49
20	Dissolved Zn	1.31	NO$_3$-N	1.42	Dissolved Cd	1.39	NO$_3$-N	1.39
21	Turbidity	1.29	Hardness	1.40	NO$_2$-N	1.37	Dissolved Cd	1.37
22	Total Cr	1.27	Total Pb	1.39	Dissolved Pb	1.35	NO$_2$-N	1.36
23	Total Cd	1.19	Conductivity	1.37	Total Cr	1.26	Turbidity	1.29
24	Dissolved Cr	1.15	Dissolved Cr	1.35	Total Pb	1.24	Total Pb	1.26
25	Dissolved Cd	1.06	NO$_2$-N	1.25	Turbidity	1.22	Total Cr	1.23
26	NO$_3$-N	0.94	Total Cr	1.19	Dissolved Cr	1.16	Dissolved Cr	1.19

To determine the effect of storm and site characteristics on first flush, statistical and graphical correlation analyses of the MFF with storm characteristics and other constituents were made. Seven parameters were selected as storm characteristics: total event rainfall; rainfall intensities (max. 5 min. and 15 min., and average); antecedent dry days (ADD); antecedent rainfall, and rainfall duration. Almost no correlations existed between MFF and storm characteristics, with the largest values of R being about 0.5. Table 7 shows the correlation among TSS, COD, DOC and the total forms of the six previously discussed metals. The table shows that the constituents are generally well correlated, suggesting that BMPs that can treat the first flush have an opportunity to remove all six metals and TSS, COD and DOC.

Table 7 Correlation among MFF$_{20}$ Ratios for TSS, COD, DOC and the total metals

	TSS	COD	DOC	Tot. Cd	Tot. Cr	Tot. Cu	Tot. Ni	Tot. Pb	Tot. Zn
TSS	1	0.23	0.14	0.13	0.45	0.18	0.20	0.57	0.16
COD	0.12	1	0.87	0.53	0.47	0.79	0.73	0.08	0.70
DOC	0.36	0.00	1	0.53	0.38	0.85	0.77	0.02	0.73
Tot. Cd	0.45	0.00	0.00	1	0.37	0.67	0.68	0.28	0.79
Tot. Cr	0.00	0.00	0.01	0.03	1	0.65	0.45	0.60	0.43
Tot. Cu	0.24	0.00	0.00	0.00	0.00	1	0.93	0.20	0.84
Tot. Ni	0.20	0.00	0.00	0.00	0.00	0.00	1	0.23	0.87
Tot. Pb	0.00	0.62	0.91	0.11	0.00	0.19	0.13	1	0.24
Tot. Zn	0.31	0.00	0.00	0.00	0.00	0.00	0.00	0.12	1

- above the diagonal: Pearson's coefficient "r"
- below the diagonal: Probability values (P-value)

Conclusions

Three years of highway runoff data were examined to determine the event mean concentrations of the pollutants normally associated with highway runoff. Strong correlations were found among many pollutants. TSS was poorly correlated to DOC, COD and TKN, but the DOC, COD and TKN were all well correlated. TSS showed a relatively good correlation to particulate metals. Among particulate metals, Cu-Ni and Cu-Zn were strongly correlated. Cr and Pb were primarily particulate-bound (72.5% and 82.6% particulate) and Cu, Ni and Zn were more associated with the dissolved phase (34.9%, 32.4% and 29.8% particulate).

Highway runoff also showed a first flush, with median MFF_{10} and MFF_{20} ratios ranging from 1.3 to 2.5 and 1.2 to 1.9, respectively. TSS, COD, DOC, TKN and NH_3-N had the greatest MFF ratios, whereas NO_2-N, dissolved Cd, total Cr, dissolved Cr and total Pb had the lowest values. The MFF ratios for total metals, COD and DOC were highly correlated except for Pb, suggesting that a BMP that treats the early runoff has an opportunity to treat high concentrations of all eight pollutants.

Acknowledgements

This study was supported in part by the Division of Environmental Analysis, California Department of Transportation through a contract with the University of California and task order 43A0073.

References

Barrett, M.E., Irish, L.B., Jr, Malina, J.F., Jr and Charbeneau, R.J. (1998). Characterization of highway runoff in Austin, Texas, area. *J. of Envir. Engrg.*, **124**(2), 131–137.

Bertrand-Krajewski, J.L., Chebbo, G. and Saget, A. (1998). Distribution of pollutant mass vs. volume in stormwater discharges and the first flush phenomenon. *Wat. Res.*, **32**(8), 2341–2356.

Furumai, H., Balmer, H. and Boller, M. (2002). Dynamic behavior of suspended pollutants and particle size distribution in highway runoff. *Wat. Sci. Tech.*, **46**(11-12), 413–418.

Gupta, K. and Saul, A.J. (1996). Specific relationships for the first flush load in combined sewer flows. *Wat. Res.*, **30**(5), 1244–1252.

Lee, J.H., Yu, M.J., Bang, K.W. and Choe, J.S. (2003). Evaluation of the methods for first flush analysis in urban watersheds. *Wat. Sci. Tech.*, **48**(10), 167–176.

Ma, J.-S., Khan, S., Li, Y.X., Kim, L.H., Ha, S., Lau, S.-L., Kayhanian, M. and Stenstrom, M.K. (2002). First flush phenomena for highways: how it can be meaningfully defined. In *Proceedings of the 9th International Conference on Urban Drainage*, Portland, Oregon.

Sansalone, J.J. and Buchberger, S.G. (1997a). Characterization of solid and metal element distributions in urban highway stormwater. *Wat. Sci. Tech.*, **36**(8-9), 155–160.

Sansalone, J.J. and Buchberger, S.G. (1997b). Partitioning and first flush of metals in urban roadway storm water. *J. of Envir. Engrg.*, **123**(2), 134–143.

Shinya, M., Tsuchinaga, T., Kitano, M., Tamada, Y. and Ishikawa, M. (2000). Characterization of heavy metals and polycyclic aromatic hydrocarbons in urban highway runoff. *Wat. Sci. Tech.*, **42**(7-8), 201–208.

Wu, J.S., Allan, C.J., Saunders, W.L. and Evett, J.B. (1998). Characterization and pollutant loading estimation for highway runoff. *J. of Envir. Engrg.*, **124**(8), 584–592.

Characteristics of particle-associated PAHs in a first flush of a highway runoff

R.K. Aryal*, H. Furumai*, F. Nakajima** and M. Boller***

*Department of Urban Engineering, University of Tokyo 7-3-1 Hongo, Bunkyo-ku, 113-8656 Tokyo, Japan
**Research Center for Advanced Science and Technology (RCAST), University of Tokyo, 4-6-1 Komaba, Meguro, Tokyo 153-8904, Japan
***Swiss Federal Institute for Environmental Science and Technology (EAWAG), Ueberlandstrasse 113, Duebendorf 8600, Switzerland

Abstract Runoff monitoring of six rainfall events was carried out in a highway, Winterthur, Switzerland focusing on first flush (runoff volume up to 2.88 mm). Six runoff events were used to investigate the characteristics of particle-associated PAHs in first flush. The fine fraction (<45 μm) had a relatively higher contribution than the coarse fraction. A significant contribution of the coarse fraction was observed at some periods when the runoff flow rapidly increased. Fluctuation of PAH content during a runoff event was significant in the coarse fraction and, in contrast, the PAH content in the fine fraction was less fluctuating. The weighted average PAH content in each event ranged from 17 to 62 μg/g in total SS, from 23 to 54 μg/g in the fine fraction and from 16 to 84 μg/g in the coarse fraction. The loading of particle-associated PAHs from the first flush of highway runoff ranged from 0.06 to 0.22 g/ha in a total of 12 PAH species.
Keywords First flush; highway runoff; PAH content; PAH profiles; suspended solids

Introduction

For the last few decades road/highway has been considered as one of the major contributing sources of diffuse pollution (Lee et al., 1995; Barret et al., 1998; Drapper et al., 2000; Furumai et al., 2002). During wet weather period, stormwater runoff carries many toxic pollutants such as polycyclic aromatic hydrocarbons (PAHs) (Pitt et al., 1995; Boxall and Maltby, 1997). The PAHs are contained in vehicle exhaust gas as well as asphalt pavement and tire rubber. Since they are hydrophobic in nature, they are mostly attached to the solid particles and deposit on the road surface (Yang et al., 1999). These PAHs are mainly discharged at an early period of runoff (Smith et al., 2000). The phenomenon in the early period of stormwater runoff, in which the concentration of pollutants is substantially higher than in the later period, is called "first flush". For good management of pollution control, it is required to understand the first flush phenomenon (Hoffman et al., 1984; Xanthopoulos and Augstin, 1992).

During runoff the solid particles of various sizes are remobilized and transported under different hydraulic conditions. One of the most important factors which control the transportation of particle-associated PAHs in runoff is particle size. Only a few studies discuss the behavior of PAHs in the first flush and the influence of different particle sizes (Krein and Schorer, 2000; Shinya et al., 2000). The wash-off behavior of size fractionated SS and associated PAHs is less known. The current research work aims to investigate particle-associated PAH content in size fractionated SS and PAH composition in total SS at the early period of runoff (first flush) from a highway in six rainfall events.

doi: 10.2166/wst.2006.058

Materials and method

Study area

Runoff monitoring of six rainfall events was carried out at an inlet point of a retention pond facility for a highway drainage system covering an area of 8.4 ha in Winterthur, Switzerland from October to December 2000. Figure 1 provides the area of sample collection and Table 1 shows the outline of the sampling area. The highway traffic volume was 25,300–73,700 vehicles per day from and into Zurich. The average number of light duty vehicles (bike, car and van) in the area was 22,200–58,700 while the average number of heavy-duty vehicles (bus and trucks) was 4,200–7,500 on weekdays and 800–1,300 at the weekend. The average daily wind speed of the area ranged from 0.5 to 4.7 m/s (average 1.85 m/s).

Sample collection

The number of total rainfall events in the area was 166 in the year 2000. The continuous rainfall record was separated into discrete rainfall events by applying an inter event time definition (IETD) of 8 hours. The initial loss from the road surface was 0.7 mm from the field monitoring data using the rainfall runoff relationship. Thus, we considered effective rainfall (able to runoff) only when the rainfall height exceeded 0.7 mm and IETD greater than 8 hrs. The total effective rainfall was 121 in year 2000. The effective rainfall events were categorized into four groups according to the rainfall height (Figure 2a). Stormwater samples of six runoff events on Oct.26, Oct.31, Nov.3, Nov.14, Nov.17 and Dec.8 in 2000 were used to study the characteristics of particle associated PAHs in a first flush. Table 2 provides the characteristics of the six events. The collected runoff samples represented moderate/strong rainfalls as well as three antecedent dry weather period (ADWP) conditions (Figure 2b).

The early phase of runoff, corresponding to the first 2.88 mm rainfall, was captured for the "first flush" phenomenon. Autosamplers were set to collect the runoff samples into 24 bottles at an interval of 10 m^3 runoff volume (0.12 mm rainfall) in the drainage area. The sampling was carried out once the flow rate exceeded 8 L/sec and was stopped at below 5 L/sec. The collected runoff samples were filtered through the glass fiber filter (Whatman GF/C with pore size 1.2 µm) and SS was recovered. The filters with SS were dried below 40 °C and kept in a desiccator for a while. Then they were preserved in refrigerator till further analysis.

The SS on four runoff events (Oct.26, Oct.31, Nov.3 and Nov.14) was fractionated into two different particle sizes, 1.2–45 µm (fine) and 45–2,000 µm (coarse) using stainless steel mesh sieves (wet sieving). In case the SS concentration was low, the SS

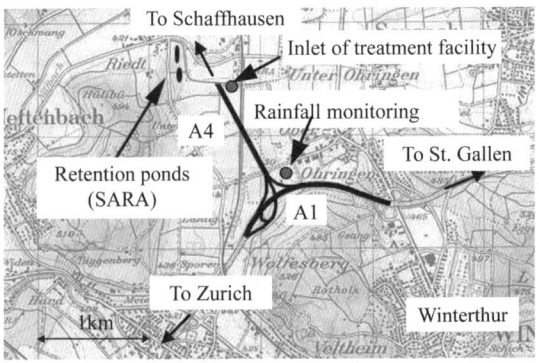

Figure 1 Sampling area

Table 1 Characteristics of highway

Highway route name	Length (km)	Area (ha)	ADT* (vehicles/day)
A1	1.8	5.5	57,500
A4	1.2	2.0	25,300
Ramp area	0.6	0.9	73,700
Total		8.4	

*ADT = Average daily traffic

samples in serial bottles were mixed together to recover a sufficient amount of suspended particles for the size fractionation.

Chemical Analysis

For PAH analysis, the SS samples were mixed with a solvent of dichloromethane and methanol (5:1 by volume). Three internal standards (phenanthrene-d10, chrysene-d12, and perylene-d12) were added into the mixture sample to correct the recovery rate of the analytes. The mixture was then applied to sonication for 30 minutes (50% cycles, 160 watt) for extraction of PAHs. The obtained extract was filtered through a glass fiber filter (GF-75, Advantec). The filtrate was evaporated to dryness below 40 °C. The dried extract was finally dissolved in dichloromethane and PAHs in the solution were analyzed using GC/MS (Hewlett Packard HP6890 GC coupled with HP 5973 mass selective detector) in SIM mode. The quantified PAHs were phenanthrene, anthracene, fluoranthene, pyrene, benzo(a)anthracene, chrysene, benzo(k)fluoranthene, benzo(b)fluoranthene, benzo(a)pyrene, indeno(1,2,3-cd)pyrene, dibenz(a,h)anthracene, and benzo(ghi)perylene. The PAH content reported in this paper is the sum of all individual target analytes.

Results and discussion
SS particle size variation in runoff

The size distribution of SS in runoff samples showed a significant fluctuation with runoff volume (Figure 3). In all four events, the fine fraction had a relatively higher contribution than the coarse fraction. Especially in the events on Oct.26 and Nov.3, the fine fraction showed its domination during the first flush period. The fine fraction contribution varied from 49% to 92% on Oct.26 and from 73 to 90% on Nov.3. In the events on Oct.31 and Nov.14, a significant contribution of the coarse fraction was observed when the flow increased rapidly. But a similar increase on Nov.3 did not give a significant rise of coarse fraction contribution. The antecedent condition (antecedent rainfall, wind speed in antecedent dry weather period, etc.) may give the difference as well as a variation of the fine/coarse ratio at the beginning of each event despite a similar flow rate.

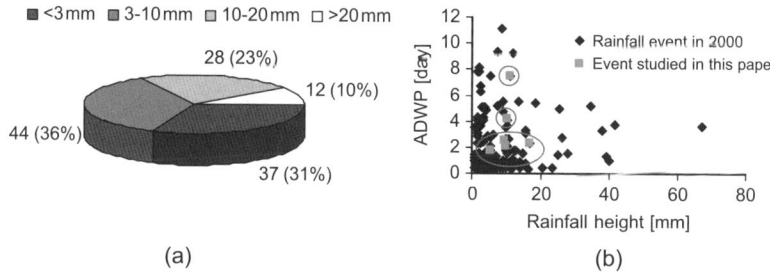

Figure 2 Characteristics of effective rainfall events (a) according to rainfall height and (b) ADWP (antecedent dry weather period) at Winterthur, Switzerland in 2000

Table 2 Characteristics of runoff events

	Whole event					First flush period	
	RH	Duration	PRV	ADWP	ADWP (2.88 mm)	RT	MRI
Oct.26	5.2	18.83	2.1	1.77	10.13	10.04	4.8
Oct.31	10.2	11.16	5.2	4.19	4.19	4.68	3.0
Nov.3	9.6	20.16	1.6	2.57	2.94	6.02	4.2
Nov.14	11.0	13.33	2.2	7.53	9.22	4.80	3.0
Nov.17	17.0	20.66	11.0	2.28	2.28	1.33	2.4
Dec.8	5.2	24.50	1.5	2.04	10.24	4.40	3.0

Note: RH = rainfall height [mm], Duration = rainfall duration [h], PRV = previous rainfall volume [mm], ADWP = antecedent dry weather period [day], ADWP (2.88 mm) = antecedent dry weather period when neglecting small rainfalls (<2.88 mm) [day], RT = runoff time for 2.88 mm [h], MRI = maximum rainfall intensity in first 2.88 mm runoff [mm/h]

Content of PAHs in total SS and fractional SS

PAH content in runoff particles is shown in Figure 4 and Table 3. Fluctuation of the PAH content in the first flush period was significant in the coarse fraction and, in contrast, the PAH content in the fine fraction was less fluctuating (Figure 4 (bottom)). The fluctuation may be caused by the particle size distribution in the coarse fraction. In this study, the fractionation was conducted only at 45 μm and no further fractionation was done. According to Yang et al. (1999), the PAHs are mostly attached to the road dust particles of about 63 and 65.5 μm. Murakami et al. (2003) statistically showed that the PAH content was high in the road dust finer than 106 μm compared with coarser dust particles (>106 μm).

The weighted average PAH content in each event ranged from 17 to 62 μg/g in total SS, from 23 to 54 μg/g in the fine fraction and from 16 to 84 μg/g in the coarse fraction. The PAH content in total SS on Oct.26 was highest among the six runoff samples.

Compared with other samples, the samples on Oct.26 and on Nov.14 had obviously high and low PAH content (Table 3). Both of the events had a long antecedent dry weather period (if small rainfalls were neglected as shown in Table 2). There are both possibilities to increase and to decrease the PAH content of runoff particles by a long

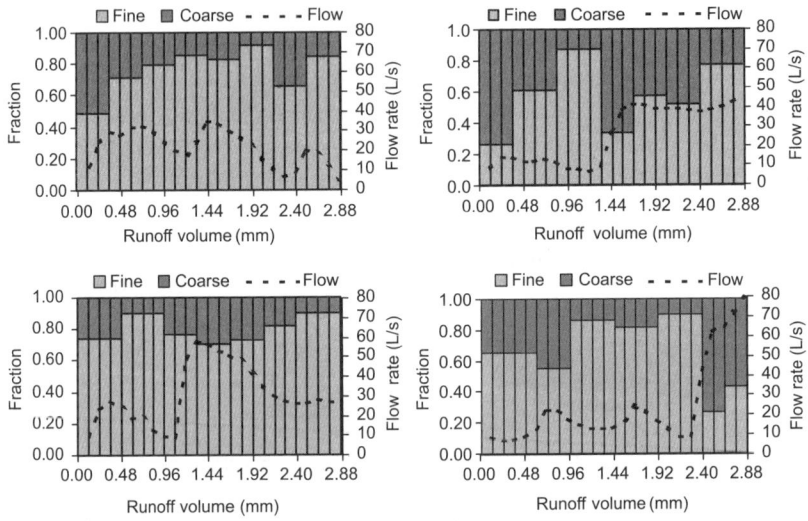

Figure 3 Fractional contribution by weight of fine and coarse particles during runoff on Oct.26 (top left), Oct.31 (top right), Nov.3 (bottom left) and Nov.14 (bottom right)

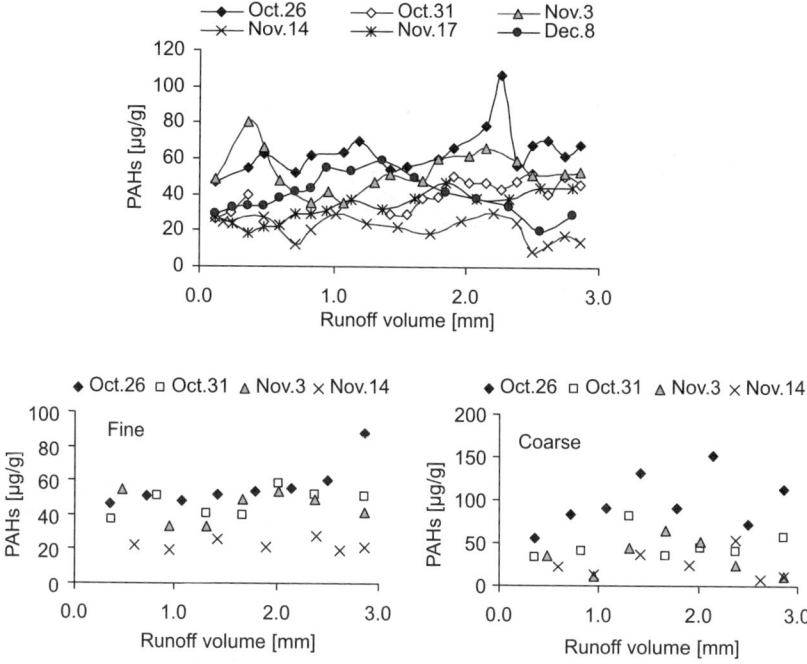

Figure 4 Variation of PAH content in total SS (top) and fractional SS (bottom)

exposure to antecedent dry conditions. Particles may be exposed repeatedly to the gas phase PAHs in a highway environment and it may increase PAH content. On the other hand, photodegradation of PAHs by sunlight during the antecedent dry weather period will decrease the observed PAH content. Wind may selectively remove fine particles which may have a specifically high/low content of PAHs. This field survey shows that the situation is too complicated and difficult to be generalized, and this means that a single/spot sampling is insufficient to understand the highway environment.

An average PAH profile (composition) of the runoff particles in each event is plotted in Figure 5. The PAH profiles were similar to each other but some difference was observed in the event on Oct.26 and on Nov.14, which had a long ADWP (2.88 mm) as already discussed. The contribution of 4-ring PAHs was high on Oct.26 and benzo(a)pyrene content was high on Nov.14.

The loading of particle-associated PAHs from first flush of highway runoff was calculated as shown in Table 4. The PAH load ranged from 0.06 to 0.22 g/ha in a total of 12 PAH species. Some major PAH species were benzo(k)fluoranthene, fluoranthene and pyrene. In this study, only the particle associated PAHs were measured and some of the low molecular weight PAHs must be present also in a dissolved form. Then this calculated load must be lower than the real load of total PAHs from the highway runoff.

Table 3 Weighted average content (μg/g) of PAHs in runoff particles

	Oct.26	Oct.31	Nov.3	Nov.14	Nov.17	Dec.8
Fine fraction	54	50	48	23		
Coarse fraction	84	40	78	16		
Total	62	44	48	17	32	37

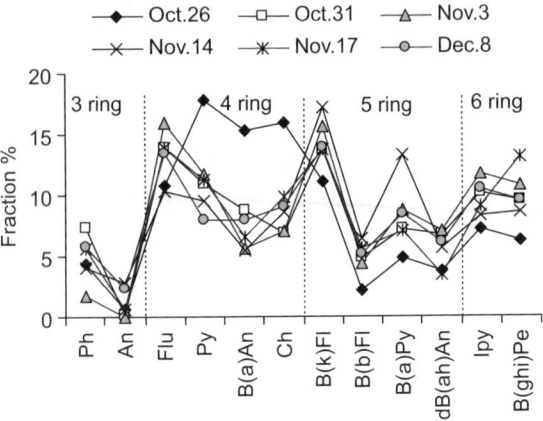

Figure 5 Average PAH profiles of six runoff events. Where, Ph-phenanthrene, An-anthracene, Flu-fluoranthene, Py-pyrene, B(a)An-benzo(a)anthracene, Ch-chrysene, B(k)Fl-benzo(k)fluoranthene, B(b)Fl-benzo(b)fluoranthene, B(a)Py-benzo(a)pyrene, Ipy-indeno(1,2,3-cd)pyrene, dB(ah)An-dibenzo(a,h)anthracene, B(ghi)Pe-benzo(ghi)perylene

Table 4 Loading (g/ha) of particle-associated PAHs from first flush of highway runoff

	Oct.26	Oct.31	Nov.3	Nov.14	Nov.17	Dec.8	Average load
Phenanthrene	0.0079	0.0093	0.0037	0.0025	0.0077	0.0097	0.0068
Anthracene	0.0012	0.0002	0.0000	0.0017	0.0011	0.0037	0.0013
Fluoranthene	0.0196	0.0161	0.0347	0.0064	0.0237	0.0235	0.0207
Pyrene	0.0305	0.0120	0.0254	0.0059	0.0143	0.0139	0.0170
Benzo(a)anthracene	0.0260	0.0095	0.0122	0.0034	0.0084	0.0134	0.0121
Chrysene	0.0268	0.0077	0.0149	0.0054	0.0126	0.0164	0.0140
Benzo(k)fluoranthene	0.0192	0.0171	0.0335	0.0105	0.0174	0.0237	0.0202
Benzo(b)fluoranthene	0.0040	0.0059	0.0091	0.0038	0.0072	0.0088	0.0065
Benzo(a)pyrene	0.0084	0.0090	0.0189	0.0080	0.0088	0.0140	0.0112
Dibenzo(a,h)anthracene	0.0062	0.0090	0.0157	0.0033	0.0043	0.0096	0.0080
Indeno(1,2,3-cd)pyrene	0.0125	0.0130	0.0254	0.0049	0.0114	0.0174	0.0141
Benzo(ghi)perylene	0.0112	0.0120	0.0233	0.0052	0.0166	0.0161	0.0141
Sum of 12 PAHs	0.17	0.12	0.22	0.06	0.13	0.17	0.15

Conclusion

A long term monitoring of a highway runoff was carried out. The particle-associated PAHs were subjected to analyses in total SS as well as in fractional SS. The following could be concluded.

- The fine fraction (<45 μm) had a relatively higher contribution than the coarse fraction. A significant contribution of the coarse fraction was observed at some periods when the runoff flow rapidly increased.
- Fluctuation of PAH content during a runoff event was significant in the coarse fraction and, in contrast, the PAH content in the fine fraction was less fluctuating.
- The weighted average PAH content in each event ranged from 17 to 62 μg/g in total SS, from 23 to 54 μg/g in the fine fraction and from 16 to 84 μg/g in the coarse fraction.
- The loading of particle-associated PAHs from first flush of highway runoff ranged from 0.06 to 0.22 g/ha in a total of 12 PAH species.

Acknowledgements

We are thankful to H. Balmer and J. Eugster (EAWAG) for their assistance in field monitoring.

References

Barret, M.E., Irich, L.B., Jr and Charbeeneau, R.J. (1998). Characterization of highway runoff in Austin, Texas Area. *J. Environ. Eng*, **124**(2), 131–137.

Boxall, A.B.A. and Maltby, L. (1997). The effects of motorway runoff on freshwater ecosystems: 3 Toxicants confirmation. *Arch. Environ. Contam. Toxicol.*, **33**(9), 9–16.

Drapper, D., Tomlinson, R. and Williams, P. (2000). Pollutant concentration in road runoff: Southeast Queensland case study. *J. Environ. Eng.*, April 313–320.

Dunbar, C.J., Lin, C.I., Vergucht, I., Wong, J. and Durant, J.L. (1995). Estimating the contributions of mobile sources of PAH to urban air using real-time PAH monitoring. *Sci. Tot. Environ.*, **279**, 1–19.

Furumai, H., Balmer, H. and Boller, M. (2002). Dynamic behavior of suspended pollutants and particle size distribution in highway runoff. *Wat. Sci. Tech.*, **46**(11), 413–418.

Hoffman, E.J., Mills, G.L., Latimer, J.S. and Quinn, J.G. (1984). Urban runoff as source of polycyclic aromatic hydrocarbons to coastal waters. *Environ. Sci. Technol.*, **18**(8), 580–587.

Krein, A. and Schorer, M. (2000). Road runoff pollution by polycyclic aromatic hydrocarbons and its contribution to river sediment. *Wat. Res.*, **34**(16), 4110–4115.

Lee, W.J., Wang, Y.F., Lin, T.C., Chen, Y.Y., Lin, W.C., Ku, C.C. and Cheng, J.T. (1995). PAHs characteristics in the ambient air of traffic-source. *Sci. Tot. Environ.*, **159**, 185–200.

Murakami, M., Nakajima, F. and Furumai, H. (2003). Distinction of Size-Fractionated Road and Roof Dust Based on PAH Contents and Profiles. *J. Japan Society on Wat. Environ.*, **26**(12), 837–842 (text in Japanese, abstract and figures/tables in English).

Pitt, R., Field, R., Lalor, M. and Brown, M. (1995). Urban stormwater toxic pollutants: assessment, sources and treatability. *Wat. Environ. Res.*, **67**(3), 260–275.

Shinya, M., Tsuchinaga, T., Kitano, M., Yamada, Y. and Ishikawa, M. (2000). Characterization of heavy metals and polycyclic aromatic hydrocarbons in urban highway runoff. *Wat. Sci. Tech.*, **42**(7–8), 201–208.

Smith, J.A., Sievers, M., Huang, S. and Yu, S.L. (2000). Occurrence and phase distribution of polycyclic aromatic hydrocarbons in urban storm-water runoff. *Wat. Sci. Technol.*, **42**(3–4), 383–388.

Vardar Nedim and Noll, K.E. (2003). Atmospheric PAHs concentrations in fine and coarse particles. *Environmental Monitoring and Assessment*, **87**, 81–92.

Xanthopoulos, C. and Augstin, A. (1992). Input and characterization of sediments in urban sewer systems. *Wat. Sci. Tech.*, **25**(8), 21–28.

Yang, H.H., Chiang, C.F., Lee, W.J., Hwang, K.P. and Wu, E.M.Y. (1999). Size distribution and dry deposition of road dusts PAHs. *Environmental International*, **25**(5), 585–597.

Water quality modeling to evaluate BMPs in rice paddies

J.H. Jeon*, C.G. Yoon**, H.S. Hwang*** and K.W. Jung**

*Research Division, Korea Environment Institute, 613-2 Bulgwang-dong, Eunpyeong-gu, Seoul 122-706, Korea (E-mail: *jihongjeon@hanmail.net*)
**Department of Environmental Science, Konkuk University, 1 Hwayang-dong, Kwangjin-gu, Seoul 143-701, Korea (E-mail: *chunyoon@konkuk.ac.kr; jungkw@konkuk.ac.kr*)
***National Institute of Environment Research, Ministry of Environment, Kyeongse-dong, Seo-gu, Inchen 404-170, Korea (E-mail: *undersun@me.go.kr*)

Abstract A water quality model applicable to rice paddies was developed using field data from 1999–2002. Use of the Dirac delta function efficiently explained the nutrient-concentration characteristics of ponded water. The model results agreed reasonably well with the observed data. The ponded-water quality was influenced primarily by fertilization; nutrient concentration was especially high during early cultivation periods. Reducing surface drainage during the fertilization period may substantially reduce nonpoint source loading from paddies. Increased weir heights and shallow irrigation methods were evaluated by the model as practical methods for reducing nutrient loading from paddies. These methods were effective in reducing surface drainage and are suggested as "best management practices" (BMPs) if applied based on site-specific paddy conditions.
Keywords Model development; paddy management; shallow irrigation; water quality; weir height

Introduction

Rice grown in paddies requires certain water-layer depths. Field water-layer depths during the growing season may vary from 500–800 mm (De Datta *et al.*, 1973) to more than 3,000 mm (Hukkeri and Sharma, 1980). In South Korea, paddy water levels are approximately 1,250 mm, and water is supplied primarily by irrigation (Chae, 1998). The hydrological and water quality characteristics of paddies differ slightly from those of other land-use types. The water retentiveness of paddies can reduce surface runoff, while paddy drain-off can produce surface runoff without a rainfall event. Therefore, the ridge height of diked rice fields and operational drain-off during the rice cultivation period can change the hydrologic pattern. Rice, like other crops, needs 16 essential elements, and all of these must be present in optimum amounts and in forms available to rice plants for proper growth. Among these elements, nitrogen, phosphorus, and potassium are most commonly applied as fertilizers, and major portions of these nutrients are used by rice plants throughout the growth cycle (Lee, 2001). However, a significant portion of these nutrients can be lost from paddies through surface drainage, seepage, and percolation This loss might result in excessive nutrient supply to receiving water bodies and eutrophication problems (Cooke *et al.*, 1993).

In South Korea, paddies cover approximately 1 million ha, and irrigation for paddy rice culture ranks first among water uses and accounts for over 50% of the nation's total water consumption. Paddies are thus important to studies of water resources and watershed management. This paper describes a field experiment performed during the 1999–2002 growing seasons to analyze water and nutrient balances in rice paddies; a water quality model was developed to evaluate best management practices (BMPs) for paddies.

doi: 10.2166/wst.2006.059

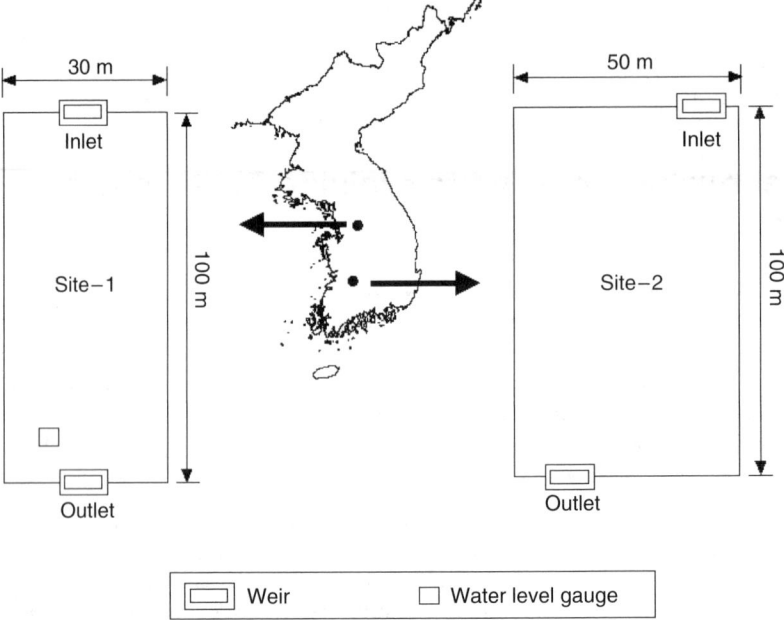

Figure 1 Study area and sampling points

Study area

The field experiment was performed at two separate sites in Korea as shown in Figure 1. Site-1 was a Konkuk University agricultural research farm in Yojoo (37°14′N, 127°33′E) irrigated with groundwater; experiments were performed here for two years (2001–2002). Site-2 was a field research farm in Jinan (35°37′N, 127°16′E) irrigated with surface water; experiments were performed here in 1999 and 2000.

Model development

The planning-level model PADDIMOD was developed to predict water and nutrient balances in rice paddies. Figure 2 illustrates the water balance concept for paddies and the model. The modeled inflow to a paddy consists of irrigation (IR_1), input from any upper paddies (IR_2), and rainfall (PR); the outflow consists of evapotranspiration (ET), infiltration (INF), and surface runoff (DR). Fertilization loading is presented as impulse loading, represented mathematically by the Dirac delta function (or impulse function) $\delta(t)$ (Chapra, 1997). The Dirac delta function can be visualized as an infinitely thin spike centered at $t = 0$ and having a specified unit area (Figure 3). The solution indicates that fertilizer is instantaneously distributed throughout the water body of the paddy, resulting in an

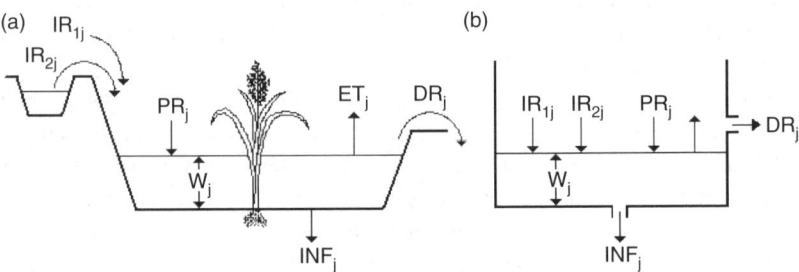

Figure 2 Water balance concept in a paddy (a) and the model (b)

initial concentration of m/V. Thereafter, the result is identical to the general solution, i.e. the concentration decreases exponentially at a rate dictated by the magnitude of λ.

Results
Model calibration and validation
The model was calibrated and validated with two independent data sets from two separate sites. The simulation results of ponded-water depth (mm) and surface runoff (mm) are shown in Figure 4; data for ponded-water depth at Site-2 were not available. The ponded-water depth generally varied with rainfall and irrigation, except during two forced drainages for fertilization (tillering and panicle). Simulated nutrient concentrations of ponded water are shown in Figure 5. Nutrient concentration was mainly influenced by fertilization, and the total nitrogen (TN) and total phosphorus (TP) concentrations reached 50 mg/L and 5 mg/L, respectively, for the basal fertilization period. This result implies that control of surface drainage especially during May and June can substantially reduce nonpoint source loading from paddies. Table 1 summarizes the statistical analyses of model fitness; average error (AE) and root mean square (RMS) values were low, while model efficiency (EF) values were high, indicating that the model simulation results matched the observed data quite well.

BMPs evaluation by PADDIMOD
The effect of paddy BMPs on the nutrient loading from Site-1 was evaluated using the PADDIMOD simulation for the conditions in 2001. Increasing weir height from 100 to 200 mm retained more ponded water and reduced surface drainage especially during May when high nutrient concentration resulted from fertilization; Table 2 shows the marked reduction in nutrient loading that resulted. The reduction of TN and TP surface loadings was about 78 and 49%, respectively, attained by simply increasing the weir height by 100 mm (Figure 6).

The effect of shallow irrigation was also simulated, and the results are summarized in Table 2. The ponded-water depths of the field experimental sites were generally maintained at 100 mm; shallow irrigation was simulated with ponded-water depths of 10–30 mm and a weir height of 100 mm.

The model predicted that shallow irrigation can also substantially reduce nutrient loading from paddies. The shallow irrigation condition reduced modeled TN and TP surface loadings by approximately 74 and 47%, respectively, and the total irrigation depth from 296 to 130 mm. Note that while the reduction in nutrient loading was substantial, total surface runoff was reduced only slightly. Surface drainage was controlled during the initial stage of paddy preparation; this control likely contributed most to the nutrient-load reduction because of the high nutrient concentration of ponded water during the basal fertilization period.

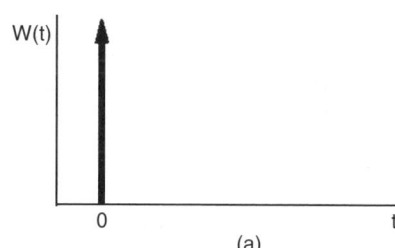 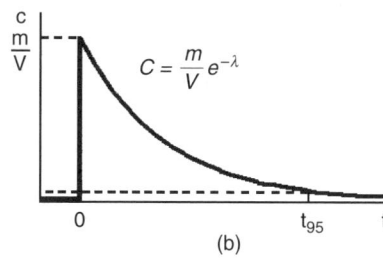

Figure 3 Loading and concentration plot of impulse loading

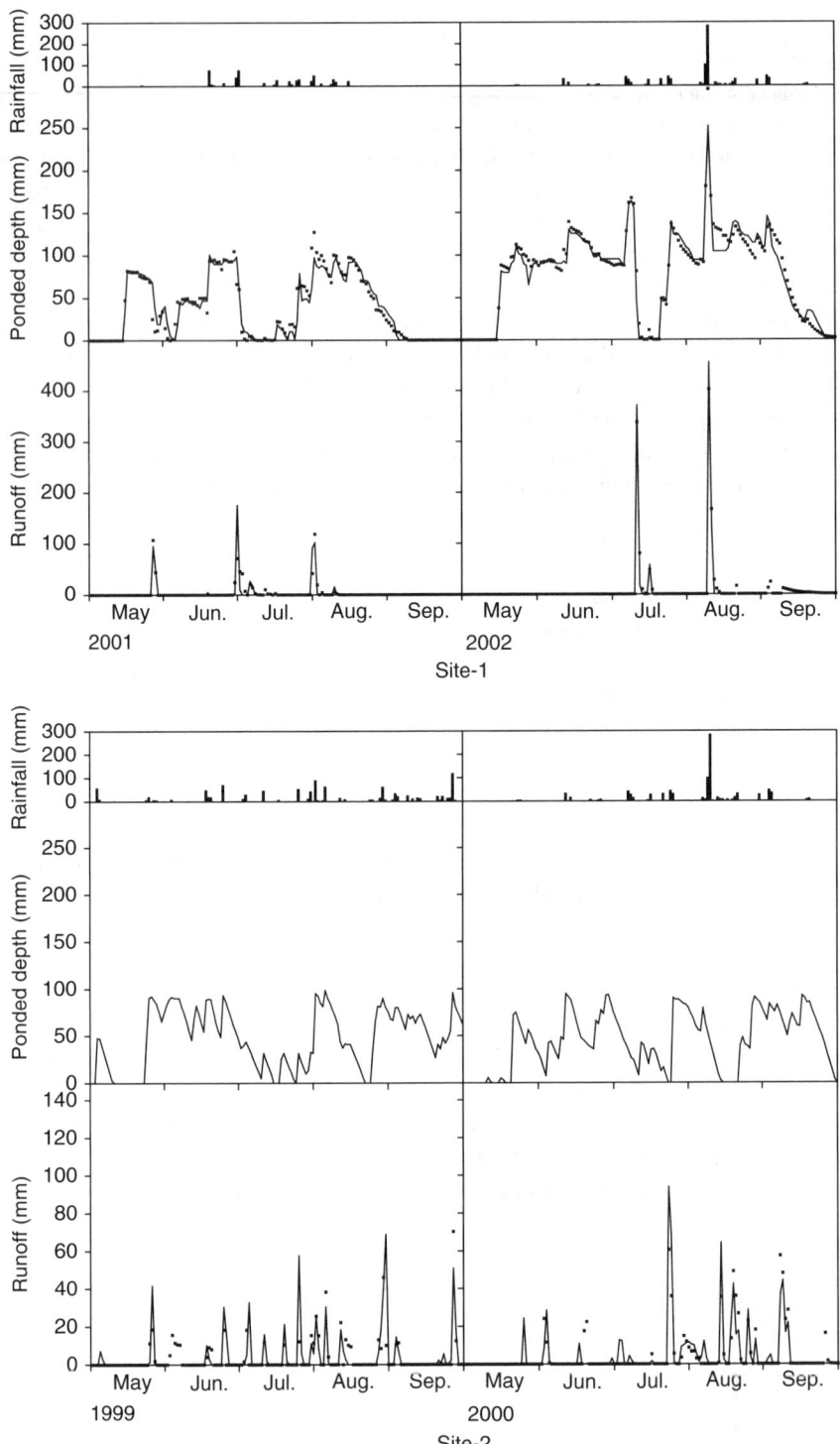

Figure 4 Observed and predicted ponded-water depth and surface runoff from the paddies

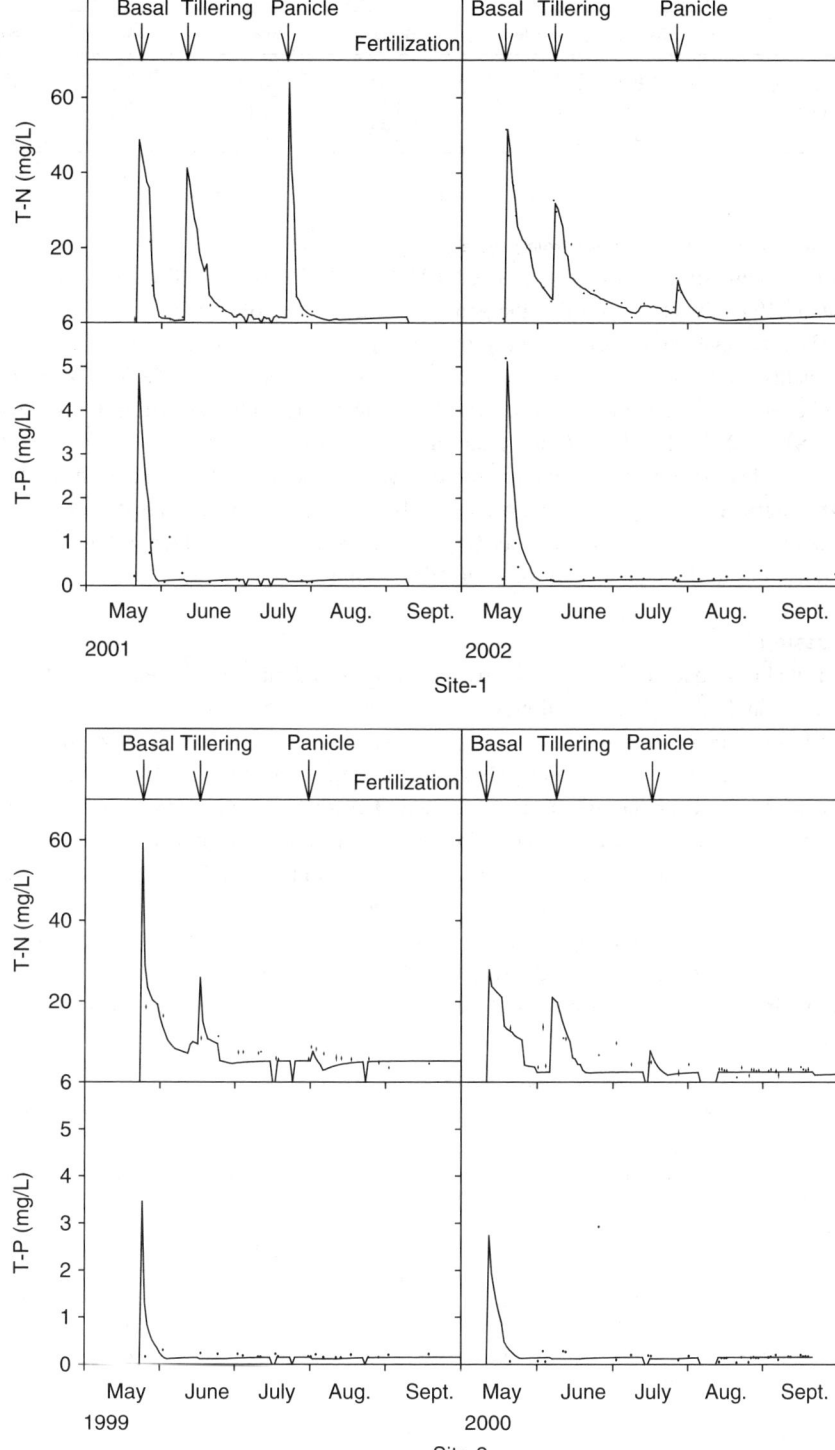

Figure 5 Observed and predicted nutrient concentrations of ponded-water in the paddies

Table 1 Results of the model fitness tests for the PADDIMOD model

	Ponded-water depth		TN		TP	
	Site-1	Site-2	Site-1	Site-2	Site-1	Site-2
AE (mm, mg/L)	0.81	–	−0.11	−0.53	−0.06	0.02
RMS (%)	11.48	–	1.99	2.52	0.24	0.38
EF	0.93	–	0.98	0.99	0.95	0.70

Characteristics of surface drainage from paddies

Two sites were simulated using the PADDIMOD model with rainfall data from the last 10 years (1994–2004) to examine the general characteristics of surface loading from paddies. The average rainfall during the growing season was 978.8 mm, and average runoff coefficients for Site-1 and Site-2 were 0.55 and 0.59, respectively (Table 3). The average runoff coefficient for Korea is 0.60; thus, the simulation result was within the expected range. Site-2 showed higher loading than Site-1; more frequent drainage during the initial stage of paddy farming might explain the difference. The simulation results showed that surface nutrient loading was influenced mainly by rainfall amount and demonstrated a correlation between these two factors (Figure 7). Data from two other studies in Korea generally followed the pattern of the simulation results.

Discussion

Water quality in paddies is influenced by fertilization, and nutrient concentrations become especially high during fertilization periods (May–June). Surface runoff from paddies depends on rainfall and forced drainage. Paddy runoff during the fertilization period drains to outside water bodies and can cause eutrophication. Runoff early in the rice culture period can occur through agricultural activities such as forced drainage or lowered weir heights. Nutrient concentration during surface runoff caused by heavy rainfall events was similar to the background concentration, as shown in Figure 5. Therefore, compared to other land-use types, it may be possible to more effectively control nutrient loading from paddies using BMPs. Increased weir height and shallow irrigation were evaluated using the calibrated and validated PADDIMOD model; these measures were found to be effective in reducing surface nutrient loading. The increased weir height reduced TN and

Table 2 The effects of increased weir height and shallow irrigation on surface loading from paddy fields

		Runoff (mm)		TN (kg/ha)		TP (kg/ha)	
		Observed	Simulated	Observed	Simulated	Observed	Simulated
Increased weir height							
2001	May	143.10	0.0	24.64	0.00	1.13	0.00
	June	182.42	182.4	0.15	1.55	0.01	0.11
	July	252.87	252.8	2.67	4.03	0.05	0.41
	Aug.	14.81	14.8	0.00	0.43	0.00	0.08
	Sept.	0.00	0.0	0.00	0.00	0.00	0.00
	Total	593.20	450.1	27.45	6.01	1.06	0.60
Shallow irrigation							
2001	May	143.10	0.00	24.64	0.00	1.13	0.00
	June	182.42	141.19	0.15	0.98	0.01	0.12
	July	252.87	281.72	2.67	5.86	0.05	0.39
	Aug.	14.81	11.30	0.00	0.42	0.00	0.04
	Sept.	0.00	0.00	0.00	0.00	0.00	0.00
	Total	593.20	434.21	27.45	7.26	1.06	0.56

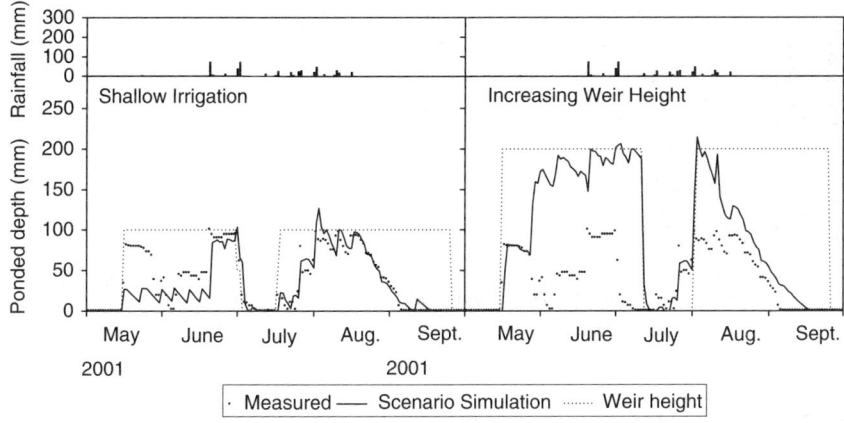

Figure 6 Simulation results for each scenario using PADDIMOD

TP surface loadings by approximately 78 and 49%, respectively. Results from shallow-irrigation modeling also suggested TN and TP surface loading reductions of approximately 74 and 53%, respectively. A three-year experimental study (Mishra et al., 1998) in India with weir heights of 6 and 30 cm at 4-cm intervals revealed that about 56.75 and 99.5% of the rainfall could be stored in 6- and 30-cm weir-height plots, respectively, without a significant impact on grain yield. Bouman and Tuong (2001) reported that reducing ponded-water depths from 5–10 cm to the level of soil saturation did not reduce land productivity, and they found that 23% water savings caused only a 6% yield reduction.

Water-saving irrigation, raising the drainage weir height in diked rice fields, and minimizing forced surface drainage are suggested measures for reducing nutrient loading from paddies. The possible benefits may include reduced irrigation water use and more efficient water resource allocation; increased storage of rainwater, with resulting flood prevention and groundwater recharge effects; and reduced nutrient loss, i.e. reduced nutrient loading and better water quality. However, these practices affect only some aspects of the overall water environment and may have other effects. For example, excessively increased weir height may create submerging damage or cold-water damage to the rice crop, and shallow irrigation requires support for participation and irrigation facilities. Therefore, other various aspects and local conditions in addition to environmental concerns need to be considered when developing practical BMPs for paddies.

Table 3 Simulation results for the past 10 years using the PADDIMOD model

	Rainfall (mm)	Runoff (mm)	Runoff rate	TN (kg/ha)	TP (kg/ha)
Groundwater (Site-1)					
Avg.	978.8	533.48	0.55	10.23	0.77
Max.	1,468.8	991.91	0.68	18.36	1.39
Min.	668.0	276.63	0.41	5.31	0.40
S.D.	276.6	244.95	0.10	4.08	0.34
Surface water (Site-2)					
Avg.	978.8	581.39	0.59	15.09	0.94
Max.	1,468.8	961.04	0.68	22.77	1.54
Min.	668.0	392.77	0.47	8.72	0.49
S.D.	276.6	204.95	0.08	5.90	0.43

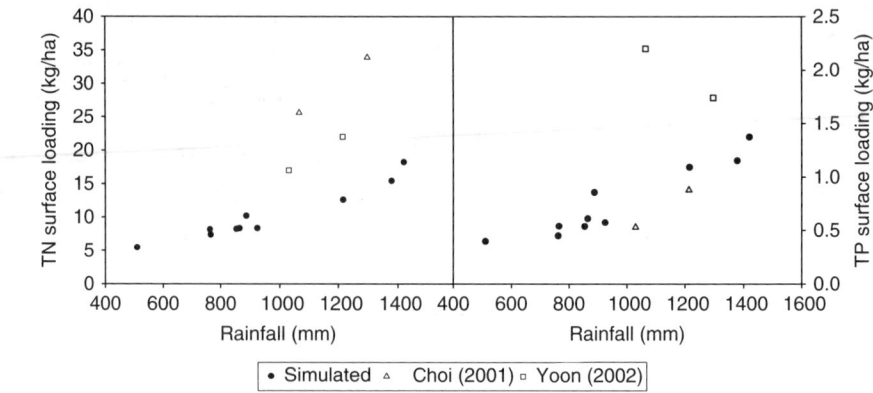

Figure 7 Relationship between rainfall amount and nutrient loading

Conclusions

The PADDIMOD model was developed using field data from two separate sites to simulate water and nutrient behavior in a paddy system. It was formulated with a few equations and simplified assumptions aimed at evaluating paddy BMPs for nonpoint source pollution control. The model produced realistic predictions and the simulated results reasonably matched the observed data. The model predicted daily ponded-water depth, surface drainage flow, and nutrient concentrations. As a simple and convenient planning model, the PADDIMOD can be used to evaluate BMPs for paddies alone or in combination with other complex watershed models.

Irrigated rice agriculture involves large amounts of water and nutrients, and substantial amounts of both inputs are lost through surface drainage. Nutrients lost by surface drainage flow into receiving water bodies and can cause eutrophication and other excessive algal growth problems. Nutrient loading from paddies occurs mainly by surface water flow, and surface drainage plays a key role in the movement of nutrients to receiving waters. Saving water by maintaining a shallow ponded-water depth and raising the drainage weir height in diked rice fields may be applicable BMPs to reduce surface drainage and associated nutrient loading. These BMPs were evaluated by the calibrated PADDIMOD model, which demonstrated their effectiveness. Model results showed that increasing the weir height from 100 to 200 mm could lead to greater water retention and reduced TN and TP surface loadings by approximately 78 and 49%, respectively. Shallow irrigation practices may also reduce TN and TP surface loadings by about 74 and 53%, respectively, and lower the total irrigation depth from 296 to 130 mm. The model demonstrated that controlling surface drainage during the initial stage may be the most critical factor in nonpoint source loading reduction from paddies because nutrient loads tend to be highest during the basal fertilization period.

References

Bouman, B.A.M. and Tuong, T.P. (2001). Field water management to save water and increase its productivity in irrigated lowland rice. *Agricultural Water Management*, **49**, 11–30.

Chae, J.C. (1998). Effect of tillage and seeding methods on percolation and irrigation requirement in rice paddy condition. *Korean Journal of Crop Science*, **43**, 264–268.

Chapra, S.C. (1997). *Surface Water-Quality Modeling*. McGraw-Hill, Inc., New York, NY, USA, pp. 47–87.

Cooke, G.D., Welch, E.B., Peterson, S.A. and Newroth, P.R. (1993). *Restoration and Management of Lakes and Reservoirs*, 2nd ed. Lewis Publishers, Boca Raton, FL, USA.

De Datta, S.K., Abilay, W.P. and Kalwar, G.N. (1973). Water stress effects in flooded tropical rice. In *Water management in Philippine Irrigation Systems: Research and Operations*. International Rice Research Institute, Los Baños, The Philippines, pp. 19–36.

Hukkeri, S.B. and Sharma, A.K. (1980). Water-use efficiency of transplanted and direct-sown rice under different water management practices. *Indian Journal of Agricultural Science*, **50**, 240–243.

Lee, B.W. (2001). Rice cultural practices in Asia. In *Rice Culture in Asia*, International Commission on Irrigation and Drainage, and Korean National Committee on Irrigation and Drainage, Seoul, Korea, pp. 36–54.

Mishra, A., Ghorai, A.K. and Singh, S.R. (1998). Rainwater, soil and nutrient conservation in rainfed rice lands in eastern India. *Agricultural Water Management*, **38**, 45–57.

Evaluation of AnnAGNPS in cold and temperate regions

S. Das, R.P. Rudra, P.K. Goel, B. Gharabaghi and N. Gupta

School of Engineering, University of Guelph, ON N1G 2W1, Canada

Abstract Identification of the pollution sources and understanding the processes related to runoff generation and pollution transportation is effective for the water quality management and selection of the Best Management Practices. The ANNualized AGricultural Non-Point Source (AnnAGNPS) model was applied to a watershed in Southern Ontario to evaluate the hydrology and sediment component from the non-point sources. The model was run for two years (1998 to 1999); one year's data was used to calibrate and the second year's data was used for validation purposes. The model has under predicted runoff amount and over predicted the sediment yield. However, the simulated runoff and sediment yield compared fairly well with the observed data indicating that the model had an acceptable performance in simulation of runoff and sediment. The study is still in progress to assess its performance for estimation of TMDL and improvements needed for the model to use under Ontario conditions.

Keywords AnnAGNPS; best management practice; hydrology; non-point source pollution; sediment

Introduction

Intensive agricultural activities are potential sources of suspended sediment, nitrogen, phosphorus and pathogens. These are the main causes of non-point source (NPS) pollution at different tributaries in Ontario. A number on studies on the water quality and quantity are being carried out in the Grand River basin in Ontario, which is a tributary to Lake Erie (GRCA, 1998; GRIC, 1982). Those studies have indicated that both point and non-point pollution sources are adding pollutants to surface and groundwater. Pollution reduction from the point sources, such as industrial and municipal wastewater discharges at the basin area has improved during the past two decades. On the other hand, developing and understanding of the role of non-point sources and their effective control in the Grand River basin is less successful. Non-point agricultural source pollution has created many problems for Ontario's environment, especially in terms of water quality (GRCA, 1998). Different parts of the Grand River and its tributaries are currently experiencing low dissolved oxygen level, high nutrient and suspended sediment concentrations and degraded drinking water quality and habitat for fish and other aquatic organisms. For the purpose of assuring clean and safe use of surface water, it is required to identify those source areas responsible for the impairment of water bodies. According to Dickinson *et al.* (1990), targeting NPS pollution sources has the potential to triple pollutant reduction, and minimize the extent of area affected negatively by restrictive land practices. Once the target for pollution sources is specified and identified, the best management practices can be applied to reduce the pollutants and contaminants at watershed outlets. Watershed based NPS models have played an effective role to perform the above jobs from the last two decades. They are the important tools in the development of management strategies to ameliorate the effects on water quality from NPS pollution. The Grand River Conservation Authority (GRCA) of Ontario in Canada has recognized the need to improve the estimation of the amounts and timing of pollution from NPS entering the Grand River. Different models are being investigated for this purpose. The AnnAGNPS is one of them and is being evaluated in the Canagagigue

doi: 10.2166/wst.2006.060

subwatershed. The main objective of this study is to examine the applicability of the AnnAGNPS model on a watershed scale in the cold and temperate region of Canada.

AnnAGNPS model

The AnnAGNPS (Cronshey and Theurer, 1998) model is a continuous simulation, daily time step, watershed scale, pollutant loading model (Frank et al., 1998) developed to simulate long-term sediment and chemical transport from ungaged agricultural watersheds. It can be used to locate the possible sources or areas that are responsible for the impairment of the water bodies. AnnAGNPS is a direct replacement for the single event model, AGNPS (Young et al., 1989a,b) but retains many of the features of AGNPS (Yuan et al., 2001). Bosch et al. (1998) described the model as flexible, accurate, and discretized. The hydrology part of the model is based on a water balance approach, which is based on a simple account of inputs and outputs of water during a day. Water inputs include rainfall, snowmelt, and irrigation water, while surface runoff, percolation, evapotranspiration, and drainage are the outputs. It is a distributed parameter model where the watershed is divided into several small scale areas named as "cells". The cells are often grouped into hydrologically similar areas. Generated runoff is routed through the in-cell watershed flow-system on a continuous basis, allowing moisture stored in the soil to be carried over from one day to the next. Soil moisture conditions are then used to calculate the SCS curve number (CN), which forms the basis of the surface and subsurface runoff quantities for that day.

The subsurface flow in AnnAGNPS considers either tile drainage or lateral subsurface flow and only occurs with the presence of an impervious layer within the soil profile. When the water table is below the drainage depth, lateral flow is calculated using Darcy's equation as described for lateral subsurface flow. AnnAGNPS utilizes the Revised Universal Soil Loss Equation (RUSLE, Renard et al., 1997) for calculating the sediment delivery to a field edge when a runoff event occurs due to rainfall, irrigation, or snowmelt. The Hydro-geomorphic Universal Soil Loss Equation (HUSLE, Theurer and Clarke, 1991) is used to estimate the total sediment yield leaving each field to the stream reach after deposition. The model calculates nitrogen (N), phosphorus (P), and organic carbon (OC) concentrations within each field, using a mass-balance basis which is dependent upon soil moisture and the environmental cycles of the concerned area. Three sets of data are required to run AnnAGNPS. They are topographic data, soil and land-use related data, and climate data. Input data preparation is organized using the tools and/or models contained in the AnnAGNPS package.

Description of the study area

The Grand River is one of the largest rivers in Southern Ontario and is a tributary to Lake Erie. The selected watershed (Canagagigue Creek) is located in the northwest part of the Grand River basin. It is a minor tributary of the Grand River. The watershed has a total area of about 150 km^2 and is situated between the latitude of 43° 36' N and 43° 42' N and longitude 80° 33' W and 80° 38'. About 80% of the land within the watershed is under agricultural activities and 10% is woodlot (Carey et al., 1983). The area upstream of the Floradale Dam with a total contributing area of about 53 km^2 is selected. This selection was primarily based on the availability of observed flow and sediment data. These data were used for calibration and validation of the model. Figure 1 displays the studied subwatershed within Canagagigue watershed. The topography of the study area is flat to gently undulating with a slight slope towards the outlet in the south. 95% of the selected subwatershed area is under agricultural practices. A major portion of the watershed has 200 to 600 mm of loam or silt loam of the Huron and Harriston series overlying

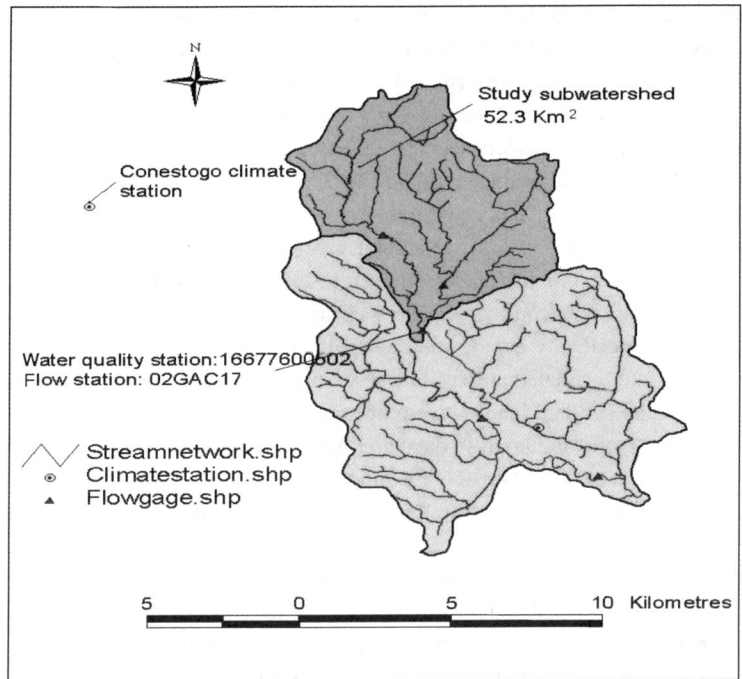

Figure 1 The location of the subwatershed inside Canagagigue watershed

a loam till (Presant and Wicklund, 1971; Hoffman et al., 1963). Loam is the predominant soil type in the central portion of the catchment.

Agricultural practices include mixed farming with predominantly dairy farming and cropping of corn, small grains, some other row crops and hay. There are limited areas used to cultivate cash crops as most crops are grown for livestock feed (So and Singer, 1982). General tillage practices within the watershed vary from conventional tillage to moldboard plough. Agricultural lands of the study area are usually ploughed in October followed by tillage operations in May. Most of the farmers apply manure to agricultural land. Generally the manure is applied two times a year in April–May and July–November.

The average annual precipitation ranges from 750–1,000 mm, of which 100–200 mm falls as snow. The average annual temperature of the area is about 6.5°C and annual evaporation is about 65% of the annual precipitation. The major part of the evaporation occurs during the summer season. There are five flow gage stations and three water quality measuring stations situated in Canagagigue creek watershed area. The current study used the observed flow data from flow gage station (No. 02GAC17) and sediment data from quality gage station (No.16677600502) respectively. Positions of the gage stations are shown in Figure 1. The observed data including hourly flow rates and daily sediments rates are available during April 1 to October 31 in every year from 1998 to 2000.

Model evaluation

Input data preparation

A Digital Elevation Model for the whole Canagagigue watershed was obtained from the Grand River Conservation Authority (GRCA). The DEM has 100 m resolution based on the contour data and digital terrain data from 1:1,000 OBMs (Ontario Base Map). Stream flow directions and the watershed and subwatershed boundaries are delineated by using the TOPAGNPS and AGFLOW modules of the AnnAGNPS Arc View interface. Cells were hydrologically determined by a user-defined critical source (CSA) area of 150 ha and a minimum source channel length (MSCL) of 250 m. Physical properties of each cell (area, length, and slopes) were determined by the AGFLOW module. Dominant soil and land use for each cell was predetermined by extrapolating the soil and land use shape files over the delineated subwatershed. The climate file was prepared based upon the historical data. The required parameters were the 2-year 24 hour rainfall, daily maximum and minimum temperatures, daily precipitation, dew point temperature, sky cover and wind speed. The Conestogo rain gauge station was selected for the historical rainfall and daily temperature data because it was the closest station to the selected area. It was difficult to obtain all the required climate parameters from one station; therefore some parameters (wind speed, sky cover, and dew point temperature) were obtained from the nearby weather station situated at the Elora research centre of the University of Guelph.

GIS layer for the soil and landuse were obtained from Ontario Ministry of Agriculture and Food (OMAF) soils and landuse database. This includes a series of county-wise geospatial soil survey data. The AnnAGNPS model has been developed in the USA where the model has the capability to extract most of the soil information from their SSURGO database provided by US Geological Survey (USGS). Therefore, the OMAF soils shape file was needed to rearrange for overlaying it with the delineated subwatershed. The entries in the attribute table for the soil shape file needed a Map Unit Symbol (MUSYM) parameter that should be identified by the model as respective soil group. Hence, the original attribute table of soil shape file was manipulated by adding the MUSYM to it. As there is no reference for this MUSYM in the Canadian conditions, MUSYM numbers were selected arbitrarily. Most of the physical properties of soils were obtained from the Canadian Soil Information System (CanSIS) website, Report No.35 of the Ontario soil survey (Hoffman *et al.*, 1963), and Report No.44 of the Ontario soil survey (Presant and Wicklund, 1971). Nine types of dominating soil have been found after extrapolating the soil shape file over the delineated watershed. Out of them, Burford soil (MUSYM 19A) has covered 31% and Brady soil (MUSYM 11A) has covered 21% of the study area.

The selected sub watershed consists of six types of dominated landuse. Each landuse type was included under a land use identifier during the input data preparation for the AnnAGNPS model. Mixed system (ID M) has covered 46%, corn system (ID C) has covered 32%, and Woodlot (ID Z) has covered 6% of the delineated study area. Most of the crop information and their management operations were collected from OMAF publication 811 (Agronomy guide for field crop). SCS curve number (CN) is one of the key hydrologic factors in obtaining accurate prediction of runoff and sediment yields (Yuan *et al.*, 2001). All the associated outputs transported through runoff are largely affected by the CN. The curve numbers for different hydrologic soil groups related to each crop operation have been selected from the National Engineering hand book section 4 (SCS, 1985).

Running AnnAGNPS

The AnnAGNPS model was run for the selected watershed for 2 years (1998- 1999) on a daily basis. The observed flow and sediment rates from April 1 to October 31 in 1998 from the respective gage stations were used to calibrate the model. Similar data from April 1 to October 31 in 1999 of the same stations were used for validation purposes. Calibration to the model input was done by adjusting the sensitive parameters according to the field estimates to make the model predictions close to the actual conditions.

Un-calibrated run. Table 1 shows the monthly output (from April 1998 to October 1998) of runoff and sediment and the percentage of deviation from the observed data. The model has under predicted the runoff as the observed runoff is always more than the

Table 1 Comparison of observed and predicted monthly runoff for the year 1998 (uncalibrated)

Month	Ppt. (mm)	Sim_runoff (mm)	Obs_runoff (mm)	Deviation (%)	Sim_sed (ton)	Obs_sed (ton)	Deviation (%)
Apr-98	42.70	2.68	8.60	−68.80	13.02	3.14	315.37
May-98	26.30	0.08	0.43	−81.98	2.29	0.01	16,090.24
Jun-98	45.80	2.24	4.44	−49.56	212.69	84.93	150.43
Jul-98	15.40	0.01	0.11	−88.11	1.19	0.38	215.44
Aug-98	86.40	10.08	20.12	−49.91	141.30	59.65	136.88
Sep-98	10.30	0.02	0.03	−38.38	0.60	0.00	19,746.03
Oct-98	18.20	0.09	0.05	65.04	2.98	0.09	3,097.11
Total	202.40	15.20	33.79	−55.01	374.07	148.21	152.40

simulated runoff. However, the daily simulated runoff amount follows the same pattern of the observed amount. The total simulated runoff for the run period is 15.2 mm where the observed runoff was 33.8 mm. AnnAGNPS doesn't account for the base flow and this may be a reason for this variation. The model has under predicted the runoff by 55% which is without the base flow separation and this could be a reasonable range for an uncalibrated run of a model. Climate data shows that 1998 was comparatively a dry year and thus the runoff amounts in May, July, September and October are comparatively low. Runoff regression analysis gives the best-fit equation $y = 1.84x$ and coefficient of determination $R^2 = 0.89$. This is in the reasonable range for an un-calibrated hydrology submodel and hence indicates that the AnnAGNPS model is capable of simulating hydrology components under watershed conditions.

Simulated sediment in 1998 indicates that AnnAGNPS has over predicted the sediment yield in all months. Specially May and September are showing a huge deviation between the simulated and observed data. Simulated sediment yield is comparatively less in these months but the observed data has shown almost negligible values. It might happen due to the problem of observed data recording. The runoff on these months is also low. Runoff amounts do not affect the soil loss results directly as the rainfall amounts are used in RUSLE to determine an erosion index value for each storm. This is because RUSLE uses this rainfall and runoff factor in the calculation of soil loss and it is only determined when AnnAGNPS predicts runoff occurring for an event (Yuan *et al.*, 2001). However, the overall error in sediment prediction is 152.4%. Sediment regression analysis gives the best-fit equation $y = 0.4x$ and coefficient of determination $R^2 = 0.89$. This result indicates that the sediment sub model did an acceptable job of predicting sediment in 1998 without calibration though there are some unexpected deviations.

Calibrated run. Though AnnAGNPS has predicted the hydrology and sediment component fairly well with some exceptions, the simulated data varies with the observed data and thus needed the calibration to bring them more close to each other. Table 2 shows the monthly simulation results and the deviation from the observed data after calibration for both runoff and sediment. It has been seen that the simulated runoff has increased and the sediment yield has been decreased by a small amount. For sediment, due to the observed data in May and September, a huge deviation between the simulated and observed value still exists. Coefficient of determination (R^2) values between the simulated and observed amount of runoff and sediment are 0.93 and 0.91 respectively.

On average, analysis shows that the hydrology component of the model consistently under predicted runoff and over predicted sediment yield. This may happen as the model doesn't have a good groundwater component and the groundwater flow is not simulated properly by the model. For the calibration phase, model simulated runoff amount has been slightly increased and the final result appeared to be in an acceptable range.

Table 2 Comparison of observed and predicted monthly runoff for the year 1998 (calibration phase)

Month	Ppt. (mm)	Sim_runoff (mm)	Obs_runoff (mm)	Deviation (%)	Sim_sed (ton)	Obs_sed (ton)	Deviation (%)
Apr-98	42.70	3.54	10.49	−66.24	11.91	3.14	279.81
May-98	26.30	0.19	0.55	−64.78	3.58	0.02	23,067.57
Jun-98	45.80	3.58	4.44	−19.21	218.97	84.93	157.81
Jul-98	15.40	0.06	0.11	−43.56	0.69	0.38	81.42
Aug-98	86.40	11.99	20.17	−40.57	130.19	59.65	118.25
Sep-98	10.30	0.15	0.03	394.88	0.74	0.00	24,549.11
Oct-98	18.20	0.54	0.07	698.41	3.95	0.09	4,144.90
Total	202.40	20.06	35.85	−44.04	370.02	148.21	149.66

The erosion sub model still over predicts sediment yield for the year 1998 after calibration. However, considering the unavailability of very precise and detailed data for watershed, results could be considered in the acceptable range.

Model validation. Model validation is considered to be a final check on the performance of the model under certain conditions. Runoff amount and sediment yield data in 1999 at the outlet of the watershed were used to validate the model. The model outputs of runoff and sediment yield using 1999 climate data were compared to the observed data and are displayed in Table 3, Figures 2 and 3. The simulated runoff is still under predicted compared to the observed amount. Simulated runoff amount is 48.88 mm whereas the observed runoff amount is found to be 55.9 mm. Simulated runoff is only higher than the observed data in September. The deviation percentage is very high in August. There was more rainfall in that month within one week (Julian day 216 to 223). A management operation was also performed (culti weed) during this period which disturbed the soil and as a result, there was less runoff. The total percentage of deviation for the whole year was 13%. There was more rainfall found in 1999 than in 1998 which resulted in more runoff in the validation phase. The best fit equation for runoff is $y = 1.153x$ and the coefficient of determination $R^2 = 0.83$. Thus it could be concluded that the model consistently under predicted the runoff. From the above statements, it can be better described that the model can simulate the runoff component in an efficient way.

The deviation of simulated sediment from the observed was 103%, which shows the model is over predicting the sediment yield in the validation phase too. The best fit equation shows the relation $y = 0.45x$ and the coefficient of determination (R^2) is 0.79. The validation results show the improvement in sediment prediction by the model. The deviation percentage did not change abruptly as was found in the year 1998 for both the un-calibrated and the calibrated run. Hence we can say that the model can simulate the sediment component effectively.

Table 3 Comparison of observed and predicted monthly runoff for the year 1999 (validation phase)

Month	Ppt. (mm)	Sim_runoff (mm)	Obs_runoff (mm)	Deviation (%)	Sim_sed (ton)	Obs_sed (ton)	Deviation (%)
Apr-99	34.90	4.87	7.09	−31.23	23.37	16.79	39.20
May-99	61.00	3.22	3.20	0.80	48.15	25.88	86.02
Jun-99	103.70	16.35	17.84	−8.37	131.57	61.72	113.16
Jul-99	62.10	8.05	7.48	7.51	36.80	22.95	60.39
Aug-99	44.40	1.59	0.18	768.50	19.99	8.26	141.97
Sep-99	62.40	4.12	2.23	85.21	64.14	34.75	84.58
Oct-99	65.50	10.68	17.91	−40.35	88.08	32.30	172.68
Total	434.00	48.88	55.93	−12.59	412.10	202.65	103.35

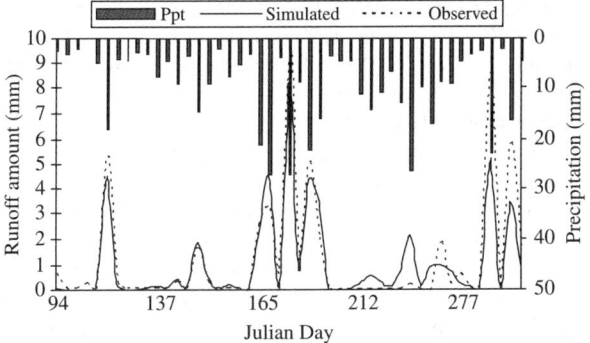

Figure 2 Comparison of simulated and observed amount of runoff (validation phase)

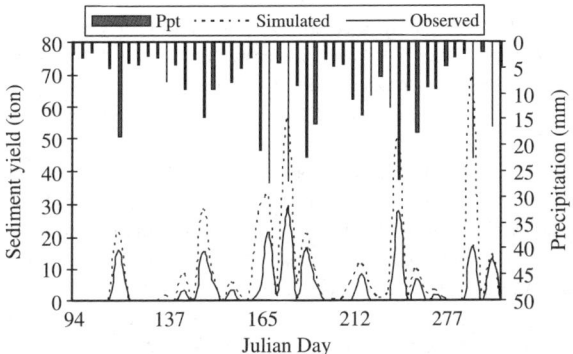

Figure 3 Comparison of simulated and observed amount of sediment yield (validation phase)

Conclusion

The AnnAGNPS model was evaluated on a watershed in Southern Ontario where the weather is cold and temperate. The model performance was calibrated and validated for the hydrology and sediment component from the non-point sources with the observed parameters. The model was applied from 1998 to 1999 on a daily basis. Depending on the availability and compilation of observed data, the model simulated hydrology and sediment component was calibrated from April 1998 to October 1998 and validated using independent data from April 1999 to October 1999. The model under predicted the simulated runoff and over predicted the sediment yield. Runoff curve number is the most sensitive parameter in the model as it has a greater impact on runoff than other parameters and runoff controls the detachment and transportation of other components from a watershed. Calibration and validation results show that the model is capable of simulating the runoff amount and sediment yield fairly well for a cold and temperate region like Ontario. Selection of the model input parameters needs careful attention especially while running for a long term period, as it is much too sensitive to input parameters.

References

Bosch, D., Theurer, F., Bingner, R., Felton, G. and Chaubey, I. (1998). Evaluation of the AnnAGNPS Water Quality Model. ASAE Paper No. 98-2195, St. Joseph, Michigan. 12 p.

Carey, J.H., Fox, M.E., Brownlee, B.G., Metcalfe, J.L., Mason, P.D. and Yerex, W.H. (1983). The Fate and effects of contaminants in Canagagigue Creek–1. Stream ecology and identification of major contaminants. Environment Canada. Scientific series No. 135. 37 p.

Cronshey, R.G. and Theurer, F.G. (1998). AnnAGNPS-Non Point Pollutant Loading Model. In *Proceedings First Federal Interagency Hydrologic Modeling Conference*. 19–23 April 1998, Las Vegas, NV.

Dickinson, W.T., Rudra, R.P. and Wall, G.J. (1990). Targeting remedial measures to control nonpoint source pollution. *Water Resources Bulletin*, **26**(3), 499–507.

GRCA (Grand River Conservation Authority) (1998). Background report on the health of the Grand River watershed 1996–97: State of the watershed report/prepared for The Grand Strategy Co-ordinating Committee by the Grand River Conservation Authority.

GRIC (Grand River Implementation Committee) (1982). Grand River Basin Water Management Study-Summary and Recommendations. 13 p.

Hoffman, D.W., Matthews, B.C. and Wicklund, R.E. (1963). Soil Survey of Wellington County, Ontario; Report No. 35 of the Ontario Soil Survey; Research Branch, Canada Department of Agriculture and The Ontario Agricultural College, 69 p. Accompanied by two coloured soil maps, scale 1:63,360.

Presant, E.W. and Wicklund, R.E. (1971). The Soils of Waterloo County; Report No. 44 of the Ontario Soil Survey; Research Branch, Canada Department of Agriculture, Department of Soil Science, University of Guelph, and The Ontario Department of Agricultural and Food, 104 p. Accompanied by soil maps, scale 1:20,000.

So, S.K. and Singer, S.N. (1982). Grand River basin water management study technical report series Report #27: Rural non-point source pollution and control, prepared for the Grand River Implementation Committee. Water Resources Branch, Ontario Ministry of the Environment, 156 p.

Soil Conservation Service (SCS) (1985). *National Engineering Handbook*, Section 4- Hydrology. Washington DC: USDA-SCS.

Theurer, F.D. and Clarke, C.D. (1991). Wash load component for sediment yield modeling. In *Proceedings of the Fifth Federal Interagency Sedimentation Conference*, March 18–21, pg. 7–1 to 7–8.

Theurer, F.D. and Cronshey, R.G. (1998). AnnAGNPS—Reach Routing Processes. In *Proceedings First Federal Interagency Hydrologic Modeling Conference*, 19–23 April, Las Vegas, NV.

Young, R.A., Onstad, C.A., Bosch, D.D. and Anderson, W.P. (1989b). AGNPS: A nonpoint-source pollution model for evaluating agricultural watersheds. *Journal of Soil and Water Conservation*, **44**(2), 168–173.

Young, R.A., Onstad, C.A., Bosch, D.D. and Anderson, W.P. (1989a). AGNPS user's guide. Verson 3.5. USDA_ARS North Central Soil Conservation Research Laboratory. Morris, MN.

Yuan, Y., Bingner, R.L. and Rebich, R.A. (2001). Evaluation of AnnAGNPS on Mississippi Delta MSEA Watersheds. *Trans. of the ASAE*, **44**(5), 1183–1190.

Nonlinear regression approach to evaluate nutrient delivery coefficient

M.S. Bae* and S.R. Ha**

*Watershed Management Research Center, National Institute of Environmental Research, Kyongseo-dong, Seo-gu, Incheon, South Korea (E-mail: mysoba@me.go.kr)
**Department of Urban Engineering, Chungbuk National University, 48 Gaeshin-dong, Heungduk-gu, Cheongju, Chungbuk, South Korea (E-mail: simplet@chungbuk.ac.kr)

Abstract Implementation of the Korean Total Maximum Daily Load Act calls for new tools to quantify nutrient losses from diffuse sources at a river basin district scale. In this study, it was elucidated that the nonlinear regression model (NRM) reduces the uncertainty of the boundary conditions of the water quality model. The NRM was proposed to analyse the delivery coefficients of surface waters and retention coefficients of pollutants. Delivery coefficient of pollution load was considered as a function of two variables: the watershed form ratio, S_h, which is a measurable geomorphologic variable and the retention coefficient, ϕ, which is an empirical constant representing the basin-wide retarding capacity of pollutant wash-off. This model was applied on the Geum River, one of the major basins in South Korea. The QUAL2E was used to simulate stream water quality using NRM. In this paper, we elucidate the possibility to use a nonlinear regression model for delivery and retention of nutrients in a drainage basin characterized as both data-rich and data-poor, and the magnitude of the nutrient loads and sources has been uncertain for a long time.
Keywords Delivery coefficient; diffuse pollution; pollution load runoff; retention coefficient

Introduction

The total maximum daily load (TMDL) Act has been established for watershed-based water quality management in Korea since 1998. To meet water quality standards mandated by the TMDL Act at a water quality monitoring station (WQMS), a comprehensive and rational pollution runoff analysis must be carried out. Prerequisites for sustainable environmental management of watersheds require basic environmental statistics gathering and quantitative assessments of the riverine loads, in particular, including estimation of the pollution delivery and retention in the drainage basin. The nutrient level and fluxes at a specific location in a public river network depend on the pollution sources in the upstream area, and transfer, retention, and loss of nutrients in the soil, ground water, and surface water network. This is a complex function of biological, physical, and chemical processes. Therefore, it is needed to analyze how these processes influence nutrient fluxes from pollution sources to river outlets over large spatial and temporal scales in modeling (Vassilijev and Stalnacke, 2005). Several models for so-called source delivery and retention have already been developed worldwide (Ha, 1989; Grimvall and Stalnacke, 1996; Ha et al., 1998; Liden et al., 1998). However, many watersheds in developing countries are often regarded as "data-poor" and characterized by high varying quantity and quality of input data. This is a typical problem in trans-boundary waters in which the richness and details of information about the watersheds' specified data may differ between provinces or countries both in terms of quantity and quality. Unfortunately, because most of the existing models require very detailed and spatially consistent input data, their applicability may be limited. Thus, more simple models are needed to address the limitations in these basins. A simple delivery coefficient (SDC), calculated by a conventional simple

doi: 10.2166/wst.2006.061

rate method, is called to either an assimilative capacity or a purification coefficient (Lee, 2000). Actually, these kinds of SDC account for the difference between the total pollution loads generated from pollution sources and total pollution loads discharged at the outlet of a stream in a specific watershed. The SDC, however, can't be used for the estimation of pollution load delivered in a watershed that lacked measurements because it is based on water quality data observed in the field. In this paper we elucidate the possibility to use a statistical model for delivery and retention of nutrients in a drainage basin characterized as both data-rich and data-poor, and the magnitude of the nutrient loads and sources has been uncertain for a long time.

Methods
Nonlinear statistical model

The SDC has been calculated by simply dividing the pollution load monitored with one discharged from a specific watershed. This relationship could be defined as:

$$P_M = P_O \times K \tag{1}$$

where P_M is the pollution load monitored at a specific water pollution monitoring site and P_O is the total pollution load discharged, which is an assumed amount of pollutants discharged from a specific watershed based on the unit loading factor method. K is a simple delivery coefficient. Because the calculation of the K in Equation (1) is impossible unless to get the data of water quality at a WQMS in each watershed, as an alternative way, it was usual to assign a mean value of the K coefficients, which have been determined from other watersheds with WQMS to the specific watershed without observed water quantity and quality data. This means that the delivery coefficient for the watershed without observed data can be estimated using an average of K values calculated from the watersheds with data observed by Equation (1). To overcome this weakness of SDC, the non-linear function defined with two variables, the watershed form ratio, S_f, which is a deterministic variable obtainable from the digital terrain map of a watershed as well as a quantitative index representing the geo-characteristic of the watershed and the retention coefficient, ϕ, which is an empirical variable was developed (Ha and Bae, 2001; Bae and Ha, 2003; Bae, 2003). Pollution loads monitored at WQMS, P_M, make a balance with the result of multiplying the pollution loads discharged from a watershed, P_O, and the innovated delivery coefficient, K. In the Equation (3), all variables are known values except the retention coefficient, ϕ. Consequently an empirical variable ϕ can be determined using the data set of known variables.

$$K = e^{-\phi \cdot S_f} \tag{2}$$

$$P_M = P_O \times e^{-\phi \cdot S_f} \tag{3}$$

where ϕ is an empirical variable and denotes the retardation effect of pollution loads in a watershed. And S_f denotes the Horton's watershed form ratio and quantifies the portion of channels in a watershed and is defined as follows:

$$S_f = \left(\sum L\right)^2 / A \tag{4}$$

where L is a sum of stream length and A is area of a specific watershed. Digital elevation model (DEM) data can be used to determine a length of waterway, the area, and average slope of a watershed.

Establishment of a reference flow rate and a standard water quality

The reference year was 1998 because the TMDL Act was established in this year in Korea. Using flow rate data for the last 10 years, a non-excess probability distribution of flow rate was analyzed and the mean of the data of Q_{275} was determined since the water quality at WQMS in the study watershed showed the worst state at the flow rate Q_{275}. As the flow rate in December 1998 was most similar to the average Q_{275}, the monthly average flow rate and BOD concentration obtained in December 1998 were used to calculate a basin-wide delivery coefficient. In addition, the total daily pollution loads discharged from all tributaries were referenced to the data released by the National Institute of Environment Research in Korea. The water quality data observed at WQMS in the Geum River were cited from the official homepage of the Ministry of the Environment, Korea.

Water quality simulation model

The feasibility analysis of the methodology innovated to determine a basin-wide delivery coefficient is carried out through a comparison of BOD concentration calculated by the innovated method with the water quality observed at WQMS. The EPA QUAL2E model was used for simulation of stream water quality. The effectiveness of introducing the innovated delivery coefficient of pollution loads, which had been taking into account the influence of stream watershed geomorphologic properties on a pollutant wash-off behavior, into the stream water quality simulation was evaluated using the root mean square errors (RMSE) method.

Results and discussion

Study watershed

The Geum River basin is located in the central part of South Korea and includes the Dacheong Reservoir, of which the total storage capacity is 1,490 million tons. It is a water supply resource for about three million people. The watershed area is about 9,910 km² and the annual precipitation is about 1,400 mm/year but more than half of it concentrates on a rainfall season from July to September. The delivery time of the storm peak is comparatively short and less than 2 days. This study watershed is divided into 121 sub-watersheds and the mean area of the watersheds is about 82 km². The mean

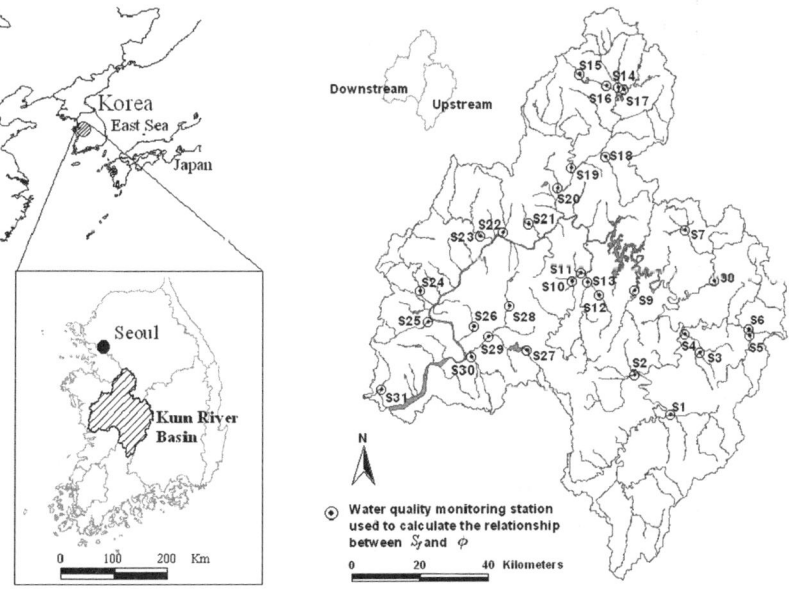

Figure 1 Study area and water quality monitoring stations

length of the waterways and the average slope of the watersheds is 28 km and 11°, respectively. A spatial distribution of WQMS used to estimate the relationship between S_f and ϕ is shown in Figure 1.

Non-linear regression model on S_f and ϕ

Non-linear regression between S_f and ϕ was derived from the data set in terms of water quality data obtained at water quality monitoring stations and the estimated values of watershed pollution loads discharged in Dec. 1998. In this analysis, taking into consideration the differences of water use in the stream reach, the Geum River was divided into the upstream reach and the downstream one on the basis of the location of the Daecheong

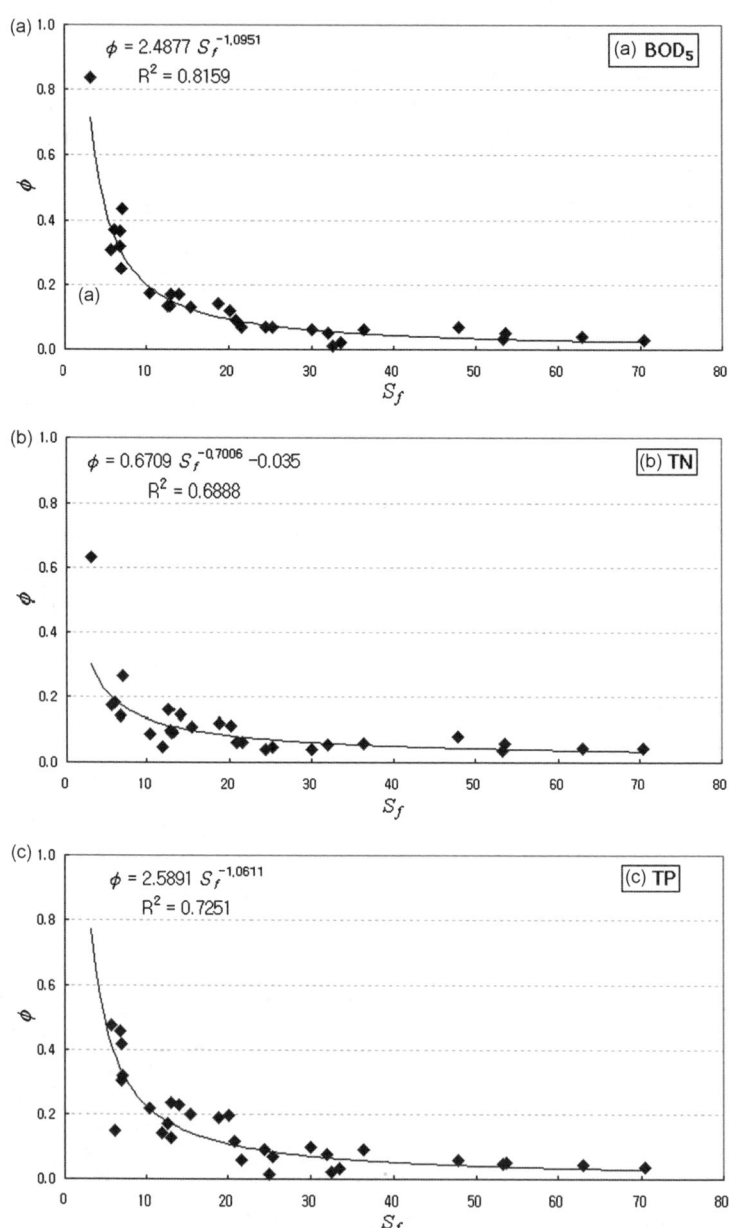

Figure 2 Relationship between S_f and ϕ on BOD, TN and TP

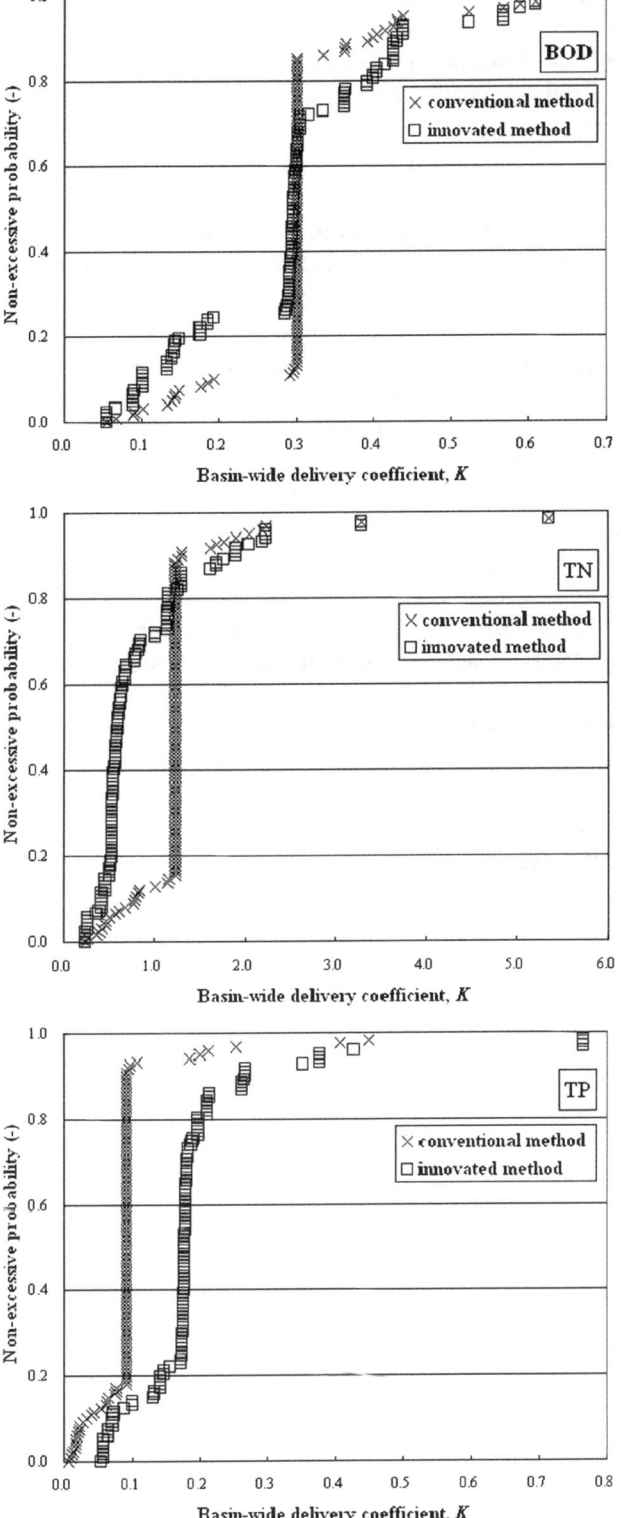

Figure 3 Comparison of non-excess probability distributions of the conventional delivery coefficient (X) and the innovated one, K ($\square$), determined by the regression on S_f vs. ϕ

Reservoir. Figure 2 shows scatter plots of the relationship between S_f and ϕ. This regression analysis was done using 31 data sets available from all WQMSs including the upstream and the downstream of Geum River. Figure 2(a) shows the scattering of BOD_5 and the correlation coefficient is about 0.82. On the other hand, Figure 2(b) and Figure 2(c) are the plots on total nitrogen and total phosphorus, respectively. And correlation coefficients on total nitrogen and total phosphorus are 0.69 and 0.73, respectively.

Determination of the innovated delivery coefficients on BOD, TN, and TP

Because it is possible to calculate S_f for any watersheds, the retention coefficient, ϕ, in the basins that lack measurement of water quantity and quality can be estimated by substituting the S_f value for the specific sub-watershed into the regression equation. Then the innovated delivery coefficient, K, in the sub-watershed without WQMS is determined by substituting two values of S_f and ϕ to Equation (2). While difference between non-excessive probability distributions of two different delivery coefficients determined using the regression model and the average K values is shown in Figure 3. This has come from the assignation of the K value averaged from all values of K determined from 31 sub-watersheds with WQMS to the other sub-watershed without WQMS in the study watershed.

Impact of the innovated basin-wide delivery coefficient on the water quality simulation

The application results of the innovated delivery coefficient to the water quality simulation using QUAL2E model showed high consistency with observed water quality concentrations.

Tables 1 and 2 show the RMSE errors between the delivered pollution loads calculated using the regression (with innovated delivery coefficient) and the pollution loads monitored at 31 WQMS sites. RMSE of pollution load delivered on BOD, TN and TP using innovated delivery coefficients are 4.27 kg/day, 17.91 kg/day and 2.53 kg/day in upstream, respectively. While RMSE using conventional delivery coefficients are 105.76 kg/day, 251.58 kg/day and 2.40 kg/day in upstream, respectively. RMSE of BOD, TN and TP concentrations with innovated delivery coefficient are 0.001 mg/L, 0.080 mg/L and 0.002 mg/L in upstream, respectively. While RMSE using conventional delivery coefficients are

Table 1 RMSE of water quality simulation using QUAL2E (upstream)

	Pollution load (kg/day)		Concentration (mg/L)	
	Innovated K*1	Conventional K*2	Innovated K	Conventional K
BOD	4.27 (0.04)*3	105.76 (1.00)	0.001 (0.001)	0.826 (1.00)
TN	17.91 (0.07)	251.58 (1.00)	0.080 (0.012)	6.676 (1.00)
TP	2.53 (1.05)	2.40 (1.00)	0.002 (1.00)	0.002 (1.00)

*1 resulted from the application of the innovated basin-wide delivery coefficient proposed in this study
*2 resulted from the application of the conventional basin-wide delivery coefficient by simple rate method
*3 numeric values in parenthesis are relative to conventional delivery coefficient

Table 2 RMSE of water quality simulation using QUAL2E (downstream)

	Pollution load (kg/day)		Concentration (mg/L)	
	Innovated K	Conventional K	Innovated K	Conventional K
BOD	38.55 (0.29)	132.98 (1.00)	0.039 (1.11)	0.035 (1.00)
TN	54.34 (0.16)	329.70 (1.00)	0.303 (0.16)	1.844 (1.00)
TP	6.44 (0.34)	19.11 (1.00)	0.001 (0.004)	0.262 (1.00)

0.826 mg/L, 6.676 mg/L and 0.002 mg/L in upstream, respectively. The simulation results from QUAL2E runs on BOD, TN and TP in upstream and downstream are shown in Figure 4 (a) and (b), respectively. As a result of water quality simulation, it was revealed that the innovated delivery coefficient resulting from the nonlinear regression model has more consistent than the conventional delivery coefficient. Following is the summary of the new findings from simulation analysis. The root mean square error for BOD (mean of observed concentrations is 1.1 mg BOD/L) was reduced from 0.82 mg BOD/L in the run case with the conventional delivery coefficient to 0.001 mg BOD/L in the run with the innovated delivery coefficient. For nutrient cases, RMSE for total nitrogen (mean is 2.57 mg TN/L) and total phosphorus (mean is 0.04 mg TP/L) are improved from 6.68 mg TN/L to 0.08 mg TN/L and from 0.0023 mg TP/L to 0.0016 mg TP/L by applying the innovated delivery coefficient to the simulation model.

Flow rate and retention coefficient

Hydrological conditions, such as rainfall intensity/duration, soil infiltration, retention time of reservoir etc., are crucial factors in diffuse pollution wash-off. The rainfall, especially, acts as a transmission mediator of pollutant from source to receiving water. In order to reveal the effects of rainfall on pollutant runoff, the relationship between S_f and ϕ was

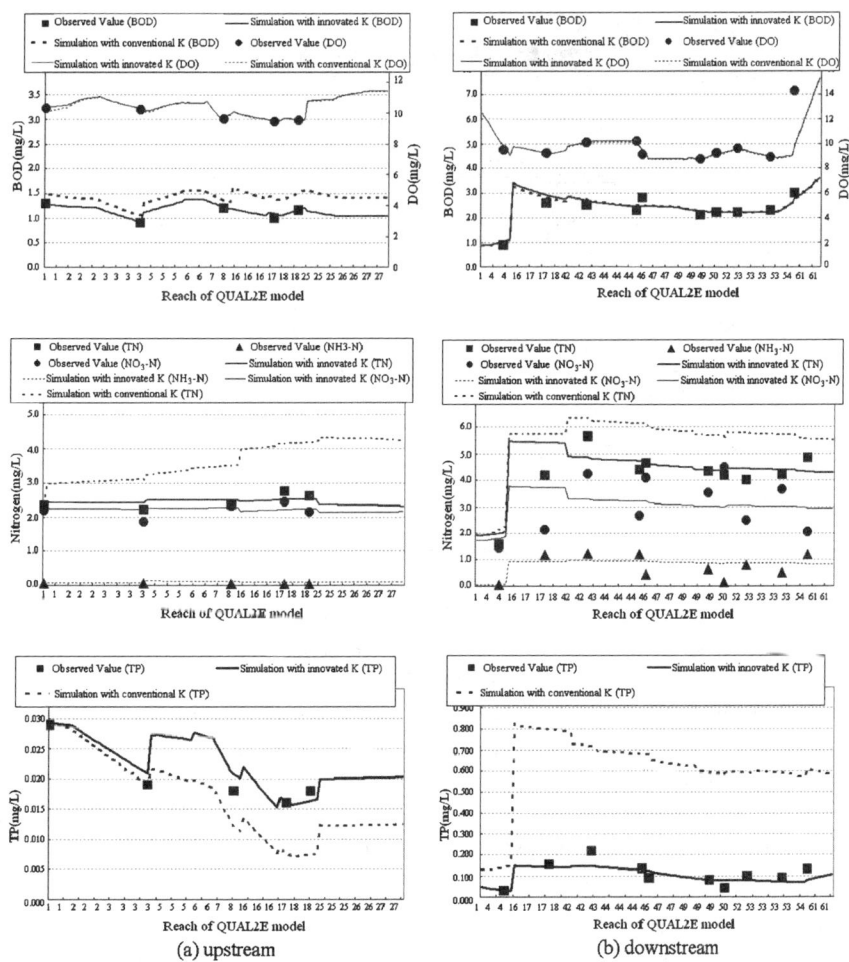

Figure 4 Consistency of simulation results on BOD and nutrients

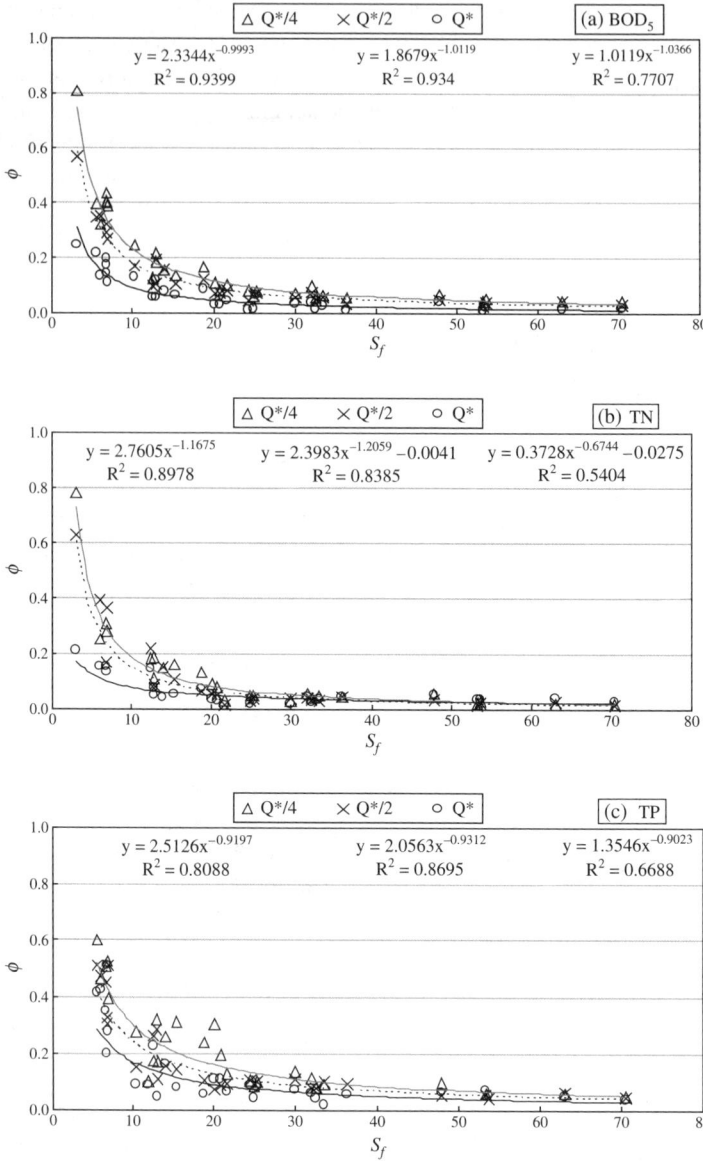

Figure 5 Variation of the flow rate and the relationship between S_f and ϕ on BOD, TN and TP

analyzed in three different periods (June, May and March, 1998) with different flow rates. Figure 5 shows the variation of the regression curve between S_f and ϕ on BOD, TN and TP based on different flow rates; Q^* is mean flow rate in June, $Q^*/2$ is mean flow rate in May when we had the most approximate flow rate to half of Q^*, and $Q^*/4$ is mean flow rate in March when we had the most approximate flow rate to a quarter of Q^*. As shown in Figure 5, it was revealed that the retention coefficient, ϕ, reduced as the flow rate increased, especially where watershed has a low S_f. This means that the variation of delivery coefficient, K, with rainfall variation can be quantified on a specific watershed.

Conclusions

In this study, the possibility to use the nonlinear regression model to calculate the delivery coefficients and the retention coefficients in terms of BOD and nutrients (TN and TP) was elucidated. The model takes into account the influence of stream watershed geomorphologic properties such as a watershed form ratio, S_f, and a basin-wide delivery retardation coefficient, ϕ, on pollution wash-off behavior. The application results of the innovated model to the stream water quality simulation done by QUAL2E model showed high consistency with the observed results. The root mean square error for BOD (mean of observed concentrations is 1.1 mg BOD/L) was reduced from 0.82 mg BOD/L in the run case with the conventional delivery coefficient to 0.001 mg BOD/L in the run with the innovated delivery coefficient. For nutrient cases, RMSE for total nitrogen (mean of observed concentrations is 2.57 mg TN/L) and total phosphorus (mean of observed concentrations is 0.04 mg TP/L) are improved from 6.68 mg TN/L to 0.08 mg TN/L and from 0.0023 mg TP/L to 0.0016 mg TP/L by applying the innovated delivery coefficient to the simulation model. As a result of this study, it was revealed that the innovated method to evaluate the delivery coefficient could cope with the limitations of the conventional method, especially in agricultural and forest areas. The accuracy improvement using this innovated model on water quality simulation in the study watershed was about 62% on average. Interestingly, it was revealed that the retention coefficient, ϕ, reduced as the flow rate increased, especially where the watershed has a low S_f. This means that the variation of delivery coefficient, K, with rainfall variation can be quantified on a specific watershed.

References

Bae, M.S. (2003). *Estimating of delivery coefficient of pollution load using geo-informatics*. PhD thesis, Department of Urban Engineering, Chungbuk National University, Korea.

Bae, M.S. and Ha, S.R. (2003). GIS-based influence analysis of geomorphological properties on pollutant wash-off in agricultural area. *Wat. Sci. Tech.*, **51**(3–4), 301–307.

Grimvall, A. and Stalnacke, P. (1996). Statistical methods for source apportionment of riverine loads of pollutants. *Environmetric*, **7**, 201–213.

Ha, S.R. (1989). *Kinematic modeling for wash-off on urban non-point source pollutants*. PhD thesis, Department of Environmental Engineering, Osaka University, Japan.

Ha, S.R. and Bae, M.S. (2001). Effects of land use and municipal wastewater treatment changes on stream water quality. *Environ. Monitor. Assess.*, **70**, 135–151.

Ha, S.R., Jung, D.I. and Yoon, C.H. (1998). A renovated model for spatial analysis of pollutant runoff loads in agricultural watershed. *Wat. Sci. Tech.*, **38**(10), 207–214.

Lee, J.H. (2000). Management of nonpoint sources in watershed with reference to Daechong Reservoir in Korea. *Journal of Environmental Impact Assessment*, **9**(3), 163–176.

Liden, R., Vassilijev, A., Loigu, E., Stalnacke, P. and Wittgen, H.B. (1998). Nitrogen source apportionment using a physically and statistically based model. *Ecological modeling*, **114**, 235–250.

Vassilijev, A. and Stalnacke, P. (2005). Statistical modeling of riverine nutrient sources and retention in the lake Peipsi drainage basin. *Wat. Sci. Tech.*, **51**(3–4), 309–317.

Comparative study of two watershed scale models to calculate diffuse phosphorus pollution

A. Kovacs

Department of Sanitary and Environmental Engineering, Budapest University of Technology and Economics, Muegyetem rakpart 3, H-1111 Budapest, Hungary (E-mail: *kovax@vkkt.bme.hu*)

Abstract The aim of this study was to compare and assess models having different principles to calculate diffuse phosphorus emissions on a selected watershed. The empirical MONERIS model and the physically based SWAT model were evaluated for comparative purposes. The approaches were applied for a sub-basin of the Hungarian Zala River watershed for five years. The calculated river loads were checked by the measured values at the catchment outlet. Due to the dissimilar results of water balance and erosion calculations, a highly different phosphorus emission was computed. It was also concluded that in the case of transport-limited watersheds, the SWAT model calculates phosphorus river loads slightly inaccurately, since it does not include the description of fate of inorganic phosphorus interacting with sediment during the channel transport. When these processes are taken into account, modeling results fit better the measured loads. The MONERIS model calculates acceptable river load by assuming very intensive in-stream retention. Additionally, the empirical method can be useful for long-term investigations as a decisions support tool for preliminary design. However, for detailed emission assessment and scenario development the physically based approach seems to be more appropriate.

Keywords MONERIS; non-point pollution; phosphorus; SWAT; watershed modeling; Zala River catchment

Introduction

Emissions from non-point sources have great importance regarding the water pollution nowadays. One of the current water quality problems related to diffuse pollution is eutrophication. Despite controlling the point sources (i.e. nutrient removal), a significant load of nutrients still reaches surface waters. Diffuse phosphorus emissions from watersheds provide nutrient supplies for the aquatic plants. Therefore knowledge of phosphorus emissions from different pathways and sources is a key issue concerning the protection of water quality and sustainable watershed management practice.

Several watershed modeling approaches are reviewed in the literature (e.g. Donigian and Huber, 1991; Novotny, 2003). They differ from each other primarily in the temporal and spatial scale, i.e. in the details of the resolution. Lumped parameter models consider the watershed as a uniform region and they do not have information on spatial differences within the watershed. In contrast, distributed parameter models divide the watersheds into smaller units. All units are individual having their own parameter set. Long-term yield models calculate the average nutrient fluxes according to a longer time period, typically several years. Dynamic models provide temporary information on the nutrient loads (time series) either as a continuous (daily time step, long-term impacts) or as an event-based (hourly or sub-hourly time-step, short-term impacts) approach. Model selection depends mainly on the aim of the examination, the characteristics of the studied water body, the availability of data, the expected accuracy and the temporal and financial costs.

This paper presents the evaluation of two different watershed models used for a Hungarian case study region, to calculate phosphorus emissions from the drainage area as well as river loads at the catchment outlet for five years (1997–2001).

doi: 10.2166/wst.2006.062

Background

Lake Balaton, which is the greatest shallow lake in the Central-Eastern European region, is sensitive to eutrophication (Somlyody and Hock, 2002). First signs of heavy eutrophication occurred during the seventies due to the increased nutrient loads reaching the high priority recreational lake (the highest algae biomass reached a value of 200 mg chl-a/m^3 (Somlyody and Hock, 2002). Sisak and Pomogyi (1994) found that the contribution of point and non-point sources during the period 1975–1987 was 55% and 45%, respectively. Since then, several investigations have been conducted to protect the water quality. Most of the wastewater treatment plants on the watershed were expanded with phosphorus removal technology. In addition, an artificial wetland area (Kis-Balaton reservoir) was constructed close to the lake inlet. It was intended to retain nutrients transported by the main influent Zala River, which carries most of the nutrient loads of the lake. Due to the control of point sources, the major portion of the phosphorus emissions is now associated with diffuse sources. Since the current phosphorus load is about two times higher than the desirable value, future water quality management of the lake needs the reduction of the non-point pollution (Somlyody and Hock, 2002). Consequently the region is a highly feasible area to study the impacts of non-point pollution.

The Zala River catchment is located in the western hilly part of Hungary. The area of the selected part of the watershed (upstream of Kis-Balaton reservoir) is 1,528 km^2. The elevation range of the region is between 90 and 325 m above the Baltic Sea. The hilly area has moderate slopes (the average value is about 6%). The average discharge at the outlet was 4.3 m^3/s during the period of 1997–2001, i.e. the average yearly runoff volume was 89 mm/a. The average phosphorus river load was 35 t/a, which means 0.23 kg/ha/a area specific flux. In the period of 1997–2001, the average annual precipitation was 651 mm. The dominant physical soil type is the loamy soil having poor to moderate hydraulic conductivity. The area is sensitive to water erosion processes (the estimated long-term average soil loss is about 6.1 t/ha/a). The majority of the watershed is an agricultural area, in particular arable land, which is 54% of the catchment area. Forests are relatively important, covering approximately one third of the area.

Methods

Two different watershed modeling tools were selected to estimate phosphorus emissions: the MONERIS model (Modeling Nutrient Emissions in River Systems, Behrendt *et al.*, 2000) and the SWAT model (Soil and Water Assessment Tool, Neitsch *et al.*, 2002).

The MONERIS model

The MONERIS model has been developed to calculate nutrient emissions entering river systems at large watershed scale (catchment area larger than 1,000 km^2). It is a lumped parameter model for the estimation of long-term averages based on mostly empirical relationships. The application of the model requires detailed statistical, sampling and literature data, default parameters and digital maps about various characteristics of the watershed. It provides a 5 year-average of water balance components, nutrient emissions and river loads at the main sub-catchment outlets. By this method the nutrient emissions of large sub-catchments into the Danube River and its main tributaries were calculated (Schreiber *et al.*, 2005).

The model does not distinguish phosphorus forms. It focuses only on total phosphorus emissions. It separates seven different pathways resulting in the total amount of phosphorus emissions. Six of these are diffuse sources: atmospheric deposition, overland flow, erosion, tile drainage, groundwater and urban systems. The seventh component is the contamination from point sources. For each pathway the appropriate water balance component (in case of erosion the sediment flux) is computed empirically and after that phosphorus concentration is determined for the water (sediment) flows. These concentrations are determined based on parameters found in the literature and mass balance calculations of the agricultural areas. Phosphorus enrichment in eroded soil is also computed empirically. Additionally, possible field retention is taken into account along the pathways (e.g. sediment deposition on surfaces). From the total emission values of sub-catchments river loads are computed using an empirical in-stream retention model.

The SWAT model

The aim of the SWAT model is to describe the fate of the water, sediment, nutrients and pesticides more accurately in the watersheds and to simulate the impact of management practices on hydrology and water quality. SWAT is physically based, thus the method requires very detailed and specific information on meteorology, topography, soil, land use and management conditions of the watersheds. The method runs on a geographical information system. SWAT is a continuous daily time-scale model; therefore it is particularly useful to simulate long-term yields and impacts. It is not designed to model detailed single-event flood routing and its impacts. It is a spatially distributed parameter model; however, the unit of the calculations is not the grid cell level, but the spatial unit called Hydrological Response Unit (HRU). This means that the grid cells with identical soil and land use types are integrated into a HRU. The number of simulated sub-basins is theoretically unlimited. Many examples of the model application can be found on the SWAT model homepage (www.brc.tamus.edu/swat).

The SWAT approach separates sub-modules for both catchment and river channel processes. In the catchment phase watershed climate and hydrology, plant growth, erosion and phosphorus transformation and transport are modeled. Six different phosphorus forms are separated, i.e. stable, active and fresh organic phosphorus, and stable, active and soluble inorganic phosphorus. The soil phosphorus cycle is modeled in detail. The main transport processes are phosphorus movement via surface runoff (soluble forms), erosion (phosphorus attached to sediment) and leaching. Enrichment of fine particles in eroded soil is also taken into account. Additionally, base flow phosphorus contents can be set to compute the subsurface contribution to the river loads. The channel phase includes water, sediment and phosphorus routing along the main river longitudinal sections, as well as fluxes from point sources. Phosphorus transformations are described with an adapted version of QUAL2-E in-stream water quality model. Mass balance of reservoirs is separately calculated.

Results and discussion

Both models were applied for the Zala River catchment for the period of 1997–2001. While only four sub-catchments were delineated for MONERIS, 40 sub-basins were appointed for SWAT. The SWAT model was calibrated to the measured discharge and water quality data at monthly time steps. The MONERIS model was applied in its original form. Emission results are compared according to the sub-basins in the MONERIS model for the period of 1997–2001 (annual average values). Dynamic results of the SWAT model at monthly time steps as well as yearly average river loads are assessed at the catchment outlet for the same time period.

Emissions of MONERIS model

The calculated area-specific phosphorus emissions and their share in the total emission according to the different pathways are presented in Figure 1. The area-specific phosphorus emissions vary in a wider range (between 0.52 and 1.23 kg/ha/a); the average for

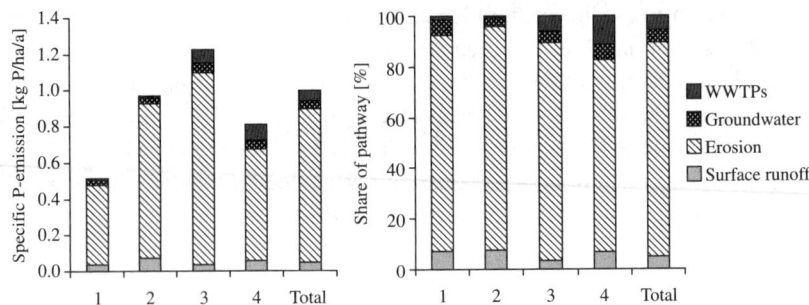

Figure 1 Annual average total phosphorus emissions according to the different pathways calculated by the MONERIS model in the Zala River catchment (1997–2001)

the total area is about 1.00 kg/ha/a, which is 153 t/a. The difference is due to the different volume of eroded soil transported into the river, which is explainable by the varying sensitivity to erosion in the watershed area. However, in sub-basin No. 1 the calculated area-specific flux is low despite the higher specific soil loss value. This is due to lower delivery ratio of sediment in this sub-catchment. Regarding the share of the sources, almost all of the total emission (about 75–85%) is caused by the erosion. Other diffuse sources have minor importance and point source emissions are also non-relevant due to the P-elimination applied in the WWTPs in this region. It is concluded that 94% of the total emitted phosphorus fluxes originate from non-point sources.

Emissions of SWAT model

Figure 2 shows the estimated area-specific phosphorus releases into the river and their proportion compared to the total phosphorus fluxes. The specific emissions have high variation among the sub-catchments ranging between 0.18 and 0.42 kg/ha/a. Emission regarding the whole catchment is 0.31 kg/ha/a, which is equal to 47 t/a. Total emitted phosphorus yield is lower by about 70% compared to MONERIS results. The reason can be found primarily in the highly different amount of sediment yield predicted (discussed later). Difference between the average enrichment ratios computed by the models is not significant. Excluding sub-basin No. 1 the area-specific values are also lower. The higher relative importance of the first sub-watershed is probably due to the decisive role of surface runoff to the river loads in areas located in the upper part of the watershed. Sub-basin No. 4 yields the least phosphorus load into the reach because of its low sensitivity to soil erosion and its low surface runoff contribution. Erosion is the primary pathway regarding phosphorus transport. Groundwater and point sources are less significant; their role is remarkable in sub-basin No. 4 only. Dominance of diffuse pollutions is obvious (90%).

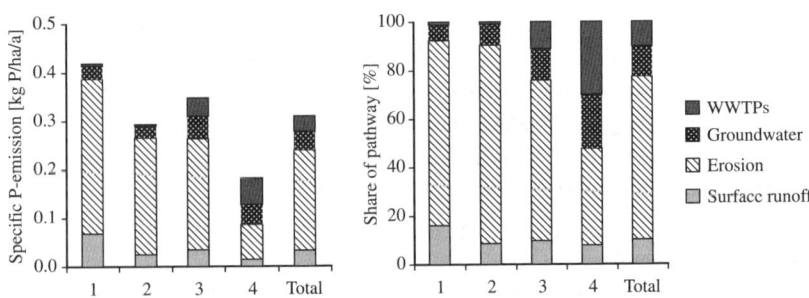

Figure 2 Annual average total phosphorus emissions according to the different pathways calculated by the SWAT model in the Zala River catchment (1997–2001)

Comparison of the river loads

Comparison of simulated river loads to each other and to the measured values at the catchment outlet is presented in Table 1. In the case of the SWAT model, monthly calibration results are shown in Figure 3. Regarding the river flow components, there are remarkable differences between the two methods. SWAT calculates quite well the total river flow based on the watershed features (Figure 3A), MONERIS uses the measured flow to calculate water balance. While MONERIS calculates lower surface water contribution and stronger base flow dominance, SWAT computes a higher amount of direct runoff (see Table 1), which is acceptable considering the topsoil with poor hydraulic conductivity. MONERIS underestimates the surface runoff volume. This is confirmed by the results of a base flow separation technique developed by Arnold et al. (1995), which determines an approximate value of 20 mm/a for the surface runoff in the studied time period. The reason for the discrepancy can be found in the extension of empirical equations for areas out of the original calibration range.

Sediment yield and load calculations also have differing results. MONERIS despite its lower surface runoff volume calculates almost four times higher sediment input than SWAT (Table 1). This inconsistency is caused by the sediment yield generation approach of MONERIS, which does not depend on the surface runoff volume and the soil and land coverage properties. It calculates empirically the sediment input from the mean soil loss and average delivery ratio determined from the catchment slope and proportion of arable lands. SWAT erosion calculation (MUSLE method) is based on surface flow amount, soil erodibility, topography and land use characters. A study on the impacts of the spatial aggregation on sediment generation found that sediment generation can vary by 44% between the coarsest and the finest watershed delineation and it decreases substantially with decreasing sub-watershed size (FitzHugh and Mackay, 2000). Consequently, further examination of the impacts of spatial scaling in the Zala River watershed is needed to clarify the possible changes in sediment generation.

Only the SWAT model contains a sediment channel routing sub-model. The calculated sediment loads approximate well the measured values (Figure 3B). The results indicate clear net sedimentation of the suspended solids (Table 1), e.g. the model computes remarkable sediment retention in the reach system (about 50%). This means, if the results of MUSLE approximate well the realistic sediment yield values, the total examined watershed is a transport-limited area, where the generated amount of eroded soil in the catchment exceeds the sediment transport capacity of the channel system (FitzHugh and Mackay,

Table 1 Comparison of annual average runoff components, sediment and phosphorus fluxes at the catchment outlet of the Zala River (1997–2001)

Component	Unit	MONERIS	SWAT
Surface runoff	mm/a	11.0	17.6
Subsurface flow	mm/a	74.3	69.9
Waste water discharge	mm/a	3.0	3.0
Total river flow	mm/a	–	90.5
Measured river flow	mm/a	89.2	
Sediment yield	t/ha/a	0.334	0.088
Sediment deposition	t/ha/a	–	0.042
Sediment load	t/ha/a	–	0.046
Measured sediment load	t/ha/a	0.045	
Phosphorus emission	kg/ha/a	0.999	0.309
Phosphorus rentention	kg/ha/a	0.817	0.022
Phosphorus load	kg/ha/a	0.182	0.288
Measured phosphorus load	kg/ha/a	0.227	

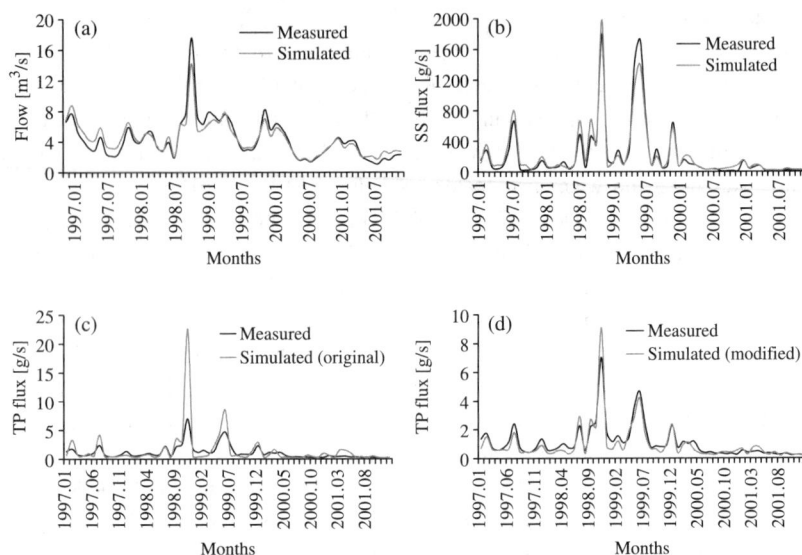

Figure 3 Monthly river flow, sediment flux and total phosphorus load values calculated by the SWAT model at the catchment outlet of the Zala River (1997–2001)

2000). In a geomorphologic sense, the stream channels are only temporary storages of the transported sediment and over a long period of time a natural river transports all sediment delivered into it from the drainage area (Novotny, 2003). Thus, long-term (more than 100 years) sediment delivery ratio for streams is 1. However, at the time step of water quality studies (less than 10 years) the deposition of sediment and the fate of settled contaminants related to it must be considered (Novotny, 2003). In many streams the sediment delivery ratio is less than 1 if the examinations concern short periods without extreme flood events. Sediment settles in the downstream reaches of the river with lower flow velocity (where the texture of sediment becomes finer), in small ponds located on the river network and primarily on the floodplains during flood events and in the reservoirs.

Dissimilarity of phosphorus emissions calculated by the two methods has already been discussed; phosphorus river load calculations have different results also. The SWAT model overestimates the observed river load (Table 1) and generates high phosphorus peaks (Figure 3C). Although sediment routing is modeled by SWAT in the river net, the fate of the inorganic phosphorus associated with the sediment particles is not tracked along the main river sections except in the reservoirs. Only organic phosphorus is accounted for in deposition in streams by a settling rate. Consequently, in watersheds where sediment retention is remarkable, SWAT is not feasible to calculate adequately the phosphorus river loads due to the lack of in-stream transport modeling of inorganic particulate phosphorus and calculation of floodplain impacts. In contrast, MONERIS gives more appropriate results for the river load compared to the measured one (Table 1). The model includes an empirical river retention model for the total phosphorus based on the specific runoff of the catchment, which results in a good fit to the measured load, though the total emissions are quite high (see Figure 2). Thus, the degree of the retention (80% of the emission) is assumed to be very intensive. As was shown by Zessner *et al.* (2005), flood events transport only a few percent of the total yearly average phosphorus loads of the Zala River, which indicates remarkable phosphorus retention in the river system. Since most of the phosphorus in soils is contained in the particulate phase and moves with eroded soil (Novotny, 2003), phosphorus can be retained in the reaches by sediment deposition. Although settled phosphorus reenters by resuspension of sediment caused by the flow turbulence and biological benthic activity, the delivery ratio for phosphorus in streams can be also less than 1 for a short time period.

To take into account the sedimentation of inorganic phosphorus attached to the sediment a simple correction was executed to improve the SWAT model performance. Assuming that the settling rate of sediment and adsorbed mineral phosphorus is approximately equal, a value for the possible river retention regarding the adsorbed inorganic phosphorus form was calculated by the ratio of daily sediment flux at the catchment outlet to the daily sediment yield from the watershed. Then, particulate inorganic phosphorus emission was multiplied with the retention ratio for each day. This slight correction resulted in smaller peaks (Figure 3D) and an additional average retention of 0.072 kg/ha/a, consequently the original value for the river load (0.288 kg/ha/a) reduced to 0.216 kg/ha/a. This value approximates better the measured one. The corrected phosphorus retention (30% of the emission) is lower than in the case of MONERIS. Based on the results it can be concluded that additional examinations are necessary to reveal the possible impacts of the in-stream retention processes regarding both sediment and phosphorus.

Conclusions

Both methods detect the dominance of diffuse sources, especially the erosion on the phosphorus pollutions of the Zala River. However, there are remarkable differences between the calculated emissions for the major sub-catchments and also for the total watershed. The MONERIS model predicts more than three times higher phosphorus emissions from the total watershed. The differences are due to the lower surface runoff volume and the huge amount of sediment yield of MONERIS. These results are explainable by the extension of the empirical equations without calibration as well as by the model application for sub-basins smaller than the originally proposed catchment size. Therefore recalibration of the empirical equations according to the local conditions and to higher resolution is needed. SWAT simulates better the water balance and sediment generation as well.

River load calculations of SWAT are slightly inaccurate due to the lack of in-stream modeling of inorganic phosphorus attached to the sediment. In watersheds where sediment deposition is significant, this shortcoming can lead to imprecise determination of phosphorus river loads, because retention primarily on the floodplains and in the river channels is not taken into account. Consequently SWAT can overestimate the observed phosphorus river loads. By a simple estimation of retention of mineral particulate phosphorus, the results fit better the measured values. Despite the high value of estimated emissions MONERIS simulates good results compared to the observed loads due to the computed very intensive river retention. Further studies are needed to clarify the accuracy of the erosion calculations and the related phosphorus emissions depending on the spatial resolution of the catchment as well as the possible river sediment and phosphorus retention.

Considering the extension problem of empirical formulas, the MONERIS model is capable of establishing the existence of the diffuse pollution problem in general. Nevertheless, regarding the accurate calculation of emission values as well as important processes (surface runoff and erosion) it has significant uncertainties. For precise emission estimations and determination of cause–effect relationships, SWAT seems to be more appropriate due to its physically based structure. However, in large basin scale, where the effects of different sub-regions are more balanced, the MONERIS model can be suitable to calculate long-term impacts of management practices. It can be useful in the phase of preliminary design of watershed management planning to evaluate the possible effects of the designed changes in the river basin management. For detailed examination of the source areas and pathways at particular temporal and spatial resolution, the SWAT model is more feasible. However, due to its high data demand, temporal and financial costs, it cannot be applicable in regions without detailed data collection. In such cases, it should be replaced even at higher resolutions with simpler methods, like the MONERIS model.

References

Arnold, J.G., Allen, P.M., Muttiah, R. and Bernhardt, G. (1995). Automated base flow separation and recession analysis techniques. *Ground Water*, **33**(6), 1010–1018.

Behrendt, H., Huber, P., Kornmilch, M., Opitz, D., Schmoll, O., Scholz, G. and Uebe, R. (2000). *Nutrient Emissions into River Basins of Germany*. UBA-Text 23/00, Berlin, Germany.

Donigian, A. and Huber, C. (1991). *Modeling of Nonpoint Source Water Quality in Urban and Non-urban Areas*. Report No. 68-03-3513. US Environmental Protection Agency, Athens, USA.

FitzHugh, T.W. and Mackay, D.S. (2000). Impacts of input parameter spatial aggregation on an agricultural nonpoint source pollution model. *Journal of Hydrology*, **236**(2000), 35–53.

Neitsch, S.L., Arnold, J.G., Kiniry, J.R., Williams, J.R. and King, K.W. (2002). *Soil and Water Assessment Tool*. TWRI Report TR-191. Texas Water Resources Institute, College Station, Texas, USA.

Novotny, V. (2003). *Diffuse Pollution and Watershed Management*. John Wiley and Sons Inc., Hoboken, USA.

Schreiber, H., Behrendt, H., Constantinescu, L.T., Cvitanic, I., Drumea, D., Jabucar, D., Juran, S., Pataki, B., Snishko, S. and Zessner, M. (2005). Nutrient emissions from diffuse and point sources into the River Danube and its main tributaries for the period of 1998–2000 – results and problems. *Wat. Sci. Tech.*, **51**(3–4), 283–290.

Sisak, I. and Pomogyi, P. (1994). Investigation of the nutrient loads of the River Zala (in Hungarian). *Hydraulic Engineering*, **4**(1994), 417–434.

Somlyody, L. and Hock, B. (2002). Water Quality and its Management. In *Strategic Issues of the Hungarian Water Resources Management* (in Hungarian), Somlyódy, L. (ed.), Hungarian Academy of Science, Budapest, Hungary, pp. 139–176.

Zessner, M., Postolache, C., Clement, A., Kovacs, A. and Strauss, P. (2005). Considerations on the influence of extreme events on the phosphorus transport from river catchments to the sea. *Wat. Sci. Tech.*, **51**(11), 193–204.

Integrating principles of nitrogen dynamics in a method to estimate leachable nitrogen under agricultural systems

M. Burkart*, D. James*, M. Liebman** and E. van Ouwerkerk***

*National Soil Tilth Laboratory, 2150 Pammel Drive, Ames, IA 50011, USA (E-mail: *burkart@nstl.gov*; *james@nstl.gov*)

**3405 Agronomy Hall, Iowa State University, Ames, IA 50011, USA (E-mail: *mliebman@iastate.edu*)

***3112 NSRIC, Iowa State University, Ames, IA 50011, USA (E-mail: *evo@iastate.edu*)

Abstract Surplus nitrogen (N) in ground and surface water is of concern in intensive agricultural regions. Surplus N leaches during lengthy periods where annual crop systems are used in temperate regions. This paper presents a model to estimate the surplus N available for leaching to ground water beneath agricultural systems and applies the model to watersheds in an intensive maize and soybean production system. The model utilizes commonly available georeferenced data on soils, crops, and livestock, making it applicable to watersheds in many regions. The model links stocks of N in soil, crops, livestock, fertilizer and the atmosphere. Nitrogen flow centers on exchange between the soil N stocks. Nitrogen mineralization rates are defined for three soil organic matter pools, crop residue, and manure based on carbon:N ratios. Nitrogen exports from the system are harvested crops, livestock and losses to the atmosphere. Application of the model in 26 Iowa watersheds finds surpluses of 18 to 43 kg-N/ha. Surpluses exceeded measured annual nitrate-N loads in regional streams by amounts equivalent to denitrification rates in groundwater. Deficits in soil N were sufficiently small to suggest that the system is in equilibrium with soils of the region.
Keywords Agriculture; leaching; nitrate; soil

Introduction

Agricultural systems rich in soluble nitrogen (N) have been shown to contaminate groundwater (Nolan *et al.*, 1997; Burkart and Stoner, 2002) and streams (Schilling and Libra, 2000; Howarth *et al.*, 2002), and contribute to hypoxic zones in coastal waters (Turner and Rabalais, 1994). Increased use of N fertilizer in the United States was cited as the prime contributor to coastal eutrophication (National Resource Council, 2000). Solutions to N contamination of aquatic systems by agriculture require a comprehensive representation of N dynamics (Goolsby *et al.*, 1999; Burkart and James, 1999). This paper presents an agricultural-N budget applied to representative watersheds in Iowa, a state that is among the most intensively cropped in the United States.

Methods and data sources

Four GIS raster databases were used to calculate spatially explicit N budgets at a 30-m resolution for analysis at cell and watershed scales. Crop data for 2001 in Iowa were obtained from the National Agricultural Statistics Service (USDA, 2002). Soils and crop-yield data were from the Iowa Soil Properties and Interpretations Database (ISPAID, ISU, 2004). Streams and watersheds were derived from the National Elevation Dataset (USGS, 1999) using methods devised by Tarboton (2002, http://moose.cee.usu.edu/taudem/taudem.html). Inorganic fertilizer use data for 1997, the most recent Census of Agriculture year, were from the US Geological Survey (Barbara C. Ruddy, written communication, 2001). Livestock-production data were from the Census of Agriculture (US Department of

Commerce, 2000). Climatic data were from the Iowa Environmental Mesonet (http://mesonet.agron.iastate.edu/agclimate/index.php).

Nitrogen budget

The N budget links soil, crop, livestock, fertilizer and atmospheric stocks to estimate surplus N (Figure 1). External flows include fertilizer and atmospheric exchanges such as volatilization, deposition, denitrification, and fixation. The model for flow among stocks is centered on exchange with soil inorganic N resulting in estimates of accumulated surplus or deficit in soil inorganic N.

Inorganic soil nitrogen

Inorganic soil N is soluble and is dominantly nitrate (NO_3) and ammonium (NH_4). NH_4 is readily nitrified to NO_3 under aerobic conditions; so inorganic N not used by crops is leachable as NO_3. Surplus N remains after removal by crop uptake, volatilization and denitrification. Inputs include inorganic fertilizer, atmospheric deposition of inorganic N and mineralized fractions of soil organic matter (SOM), crop residue and manure. Manure and residue N that is not mineralized is immobilized as SOM.

Inorganic fertilizer sources include ammonia, which dominates, and applications of urea, ammonium nitrate, and ammonium sulfate. Adjustments were made to N-fertilizer applications for gaseous losses during application, primarily as NH_4 (Table 1, modified from Meisinger and Randall, 1991). The maize application rate (average: 162 kg/ha) was calculated by subtracting N applied to soybean (5 kg/ha) and oat (45 kg/ha) from each county total and dividing by the area in maize. Application of N to soybean is a by-product of diammonium or monoammonium phosphate.

Redeposition of locally derived atmospheric ammonia and ammonium (NHx) represents the return of local atmospheric emissions to the soil. Local emissions include volatilized manure, inorganic fertilizer, mineralized soil organic matter, and emissions during crop senescence, all of which are described below. Measures of redeposition are not available in this region, but a review of NHx dynamics and movement (Ferm, 1998) showed 50% of the annual volatilized NH_4 was deposited within 50 km of the source. Farther deposition was halved every 400 km. Redeposition of locally derived NHx was simplified to 75% of the local emissions assuming emissions were similar within a 400 km radius.

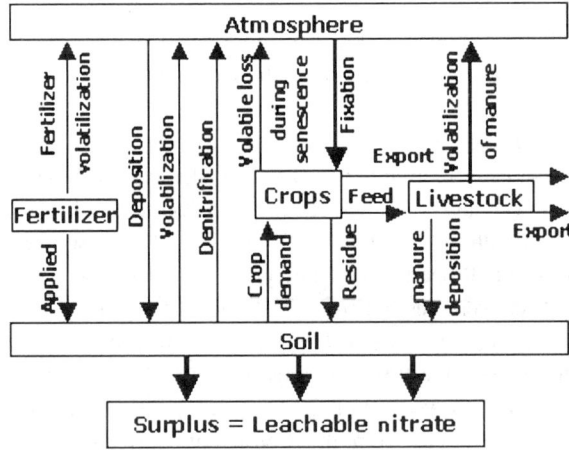

Figure 1 Stocks and flows in N budget

Table 1 N-fertilizer lost on application

Fertilizer type	Soil pH ≥ 7.0			Soil pH < 7.0		
	Urea & N solutions	Ammonium Nitrate	Anhydrous ammonia	Urea & N solutions	Ammonium Nitrate	Anhydrous ammonia
N lost (%)	7.5	7.5	1	2	0	0

Mineralization of SOM was estimated by applying mineralization rates (Table 2) to soil and climatic data. Mineralization is a biological transformation of organic N into inorganic forms. Annually, 5% of mineralized SOM is lost to the atmosphere (Parton et al., 1987). SOM constitutes the principal stock in which N is stored and exchanged. Consequently, SOM is a major source of mobile or inorganic N. SOM is partitioned into three pools (Jenkinson and Raynor, 1977; Van Veen and Paul, 1981). The distribution of N among these pools was calculated from carbon content and C:N ratios in each pool (Parton et al., 1987) similar to those incorporated in the CENTURY model (Parton et al., 1994). Each pool has a residence-time domain and mineralization rate constant determined by the microbial availability of constituent organic compounds (Table 2). The SOM in the upper 18 cm of soils was used to estimate the mass in each pool using the fractions shown in Table 2.

Mineralization varies with temperature and moisture that regulate micro-organism activity. Mineralization rates were modified using the concepts of Jenkinson (1990) and Van Veen and Paul (1981) (Eq. 1).

$$M_{om} = (1 - e^{-kt}) * SOM * T * W, \qquad (1)$$

where: M_{om} = mass of mineralized N (kg/ha/yr); k = mineralization rate constant (1/yr); t = time (yr); SOM = soil organic matter (kg/ha); T = temperature stress factor; and W = water stress factor. Mineralization rates double with each 10 °C increase in soil temperature to a maximum near 30 °C (Jenkinson, 1990; Van Veen and Paul, 1981). Temperature was accounted for using air temperature as a proxy for soil temperatures (Eq. 2).

$$T = 47.9/(1 + e^{106/(MAT+18.3)})(MAT) \qquad (2)$$

where MAT = mean annual temp (°C). Nitrogen mineralization is maximum when soil moisture is near field capacity (Jenkinson, 1990) and W is 1.0. Prolonged saturation results in anoxia, limiting mineralization. Moisture below the wilting point reduces microbial activity and mineralization rates. Maximum mineralization rates operate under soil-moisture deficits producing water tension as small as -100 kPa (Jenkinson, 1990; Linn and Doran, 1984; Van Veen and Paul, 1981). Midwest conditions approximate field capacity during the non-growing season and average growing-season soil-water tension is ≥ -100 kPa except during rare, very dry years (Baker et al., 1979). Soil water content was assumed to support maximum mineralization rates and the water stress factor (W) was set to 1.0.

Table 2 Carbon and nitrogen mass and fractions in soil organic matter

Soil organic matter (SOM) pool	Carbon (Mg/ha)[1]	C:N[2]	N (kg/ha)	Fraction of total soil N (%)	Residence time (yr)[2]	Mineralization rate constant (1/yr)
Active (ASOM)	0.28	8	35	1.6	1.5	67
Slow (SSOM)	11.3	11	1,027	47.3	25	4
Passive (PSOM)	12.2	11	1,109	51.1	1,000	0.001

[1] Jenkinson and Rayner, 1977; [2] Parton et al., 1987

Mineralization of crop residue was estimated by partitioning residues (Table 3) into the active soil organic matter (ASOM) and slow soil organic matter (SSOM). Residue mineralization rates vary by crop type and are related to C:N ratios. Crop residues with large C:N ratios mineralize more slowly and those with small C:N ratios more rapidly (Parr and Papendick, 1978). Mineralization rates for legume residues are consistently larger than those for other crops. Soybean residue mineralizes at rates approximately 1.5 times that of non-legumes (Buyanovsky and Wagner, 1997). An approximation of this difference allocates 100% of legume residue to the ASOM, 70% of non-legume residue to ASOM and 30% to SSOM. Mineralized N is calculated using Eq. 1. The remaining organic N from residues is immobilized in ASOM or SSOM.

Mineralized manure-N was calculated using county livestock inventory data (US Department of Commerce, 2000) and estimates of the N content of manure by livestock class (Lander et al., 1998). Values for excreted N were adjusted for volatilization of NH_x and amine during storage (Lander et al., 1998; Midwest Planning Service, 1985) and application (Schepers and Mosier, 1991). Fractions of net manure-N applied were mineralized (Table 4) or immobilized in the SSOM (Schepers and Mosier, 1991). Adjusted values were distributed uniformly to all 30-m cells where maize was grown.

Crop nitrogen

The stock of N associated with crops requires inputs from fixation and soil inorganic N to balance removal as harvest, crop residue, and gases lost during senescence. Inorganic soil N is derived from inorganic fertilizer, atmospheric deposition, and mineralized fractions of soil organic matter, crop residue, and manure.

Symbiotic fixation converts atmospheric N_2 into plant tissue N through bacteria in legume root nodules. The estimate of annual N fixed by alfalfa was 52% of the N yield (Heichel et al., 1984). The fixation rate used for soybean plant mass was 56% (Patterson and LaRue, 1983). The fixation rate used for pasture cells and legume hay was 70%, an average of values reported by Heichel and Henjum (1991). The ratio of annually fixed N released directly to soil through alfalfa root turnover to that found in the plant mass is 0.47 (Goins and Russelle, 1996 and Heichel et al., 1984). Consequently, fixed N for alfalfa was increased by 47%. Soybean was found to release more than 10% of the total fixed N directly to soil from roots (Brophy and Heichel, 1989) requiring increase in total soybean fixation by this fraction. These rates were applied to all crop components including harvest (Table 5), residue (Table 3) and loss during senescence.

Crop-harvest N contents (Table 5) were applied to yields (ISU, 2004) in each 30-m cell (USDA, 2002). The net export of crop N from a watershed was calculated by subtracting N required for livestock feed from the harvested N. Separating livestock N requirements and net exported harvest-N allows comparison of exported crop and livestock N, an important distinction for evaluating integrated systems.

Crop residue constitutes a mass of N that can be much larger than harvested N (Table 3) for crops common to the region. Only below ground silage maize and forage sorghum biomass was included in estimates, assuming the above ground biomass was harvested. Similarly, only alfalfa roots and crowns were included in residue, on one-third of the area to simulate the three-year rotation commonly used in the region. Estimates of total residue N were based on an above- to below-ground N-content ratio of 1.0, except maize and small grains with a ratio of 1.1 and soybean with a ratio of 1.2 (Buyanovsky and Wagner, 1986). Permanent hay or pasture was assumed to have no residue until it is killed.

Volatile loss of N during crop senescence is the process by which N flows to the atmosphere upon flowering until the crop matures or is harvested. A conservative rate of 25 kg N/ha is used in this analysis for loss during senescence from maize, soybean,

Table 3 Nitrogen in crop residues

Crop	Maize[1]	Maize silage[1]	Soybean[1]	Grain sorghum[2]	Forage sorghum[2]	Alfalfa[3]	Oat[4]	Wheat[4]	Barley[4]	Alfalfa/grass[5]
N (kg/ha)	92	46	75	53	26	37	31	31	52	14.8

[1]Buyanovsky and Wagner (1997); [2]Power and Legg (1978); [3]Shaeffer et al. (1991); [4]Narasimhalu et al. (1998); [5]Duru et al. (1997)

Table 4 Manure nitrogen by animal type

Animal class	Beef cows	Milk cows	Heifers	Other bovines	Breeding hogs	Other hogs	Layers*	Broilers*	Turkeys*	Sheep
Excreted (kg yr^{-1})	57	100	18	32	14	8	56	40	168	7
Applied (kg yr^{-1})	17	27	4	7	3	2	37	24	87	1
Mineralized (%)	30	40	40	30	90	90	90	75	75	75
Immobilized (%)	70	60	60	70	10	10	10	25	25	25

*values for 100 birds

Table 5 Nitrogen in crop harvest (Meisinger and Randall (1991))

Crop	Maize grain	Maize silage	Soybean	Wheat grain	Barley	Oat grain	Small grain	Alfalfa	Tame hay	Wild hay
N (kg/t)	13.2	3.6	55.6	18.5	18.4	19.3	6.0	25.4	15.4	11.0

and wheat. This rate was provided by a consensus of Iowa State University staff, but also falls in the ranges cited by Wetselaar and Farquhar (1980) and Francis *et al.* (1993). An annual loss of 2 kg N/ha was used for alfalfa (Dabney and Bouldin, 1985). All other crops were assumed to lose N through this process at a rate of 22 kg/ha (Schepers and Mosier, 1991) except pasture which was directly consumed before senescence (Sutton *et al.*, 2001).

Livestock nitrogen

The flow of N from crops into livestock for each county was set equal to the sum of N exported in the animal and excreted N. Calculating the N from crop harvest needed to sustain local livestock provides the basis for distinguishing N exported as livestock from that exported directly as crops.

Livestock export of N from watersheds was calculated from market-weight livestock-N content (Table 6). Data on the number of animals sold and milk produced were from the Census of Agriculture (US Department of Commerce, 2000). Animal production was distributed proportionally among all cropland in each watershed.

Atmospheric nitrogen

The atmospheric N stock is accounted for in two pools. One pool accumulates NH_x from volatilized fractions of SOM, inorganic fertilizer, and manure, as well as loss during crop senescence. The other pool represents elemental N that flows to plants through fixation or from denitrification as N_2 and NO_x.

Soil denitrification results from microbial reduction of NO_3 to N_2 and NO_x. Low oxygen levels occur under conditions of high soil water content. Consequently, poorly drained soils provide greater potential for denitrification than well-drained soils. Rates based on soil drainage classifications (Meisinger and Randall, 1991) were applied to the sum of NO_3 derived from fertilizer, redeposition of atmospheric N, and mineralization of SOM, crop residue, and manure after subtracting N flows to crops. Differential rates (Table 7) were applied using soil drainage classes and SOM content.

Immobilized nitrogen

Immobilization of N from crop residues and manure is accounted for in addition to the SOM stock. Immobilized residue-N is the organic N from all crop residues that is not mineralized. This includes 33% of legume and 23% of non-legume residue-N that flows to the ASOM pool (Table 2) and 29% of non-legume residue-N added to the SSOM pool. A fraction of manure-N is immobilized as SSOM at rates specific to each livestock type (Table 4) with a range of 10% for hogs and layers to 70% for cattle.

Results

The primary variable of interest is the surplus or potentially leachable N. Surplus N was the inorganic soil-N stock not accounted for in crop uptake or losses to the atmosphere

Table 6 Nitrogen in livestock

Livestock type	Swine	Feeder pig	Broiler	Pullet	Layer	Turkey	Cattle & cows
Live weight (kg)	113	23.6	2.27	1.36	1.81	10.4	544
N (% live weight)	2.14 [1]	2.14 [1]	2.53 [2,3,4]	3.3 [4,5]	3.3 [6]	3.3 [6]	2.31 [7]
N mass (kg/animal)	2.41	2.41	0.06	0.05	0.06	0.34	12.57

[1] Mahan and Shields, 1998; [2] Stilborn *et al.*, 1994; [3] Szakall *et al.*, 1998; [4] Scott *et al.*, 1982; [5] Oju *et al.*, 1988; [6] Middleton and Ferket, 2001; [7] Fox and Black, 1984

Table 7 Soil denitrification rates

SOM Content (%)	Soil drainage classification				
	Excessively, somewhat excessively	Well	Moderately well	Somewhat poorly	Poorly, very poorly
		Inorganic N denitrified (%)			
<2	3	6	9	13	20
2–5	6	12	13	17	30
>5	8	13	20	25	40

from this stock as expressed in Eq. 3.

$$E = F + M_s + M_r + M_m + A - C - V - S - D, \qquad (3)$$

where E = surplus inorganic soil N; F = fertilizer application; M_s = mineralized SOM; M_r = mineralized crop residue; M_m = mineralized manure; A = atmospheric redeposition; C = crop uptake; V = volatilized soil inorganic N; S = volatile N loss during crop senescence; and D = denitrified N.

Another variable of interest is the annual change in soil organic N (SON) after accounting for mineralization of SOM and immobilization of residues and manure. This variable shows where depletion or accumulation of SON may occur and was calculated using Eq. 4.

$$\Delta S_n = I_r + I_m - M_{asom} - M_{ssom} - M_{psom}, \qquad (4)$$

where ΔS_n = change in soil N; I_r = immobilized residue; I_m = immobilized manure; M_{asom} = mineralized ASOM; M_{ssom} = mineralized SSOM; and M_{ssom} = mineralized PSOM.

The distribution of surplus N among watersheds in the region ranges from 18 to almost 43 kg/ha (Figure 2A). The largest surpluses were found in headwater watersheds

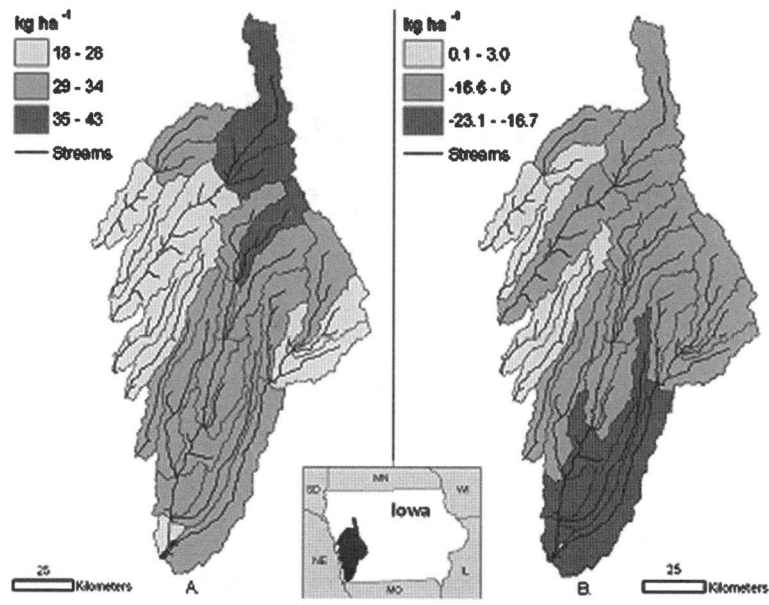

Figure 2 Nitrogen budget results in watersheds of western Iowa. A. Surplus nitrogen. B. Net change in soil organic nitrogen

of most major streams. The narrow range of surplus N reflects the dominance of the maize and soybean rotation on 71% of land in the region. The watersheds with the largest surpluses include broad drainage divides where gentler slopes accommodate the maize–soybean rotation and smaller areas in pasture on steep slopes.

Surplus N was larger than measured annual nitrate-N loads in streams of the region (Figure 3). This difference can readily be explained by denitrification in groundwater (Tomer and Burkart, 2003; Cambardella *et al.*, 1999).

The distribution of budget-components (Figure 3) in watersheds shows that fertilizer is the largest source of inorganic N. Mineralized soil, fixation, and atmospheric redeposition provide similar quantities of inorganic N. Manure contributions were not substantial and were applied only to maize cells. Livestock harvest constitutes the largest removal of inorganic N, although crop harvest removes only a slightly smaller amount. When examined at the 30-m cell scale, there were substantial differences in surpluses between the two dominant crops (Figure 4). The large ranges of surplus N reflect the variability in mineralized SOM. Soybean, however, consistently produces N deficits approximating the surpluses found with maize. Inorganic fertilizer applied to maize contributes substantially to the surplus in maize cells.

Soil organic N (ΔS_n, Eq. 4) among watersheds in the region (Figure 2B) is being accumulated in four watersheds, but depleted in most at rates up to 25 kg/ha. Larger deficits occur in watersheds where larger fractions of the drainage areas have high SOM and smaller slopes. The distribution of net change in SON (Figure 5) shows a clear distinction between maize and soybean. More maize residue is immobilized than soybean and

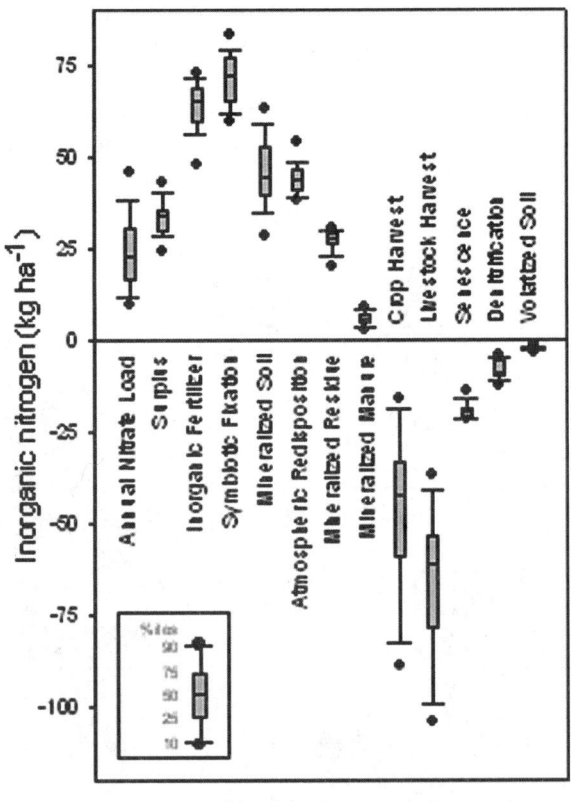

Figure 3 Distribution of inorganic nitrogen among budget components in watersheds

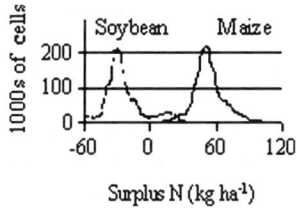

Figure 4 Surplus nitrogen by cell for maize and soybean

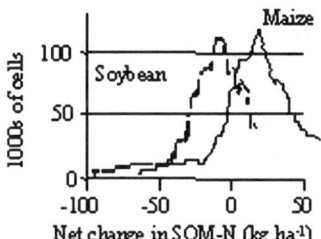

Figure 5 Net change in SOM-N by cell for maize and soybean

manure was applied only to maize, explaining much of the accumulation of SON. With minor exceptions, all soybean cells had depleted soil N. Some of this loss is balanced during the following year under maize. Variability among the cells within crop type reflects variable SOM; thus variability in mineralized SON and denitrification. The effect of these two crops on SON is extremely small values when compared to 75 to 300 Mg/ha of SON in the region. The similarity of the two curves may be an artifact of universal N-fertilizer values for each crop and soybean fixation.

Conclusions

The N budget model presented here can be applied to many intensive agricultural regions. The minimum data required includes soil characteristics and crop and livestock distribution. Climatic data and soil moisture will be required in regions where soil moisture is frequently < -100 kPa or where saturated surface soils are common during the growing season. Better spatial resolution of livestock inventories than that used here will improve the model results.

This N budget was applied to watersheds draining western Iowa, where more than 71% of the land is occupied by maize or soybean. All watersheds yielded surplus or leachable N ranging from 18 to 43 kg N/ha, with variability attributable to mineralized SON and the fraction of land occupied by maize and soybean. The change of SON in all watersheds was extremely small relative to the mass of SON, indicating an apparent equilibrium between the agricultural system and soils. Surplus values exceeded measured annual nitrate-N loads in regional streams by amounts equivalent to denitrification rates. Variability in annual SON changes reflects differences in mineralized SOM transfers to the inorganic N pool that also generate larger denitrification rates.

References

Baker, D.G., Nelson, W.W. and Kuehnast, E.L. (1979). *Climate of Minnesota: Part XII – The Hydrologic Cycle and Soil Water.* Tech. Bull. 322. Ag. Exper. Stat., Univ. Minnesota, St. Paul. 24 p.

Brophy, L.S. and Heichel, G.H. (1989). Nitrogen release from roots of alfalfa and soybean grown in sand culture. *Plant and Soil*, **116**, 77–84.

Burkart, M.R. and James, D.E. (1999). Agricultural-nitrogen contributions to hypoxia in the Gulf of Mexico. *Jour. Environ. Qual.*, **28**(3), 850–859.

Burkart, M.R. and Stoner, J.D. (2002). Nitrate in aquifers beneath agricultural systems. *Wat. Sci. Tech.*, **45**(9), 19–29.

Buyanovsky, G.A. and Wagner, G.H. (1986). Post-harvest residue input to cropland. *Plant and Soil*, **93**, 57–65.

Buyanovsky, G.A. and Wagner, G.H. (1997). Crop residue input to soil organic matter on Sanborn Field. In *Soil Organic Matter in Temperate Agroecosystems: Long-Term Experiments in North America*, Paustian, P. and Elliott, C. (eds), CRC Press, New York, pp. 73–83.

Cambardella, C.A., Moorman, T.B., Jaynes, D.B., Hatfield, J.L., Parkin, T.B., Simpkins, W.W. and Karlen, D.L. (1999). Water quality in Walnut Creek Watershed: Nitrate-nitrogen in soils, subsurface drainage water, and shallow groundwater. *J. Environ. Qual.*, **28**, 25–34.

Dabney, S.M. and Bouldin, D.R. (1985). Fluxes of ammonia over an alfalfa field. *Agron. J.*, **77**, 572–578.

Duru, M., Lemaire, G. and Cruze, P. (1997). Grasslands. In *Diagnosis of the Nitrogen Status in Crops*, Lemaire, G. (ed.), Springer, Berlin, pp. 59–72.

Fox, D.G. and Black, J.R. (1984). A system for predicting body composition and performance of growing cattle. *J. Anim. Sci.*, **58**(3), 725–739.

Ferm, M. (1998). Atmospheric ammonia and ammonium transport in Europe and critical loads: a review. *Nutrient Cycling in Agroecosystems*, **51**, 5–17.

Francis, D.D., Schepers, J.S. and Vigil, M.F. (1993). Post-anthesis nitrogen loss from corn. *Agron. Jour.*, **85**, 659–663.

Goins, G.D. and Russelle, M.P. (1996). Fine root demography in alfalfa (*Medicago sativa* L.). *Plant and Soil*, **185**, 281–291.

Goolsby, D.A., Battaglin, W.A., Lawrence, G.B., Artz, R.S., Aulenbach, B.T., Hooper, R.P., Keeney, D.R. and Stensland, G.J. (1999). Flux and Sources of Nutrients in the Mississippi-Atchafalaya River Basin, Topic 3 Report. In *Hypoxia in the Gulf of Mexico*, White House Office of Sci. and Tech. Policy Comm. on Environ. and Nat. Res. http://www.nos.noaa.gov/products/pubs_hypox.html.

Heichel, G.H. and Henjum, K.I. (1991). Dinitrogen fixation, nitrogen transfer, and productivity of forage legume-grass communities. *Crop Sci.*, **31**, 202–208.

Heichel, G.H., Barnes, G.J., Vance, C.P. and Henjum, K.I. (1984). N_2 Fixation, and N and dry matter partitioning during a 4-year alfalfa stand. *Crop Sci.*, **24**, 811–815.

Howarth, R.W., Sharpley, A. and Walker, D. (2002). Sources of nutrient pollution to coastal waters in the United States: Implications for achieving coastal water quality goals. *Estuaries*, **25**(4B), 656–676.

ISU (2004). Iowa Soil Properties And Interpretations Database (ISPAID). Iowa State University Extension, March 2004. http://extension.agron.iastate.edu/soils/PDFs/ispaid.pdf

Jenkinson, D.S. (1990). The turnover of organic carbon and nitrogen. *Phil. Trans. R. Soc. Lond.*, **329**, 361–368.

Jenkinson, D.S. and Raynor, J.H. (1977). The turnover of soil organic matter in some of the Rothamsted classical experiments. *Soil Science*, **123**, 298–305.

Linn, D.M. and Doran, J.W. (1984). Effect of water-filled pore space on carbon dioxide and nitrous oxide production in tilled and nontilled soils. *Soil Sci. Soc. Am. J.*, **48**, 1267–1272.

Lander, C.H., Moffitt, D. and Alt, K. (1998). *Nutrients available from livestock manure relative to crop growth requirements*. US Department of Agriculture, NRCS, Resource Assessment and Strategic Planning Working Paper 98–1. http://www.nhq.nrce.usda.gov/land/pubs/nlweb.html.

Mahan, D.C. and Shields, R.G. (1998). Macro- and micromineral composition of pigs from birth to 145 kilograms of body weight. *J. Anim. Sci.*, **76**, 506–512.

Meisinger, J.J. and Randall, G.W. (1991). Estimating nitrogen budgets for soil-crop systems. In *Managing Nitrogen for Groundwater Quality and Farm Profitability*. Follett, R.F., Keeney D.R. and Cruse R.M. (eds) Soil Sci. Soc. Amer. Madison, Wisconsin. 357 p.

Middleton, T.F. and Ferket, P.R. (2001). Effect of level of acidification by phosphatic acid, storage temperature, and length of storage on the chemical and biological stability of ground poultry mortality carcasses. *Poul. Sci.*, **80**, 1144–1153.

Midwest Planning Service-Livestock Waste Subcommittee (1985). *Livestock waste facilities handbook*. Midwest Planning Serv. Rep. MWPS-18 (2nd edition), Iowa State Univ., Ames.

Narasimhalu, P., Kong, D. and Choo, T.M. (1998). Straw yields and nutrients of seventy-five Canadian barley cultivars. *Canadian Journ. Animal Sci.*, **78**(1), 127–134.

National Research Council (2000). *Clean Coastal Waters: Understanding and Reducing the Effects of Nutrient Pollution.* National Academy Press, Washington, DC. http://stills.nap.edu/books/0309069483/html/.

Nolan, B.T., Ruddy, B.C., Hitt, K.J. and Helsel, D.R. (1997). Risk of nitrate in groundwaters of the United States – A national perspective. *Environ. Sci. Technol.*, **31**, 2229–2236.

Oju, E.M., Waibel, P.E. and Noll, S.L. (1998). Early protein undernutrition and subsequent realimentation in turkeys. 1. Effect of performance and body composition. *Poul. Sci.*, **67**, 1750–1759.

Parr, J.F. and Papendick, R.I. (1978). Factors affecting the decomposition of crop residues by microorganisms. In *Crop Residue Management*, ASA Spec. Pub. **31**, 101–130.

Parton, W.J., Schimel, D.S., Cole, C.V. and Ojima, D.S. (1987). Analysis of factors controlling organic matter levels in Great Plains grasslands. *Soil Sci. Soc. Am. Jour.*, **51**, 1173–1179.

Parton, W.J., Ojima, D.S., Cole, C.V. and Schimel, D.S. (1994). A general model for soil organic matter dynamics: sensitivity to litter chemistry, texture, and management. In *Quantitative Modeling of Soil Forming Processes*, SSSA Special Publication 39, Madison, WI, pp. 147–167.

Patterson, T.G. and LaRue, T.A. (1983). Nitrogen fixation by soybeans: seasonal and cultivar effects, and comparison of estimates. *Crop Science*, **23**, 488–492.

Power, J.F. and Legg, J.O. (1978). Effect of crop residues on the soil chemical environment and nutrient availability. In *Crop Residue Management Systems*. W.R. Oschwald (ed.). Special Publication No. 31. American Society of Agronomy. Madison, WI, pp. 85–100.

Schilling, K.E. and Libra, R.D. (2000). The relationship of nitrate concentrations in streams to row crop land use in Iowa. *Jour. Environ. Qual.*, **29**(6), 1846–1851.

Schepers, J.S. and Mosier, A.R. (1991). Accounting for nitrogen in nonequilibrium soil-crop systems. pp. 125–128. In *Managing Nitrogen for Groundwater Quality and Farm Profitability*. Follett, R.F., Keeney, D.R. and Cruse R.M. (eds). Soil Science Society of America, Madison, Wisconsin. 357 p.

Scott, M.L., Nesheim, M.C. and Young, R.J. (1982). *Nutrition of the chicken*, 3rd edition, M.L. Scott & Associates, Ithaca, NY, pp. 279.

Shaeffer, C.C., Russelle, M.P., Heichel, G.H., Hall, M.H. and Thicke, F.E. (1991). Nonharvested forage legumes: nitrogen and dry crop matter yields and effects on a subsequent corn crop. *J. Prod. Agric.*, **4**, 520–525.

Stilborn, H.L., Moran, E.T., Jr, Gous, R.M. and Harrison, M.D. (1994). Experimental data for evaluating broiler models. *J. Appl. Poul. Res.*, **3**(4), 379–390.

Sutton, M., Milford, C., Nemitz, E., Theobald, M., Hill, P., Fowler, D., Schjoerring, J., Mattson, M., Nielsen, K., Husted, S., Erisman, J., Otjes, R., Hensen, A., Mosquera, J., Cellier, P., Loubet, B., David, M., Genermont, S., Neftel, A., Blatter, A., Herrmann, B., Jones, S., Horvath, L., Fuhrer, E., Mantzanas, K., Koukoura, Z., Gallagher, M., Williams, P., Flynn, M. and Reido, M. (2001). Biosphere-atmosphere interactions of ammonia with grasslands: Experimental strategy and results from a new European initiative. *Plant and Soil*, **228**, 131–145.

Szakall, I., Fekete, S., Andrasofszky, E., Romvari, R. and Szita, G. (1998). Relationship of dietary fat and lysine level with body composition in broiler chickens. *Acta Vet. Hungarica*, **46**, 243–257.

Tomer, M.D. and Burkart, M.R. (2003). Long-term effects of nitrogen fertilizer use on ground water nitrate in two small watersheds. *J. Environ. Qual.*, **32**, 2158–2171.

Turner, R.E. and Rabalais, N.N. (1994). Coastal eutrophication near the Mississippi River delta. *Nature*, **368**, 619–621.

USDA (2002). *National Agriculture Statistics Service 1:100,000 2001 Cropland Data Layer for Iowa*, http://www.nass.usda.gov/research/Cropland/metadata/metadata_ia01.htm.

US Department of Commerce (2000). *1997 Census of Agriculture. Summary and Country Level Data.* US Dept. Comm., Bureau of the Census. Geographic Area Series 1b. Washington, DC. CD-ROM.

US Geological Survey (USGS) (1999). *National Elevation Dataset.* EROS Data Center, 1999.

Van Veen, J.A. and Paul, E.A. (1981). Organic carbon dynamics in grassland soils. 1. Background information and computer simulation. *Canadian J. Soil Sci.*, **61**, 185–201.

Wetselaar, R. and Farquhar, G.D. (1980). Nitrogen losses from tops of plants. *Adv. Agron.*, **33**, 263–302.

Indicator of risk of water contamination by phosphorus from Canadian agricultural land

E. van Bochove*, G. Thériault*, F. Dechmi*, A.N. Rousseau**, R. Quilbé**, M.-L. Leclerc* and N. Goussard**

*Agriculture and Agri-Food Canada, Soils and Crops Research and Development Centre, 2560 Hochelaga Boulevard, Sainte-Foy, Quebec, G1V 2J3, Canada (E-mail: *vanbochovee@agr.gc.ca; theriaultg@agr.gc.ca; dechmif@agr.gc.ca; leclercml@agr.gc.ca; goussardn@agr.gc.ca*)

**Institut national de la recherche scientifique, Eau, Terre et Environnement, 490 de la Couronne, Québec, G1K 9A9, Canada (E-mail: *alain_rousseau@ete.inrs.ca; Renaud_QUILBE@ete.inrs.ca*)

Abstract The indicator of risk of water contamination by phosphorus (IROWC_P) is designed to estimate where the risk of water P contamination by agriculture is high, and how this risk is changing over time based on the five-year period of data Census frequency. Firstly developed for the province of Quebec (2000), this paper presents an improved version of IROWC_P (intended to be released in 2008), which will be extended to all watersheds and Soil Landscape of Canada (SLC) polygons (scale 1:1, 000, 000) with more than 5% of agriculture. There are three objectives: (i) create a soil phosphorus saturation database for dominant and subdominant soil series of SLC polygons – the soil P saturation values are estimated by the ratio of soil test P to soil P sorption capacity; (ii) calculate an annual P balance considering crop residue P, manure P, and inorganic fertilizer P – agricultural and manure management practices will also be considered; and (iii) develop a transport-hydrology component including P transport estimation by runoff mechanisms (water balance factor, topographic index) and soil erosion, and the area connectivity to water (artificial drainage, soil macropores, and surface water bodies).
Keywords Agriculture; hydrology; indicator; phosphorus; soil; water contamination

Introduction

Nutrient management in agricultural production is of critical importance from the standpoint of environmental protection and agro-ecosystem sustainability. Excessive amounts of P in surface fresh water contribute to eutrophication of rivers and lakes. The contamination of water is of greatest concern in areas where soil tests show high P levels, ability of soils to retain P is low, susceptibility to runoff is high, macropore flow is high, soil erosion is high, and where connectivity to surface water and artificial drainage is dense.

Phosphorus, a major nutrient used in agricultural activities, is getting more and more attention in Canada. In the provinces of Quebec and Ontario, phosphorus is currently part of nutrient management plans developed for farming operations in order to reduce the risk of contamination of adjacent surface water bodies. Moreover, other provinces are considering the possibility to come up with similar management plans.

It is important to demonstrate to the Canadian public and to the Canadian economic partners that a national indicator of risk of water contamination by P is being developed on current scientific knowledge and that it could be efficiently used for governmental policies. The relevance to develop a national indicator of risk of water contamination by P for Canada is to identify critical areas across the country where more prospecting is required to protect surface water. Once identified, these areas would be investigated carefully at the operational management watershed scale. Stakeholders could then enforce management plans and beneficial management practices at the farm level in critical areas following the concept of strict regulations for voluntary adoption. Rather than assuming that inappropriate farm management is responsible for water quality problems, the underlying causes of the symptoms (regional and global economic pressures and constraints) must be also addressed (Sharpley et al., 2001).

An indicator of risk of water contamination by phosphorus (IROWC_P) was developed during the Agri-Environmental Indicator Project (McRae et al., 2000) at the Soil Landscapes of Canada (SLC) polygon scale on the basis of a phosphorus index (PI) developed by Lemunyon and Gilbert (1993) and USDA-NRCS (1994). The first version of IROWC_P (Bolinder et al., 2000) was applied only for Quebec because reliable site information was available from provincial soil surveys (Tabi et al., 1990), particularly the degree of soil P saturation (DSPS) and soil test P (STP). Although some soil surveys across the country included P data, no soil P-status databases were available for other Canadian provinces. Therefore, as a primary requirement to estimate risk of water contamination by P, a national DSPS database for agricultural soils will be created. The modified IROWC_P will also include a transport–hydrology component that will improve estimation of the risk of P transport to water by soil erosion, infiltration and surface runoff to water by accounting for the connectivity of agricultural land to drainage areas at a daily interval. The scope of this paper is to present the current state of development of a national indicator of risk of water contamination by P intended to cover all Canadian agricultural land.

Methodology

IROWC_P formulation and data sources

The national IROWC_P modified version will be built on the first version of IROWC_P (Bolinder et al., 2000) and adapted to Canadian soil and climate characteristics. The modified IROWC_P will be calculated at both SLC polygon and watershed levels using existing Census of Agriculture, farm environmental management surveys, hydrology, and climate databases.

Agriculture and Agri-Food Canada (AAFC) in collaboration with Statistics Canada – Agriculture Division have developed an "area-weighting" process for re-allocating Census of Agriculture data from census polygon based geographies to other "target" polygon-based geographies such as Soil Landscapes of Canada and Drainage Area (Watershed) spatial frameworks (Canadian Soil Information System, 2004). The SLCs are based on existing soil survey maps which have been recompiled at a 1:1 million scale. Each area (or polygon) on the map is described by a standard set of attributes. The full array of attributes that describes a distinct type of soil and its associated landscape, such as surface form, slope, water table depth, permafrost and lakes, is called a soil landscape. SLC polygons may contain one or more distinct soil landscape components and may also contain small but highly contrasting inclusion components. The location of these components within the polygon is not defined. SLCs were originally conceived as a standardized database consisting of major attributes relevant to plant growth, land management, and soil degradation. These data have since turned out to be a useful framework to support other databases.

The modified IROWC_P includes the soil P-status component (PS), the annual P-balance component (PB) and the P transport-hydrology component (PT_H). The different IROWC_P subcomponents will be weighted to estimate their relative importance for P transfer to water and rated by their corresponding P class values. Finally, the three component values will be combined following equation 1 to estimate the risk of water contamination by P. IROWC_P values will be associated to five vulnerability classes (very low, low, medium, high and very high) to obtain a corresponding magnitude of risk for each polygon.

IROWC_P = (PS + PB)PT_H (1)

Phosphorus status component

The degree of soil P saturation (DSPS) characterizes the actual P status of the soil and its long-term capacity to retain P. The DSPS is defined as the ratio of the soil-test P (STP) (Sims, 1993) to the inherent soil characteristic, the P sorption capacity (PSC) (Bache and Williams, 1971; Syers et al., 1973). In the first version of IROWC_P, the PSC was estimated by the Mehlich-3 extractable Al and consequently the DSPS was calculated as the ratio of Mehlich-3 extractable P to Mehlich-3 extractable Al (Beauchemin and Simard, 1999; Khiari et al., 2000). This modified version of IROWC_P will calculate a DSPS from two newly created databases: a STP and a PSC database, both at the SLC polygon and watershed levels.

Creation of the STP database. In Canada, the majority of soil and plant analyses are performed by private laboratories. Their STP databases are generally referenced on the basis of a producer, a producer association or a fertilizer dealer address level. Two current pathways of STP analysis exist between farms and soil laboratories (Figure 1). The development of the modified IROWC_P will need collaboration between the federal government and the agriculture industry through a third party to permit access to the existing STP databases. Therefore, an agreement between AAFC and a national fertilizer institute will be proposed to obtain a five-year period STP data aggregated at the SLC polygon and watershed levels.

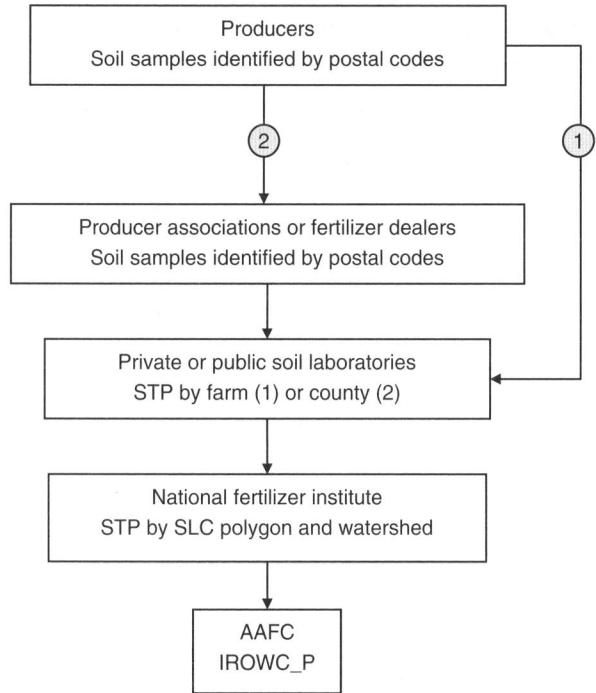

Figure 1 Flow chart of STP data aggregation at the SLC polygon and watershed levels; 1 and 2 are current pathways of soil sample and data transmission

Creation of a PSC database. To create the PSC database, a survey of the dominant soil series (A, B horizons) characterizing the agricultural SLC polygons in each province will be conducted. This survey is necessary as no agricultural soil PSC database is currently available in Canada. The soils will be analysed for their PSC and other related soil properties (pH, texture, OM, Ca, Fe and Al contents). Multivariate analysis will be performed (Leclerc *et al.*, 2001) to create soil groups presenting similar characteristics. These groups will then be used to classify each agricultural SLC polygon (dominant soil series) into a class of PSC (very low to very high). Thereafter, the first subdominant soil series of each polygon will be allocated to a PSC class using the related soil properties information available from existing soil surveys and SLC databases (Figure 2).

Phosphorus balance component

The modified IROWC_P will use the same P-balance component as in the first version of IROWC_P (Bolinder *et al.*, 2000). The current three subcomponents of P-balance are:

Mineral Fertilizer Phosphorus. This subcomponent is currently estimated from crop fertilizer P recommendation rates using information on the status of current STP levels. These estimations will be weighed against values derived from provincial summaries (total dollars spent on fertilizers and lime; the quantity of nutrient sold) and the following attribute of Census of Agriculture: "dollars spent on fertilizer and lime" at the SLC polygon level. As previously described, a STP value will be estimated for each agricultural SLC polygon and watershed. The Census of Agriculture will be the source of information for the area of production of the main crops.

Manure Phosphorus. This subcomponent is currently estimated from the Census of Agriculture database information. Data on the number of animals within different categories, manure production coefficients and manure P coefficients for each animal category will be used to calculate an estimation of the phosphorus coming from manure for a given SLC polygon. Several changes in management practices (animal feeding,

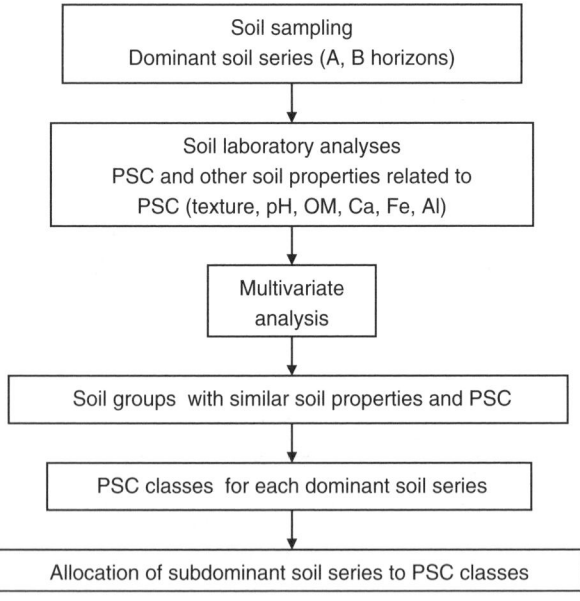

Figure 2 Different steps to allocate dominant and subdominant soil series of each SLC polygon to PSC classes

water utilization, etc.) have occurred in the past few years and have caused noteworthy changes in quantities of nutrients excreted and applied to agricultural soils. Consequently, the basic coefficients used in the calculations are under revision.

Crop residue Phosphorus. This subcomponent is currently estimated from the Census of Agriculture database and Provincial Census information: phosphorus uptake and phosphorus harvest coefficients. Only the major annual crops and hay categories for each province are considered. This subcomponent estimates values of exported phosphorus out of the SLC polygon or watershed at harvest, and also estimates values of crop residue phosphorus remaining on the agricultural soil after harvest.

Phosphorus transport–hydrology component

Under Canadian conditions, snowmelt runoff, rainfall runoff (Hortonian and saturation excess runoff), and subsurface flow represent the dominant hydrologic processes governing P transfer to surface water. The importance of each process is variable in time and space and depends on various factors such as drainage conditions, meteorological conditions, cropping and tillage practices, soil characteristics and topography (Figure 3).

In the revised IROWC_P, the risk of contamination is estimated on the basis of the transport–hydrology component *(PT_H)*, the particulate and dissolved sources of P *(PP and DP)* as well as other factors accounting for hydrological connectivity between sources and water bodies (Figure 4).

At the SLC polygon level, *PT_H* may be viewed as a function of erosion (*E*), surface runoff (*R*), infiltration (*I*), topography (*TI*), preferential flow (*PF*), tile drainage (*TD*), and surface drainage density (*SD*):

$$PT\_H = f(E, R, I, TI, PF, TD, SD) \tag{2}$$

Although we do not know a priori the form of the function introduced in (2), as a first approximation, we propose, on an annual basis, the following arrangement:

$$PT\_H = (SD)(E) + (TI + SD)(R) + (TD + PF)(I) \tag{3}$$

The risk of *PP* contamination increases with the product of *SD* and *E*; meanwhile, the risk of *DP* contamination increases with the product of the two water balance surplus factors

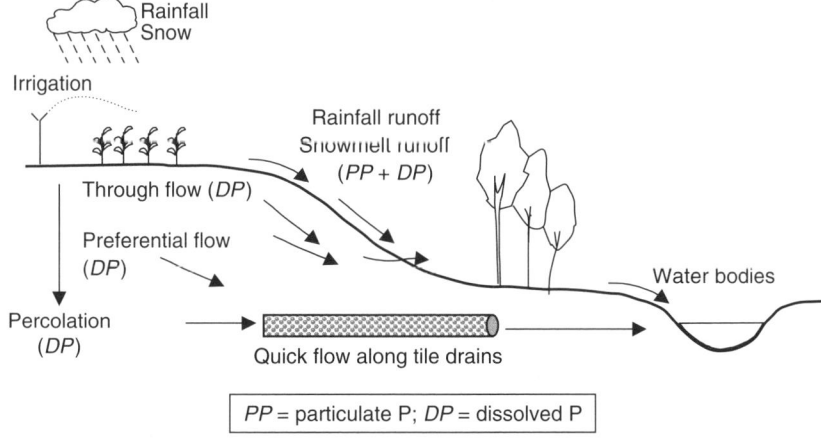

Figure 3 Phosphorus transport pathways

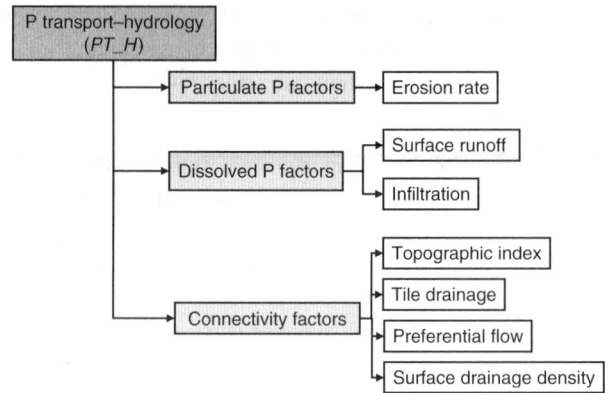

Figure 4 Phosphorus transport–hydrology factors

(R and I) and the hydrological connectivity factors governing surface runoff ($TI + SD$) and subsurface runoff ($TD + PF$) respectively without generating sediment loss.

Soil erosion rate, E, will be calculated using the Revised Universal Soil Loss Equation for Application in Canada (RUSLEFAC, 1997). This model predicts the amount of sediment loss (given in tons of soil loss per hectare and per year) in overland flow and accounts for winter and non-winter conditions in different Canadian regions (Prairie region, Eastern Canada and British Columbia). The methodology to use the RUSELFAC equation at the SLC polygon level is actually under investigation to be improved and included in the Risk of water erosion indicator (van Vliet, 2004).

A soil water balance approach is used to quantify excess soil water that runs off cropland as surface and/or subsurface flow using the Versatile Soil Moisture Budget model modified by Akinremi et al. (1996). This model uses the SCS-CN method (USDA-SCS, 1972) to partition R and I. The soil water balance calculation will be performed on a daily time step, thus providing upon summation an annual water balance. A seasonal partitioning of R will be considered to account separately for snowmelt runoff and rainfall runoff. Climatic variations between years will be considered as well.

The hydrological connectivity between P sources and water bodies is modelled using: (i) the topographic index, TI, of Kirby (1975) which accounts for the propensity of watershed area for saturation excess runoff; (ii) tile drainage, TD, and preferential flow, PF, because significant quantities of P could be exported from saturated-P soils intensively drained and through soil macropores (Sharpley et al., 1994; Hansen et al., 2002); and (iii) surface drainage density, SD, which accounts for ditches, brooks, rivers, ponds, lakes and wetlands. All these factors will be weighted with research and development work as well as already available national databases.

Discussion

The feasibility to create a national STP database depends on the data availability from the soil laboratories and the establishment of regional rating value tables. Canada has a variety of acidic and calcareous agricultural soils. Therefore, different STP methods of analysis are adapted to intrinsic characteristics: Mehlich 3, Olsen, Kelowna. Accordingly, the STP rating values will be discussed within Provincial P-expert commitees and adapted to the regional conditions and function of the crop P requirements.

In the calculation process of mineral fertilizer P inputs, the only nationwide available data giving some indirect information on the quantity of fertilizers applied on agricultural soils is the Census of Agriculture attribute "dollars spent on fertilizer and lime". Using this data in the ratio method with data coming from provincial summaries may lead to serious deficiencies. For example, a distortion by the nutrient costs encountered by specialty production can lead to an unreasonably large quantity of fertilizers used in more common productions. Also, the same proportion of nutrients is being allocated to the polygon regardless of the crop mix. As previously mentioned, several changes in management practices (animal feeding, water utilization, application techniques, etc.) have occurred in the past few years and caused noteworthy changes in quantities of nutrients excreted and applied to agricultural soils. The basic coefficients are dynamic and one should make sure that they reflect the reality as much as possible.

The development of the P transport–hydrology component should improve estimation of risk of water contamination by accounting for the main P transport processes, climate, and the connectivity of agricultural land to water bodies. Also its formulation defines separately erosion, runoff and infiltration, which control the P transport, in order to weigh their relative impacts as a function of the different agro-climatic regions. Furthermore, these transport processes are modulated by relevant factors that define the hydrological connectivity to surface and ground waters. The transport–hydrology component is also developed with the objective to estimate the risk of water contamination by other contaminants from agricultural activities such as nitrogen, pesticides and pathogens. The four indicators of risk of water contamination (IROWCs) are simultaneously developed under the National Agri-Environmental Health Analysis and Reporting Program (NAHARP) with the objective to use the same water budget model, the same national hydrology databases and the same soil erosion model.

Conclusions

The first version of IROWC_P for Quebec will be updated with 2001 Census data and SLC v3.0 by the end of 2004. The modified IROWC_P will be completed in 2008 and will be validated at the watershed level with local or regional water quality data in some catchment areas across Canada.

Acknowledgements

We thank our colleagues of IROWC-N, C. Drury, R. De Jong, B.J. Yang, and T. Huffman, IROWC-pathogens, E. Topp, and IROWC-pesticides, A.J. Cessna, A. Farenhorst, and R. McQueen for their comments and collaboration on the development of the transport–hydrology component. We acknowledge the staff from the National Agri-Environmental Health Analysis and Reporting Program of Agriculture and Agri-Food Canada: A. Lefèbvre and W. Eilers.

References

Akinremi, O.O., McGinn, S.M. and Barr, A.G. (1996). Simulation of soil moisture and other components of the hydrological cycle using a water budget approach. *Can. J. Soil. Sci.*, **75**, 133–142.
Bache, B.W. and Williams, E.G. (1971). A phosphate sorption index for soils. *J. Soil Sc.*, **22**, 289–301.
Beauchemin, S. and Simard, R.R. (1999). Soil phosphorus saturation degree: review of some indices and their suitability for P management in Quebec, Canada. *Can. J. Soil Sci.*, **79**, 615–625.
Bolinder, M.A., Simard, R.R., Beauchemin, S. and Macdonald, K.B. (2000). Indicator of risk of water contamination by P for soil landscape of Canada polygons. *Can. J. Soil Sci.*, **80**, 153–163.
Canadian Soil Information System (2004). http://sis.agr.ca/cansis/.
Hansen, N.C., Daniel, T.C., Sharpley, A.N. and Lemunyon, J.L. (2002). The fate and transport of phosphorus in agricultural systems. *Journal of Soil and Water Conservation*, **57**, 408–416.

Khiari, L., Parent, L.E., Pellerin, A., Alimi, A.R.A., Tremblay, C., Simard, R.R. and Fortin, J. (2000). An Agri-environmental phosphorus saturation index for acid coarse-textured soils. *J. Environ. Qual.*, **29**, 1561–1567.

Kirby, M.J. (1975). Hydrograph modelling strategies. In *Processes in Physical and Human Geography*, Peel, R., Chisholm, M. and Haggett, P. (eds), Heinemann, London, pp. 69–90.

Leclerc, M.L., Nolin, M.C., Cluis, D. and Simard, R.R. (2001). Grouping soils of the Montreal Lowlands (Quebec) according to fertility and P sorption and desorption characteristics. *Can. J. Soil Sci.*, **81**, 71–83.

Lemunyon, J.L. and Gilbert, R.G. (1993). The concept and need for a phosphorus assessment tool. *Journal of Productive Agriculture*, **6**, 483–486.

McRae, T., Smith, C.A.S., and Gregorich, L.J. (2000). Environmental Sustainability of Canadian Agriculture: Report of the Agri-Environmental Indicator Project. Agriculture and Agri-Food Canada, Ottawa, Ontario. (internet: http://www.agr.gc.ca/policy/environment/pdfs/aei/summary.pdf).

Revised Universal Soil Loss Equation For Application in Canada (RUSLEFAC) (1997). *A Handbook for Estimating Soil Loss from Water Erosion in Canada,* Draft copy, Wall, G.J., Coote, D.R., Pringle, E.A. and Shelton, I.J. (eds), Agriculture and Agri-Food Canada, Research Branch, Centre for Land and Biological Resources Ottawa, Ontario, pp. 53.

Sharpley, A.N., Chapra, S.C., Wedepohl, R., Sims, J.T., Daniel, T.C. and Reddy, K.R. (1994). Managing agricultural phosphorus for protection of surface water: Issues and options. *J. Environ. Qual.*, **23**, 437–451.

Sharpley, A.N., Kleinman, P. and McDowell, R. (2001). Innovative management of agricultural phosphorus to protect soil and water resources. *Commun. Soil Sci. Plant Anal.*, **32**, 1071–1100.

Sims, J.T. (1993). Environmental soil testing for phosphorus. *J. Prod. Agric.*, **6**, 501–507.

Syers, J.K., Browman, M.G., Smillie, G.W. and Corey, R.B. (1973). Phosphate sorption by soils evaluated by Langmuir adsorption equation. *Soil Sci. Soc. Am. Proc.*, **37**, 358–363.

Tabi, M., Tardif, L., Carrier, D., Laflamme, G. and Rompré, M. (1990). Inventaire des problèmes de dégradation des sols agricoles du Québec. Rapport synthèse. MAPAQ and Agriculture Canada, pp. 71.

United States Department of Agriculture – Natural Resources Conservation Service (1994). A *phosphorus assessment tool* Technical note # 1901, National Resources Conservation Service, US Department of Agriculture, Washington DC.

USDA-SCS (1972). Hydrology, section 4, Chapter 10, *Soil Conservation Service National Engineering Handbook*, Washington, DC., pp. 105–106.

van Vliet, L.J.P. (2004). Risk of water erosion indicator. Summary report, Annex C, Agriculture and Agri-Food Canada. In *National Agri-Environmental Health Analysis and Reporting Program 2004 Technical Workshop*, January 15–16, Ottawa, Ontario.

Estimation of particulate nutrient load using turbidity meter

K. Yamamoto* and T. Suetsugi**

*Ariake Sea Research Project, Saga University, Honjo 1, Saga-city, Saga 840-8502, Japan
(E-mail: *ko1yama@attglobal.net*)
**National Institute for Land and Infrastructure Management, 1 Asahi, Tsukuba, Ibaraki 305-0804, Japan
(E-mail: *suetsugi-t92fv@nilim.go.jp*)

Abstract The "Nutrient Load Hysteresis Coefficient" was proposed to evaluate the hysteresis of the nutrient loads to flow rate quantitatively. This could classify the runoff patterns of nutrient load into 15 patterns. Linear relationships between the turbidity and the concentrations of particulate nutrients were observed. It was clarified that the linearity was caused by the influence of the particle size on turbidity output and accumulation of nutrients on smaller particles (diameter < 23 μm). The *L-Q-Turb* method, which is a new method for the estimation of runoff loads of nutrients using a regression curve between the turbidity and the concentrations of particulate nutrients, was developed. This method could raise the precision of the estimation of nutrient loads even if they had strong hysteresis to flow rate. For example, as for the runoff load of total phosphorus load on flood events in a total of eight cases, the averaged error of estimation of total phosphorus load by the *L-Q-Turb* method was 11%, whereas the averaged estimation error by the regression curve between flow rate and nutrient load was 28%.
Keywords *L-Q* equation; *L-Q*-Turbidity method; nutrient load; turbidity

Introduction

To control and model water quality in lakes and other closed water areas in relation to eutrophication, it is important to accurately estimate the total input of nutrients and other chemical components from their basin. It is important that nutrient loads from the nonpoint sources are often discharged in the storm period and occupy over 50% of total discharge for a year (Tachibana *et al.*, 2001). For measuring the nutrient loads in the storm period, it is often the case that water samples are collected over 24 times per day. Therefore, it is almost impossible to measure the discharge of nutrients from the basin in every storm. That is why the relationships of nutrient load and flow rate such as the *L-Q* regression curve were developed and adapted to many rivers. However, the *L-Q* regression curve has difficulty in estimating the nutrient loads when they do not depend on the river flow rate. For example, rapid increase of nutrient loads by the suspension of mud by human activity or inflow of turbid water in ordinary water level period cannot be estimated by the *L-Q* equation. At present, *in situ* automatic analyzers of nutrient concentration are developed and released by manufacturers. It is ideal to measure the nutrient concentration for hours, but it is difficult to install them in all rivers due to its cost. The major reason for the fluctuation of total nutrient concentrations in storm periods is the fluctuation of particulate nutrient concentration. Therefore, if we could measure the particulate nutrient concentrations by any method, the estimation of total nutrient load would be more accurate. We studied measuring particulate nutrient concentrations by turbidity meter and presented a new estimation method for nutrient loads.

doi: 10.2166/wst.2006.065

Methods

Study field, observed period and chemical analysis

We studied nine time series of storm discharge in six rivers in Japan (Table 1). Flow rate and turbidity were measured *in situ*. We collected water samples and analyzed particle size distribution by laser scattered particle analyzer (SALD-3000, Shimadzu Co., Ltd.), Total Nitrogen (TN), Dissolved Nitrogen (DN), nitrate-nitrogen (NO_3^--N), nitrite-nitrogen (NO_2^--N), ammonium-nitrogen (NH_4^+-N), Total Phosphorus (TP), Dissolved Total Phosphorus (DP) and Dissolved Reactive Phosphorus (DRP) by auto analyzer (TRAACS, BRAN + LUEBBE) and Total Organic Carbon (TOC) and Dissolved Total Organic Carbon (DOC) by TOC meter (TOC-5000A, Shimadzu Co., Ltd.). Particulate components were calculated by the difference between the concentrations of total components and dissolved components.

Instruments

A turbidity meter (ATU 5-8M, Alec Electronics Co., Ltd.) was installed in the river. We set the turbidity meter 50 cm above the riverbed. The type of the turbidity meter was that of measuring backscatter of infrared (wavelength = 880 nm in air, and 660 nm in water) from suspended solid. A portable turbidity meter (AAQ-1183, Alec Electronics Co., Ltd.) was also deployed to measure the turbidity of the collected water. To measure the turbidity of the collected water, water samples were filled into the matte-black bucket.

Nutrient Load Hysteresis Coefficient (NLHC)

In the storm period, the loads of chemical components fluctuate with flow rate. In many cases, the load of chemical components increases with the increase of flow rate.

It is recognized that particulate component loads often loop in the plot of flow rate and load. Higher concentration of particulate components was often observed in the phase of increasing rather than decreasing, although their flow rates were the same. Figure 1 shows the dependence and hysteresis of the nutrient load to the flow rate. Though NO_3^--N load could be expressed as a one-valued function of flow rate, particulate phosphorus load could not. Therefore, we proposed the hysteresis coefficient of load of chemical components, namely Nutrient Load Hysteresis Coefficient (NLHC), to evaluate the hysteresis of the nutrient load to flow rate as follows.

When

$$\sum_{k=1}^{n-1}(L_k + L_{k+1})(Q_{k+1} - Q_k) \geq 0 \qquad (1)$$

then the hysteresis coefficient H is defined as

$$H = \frac{\sum_{k=1}^{n-1}(L_k + L_{k+1})(Q_{k+1} - Q_k)}{\sum_{k=1}^{M-1}(L_k + L_{k+1})(Q_{k+1} - Q_k)} \qquad (2)$$

and when

$$\sum_{k=1}^{n-1}(L_k + L_{k+1})(Q_{k+1} - Q_k) < 0 \qquad (3)$$

then the hysteresis coefficient H is defined as

$$H = -\frac{\sum_{k=1}^{n-1}(L_k + L_{k+1})(Q_{k+1} - Q_k)}{\sum_{k=M}^{n-1}(L_k + L_{k+1})(Q_{k+1} - Q_k)} \qquad (4)$$

Table 1 Sampling stations and sampling periods

Sampling station	River	Watershed (km²)	Land use	Sampling period	Rainfall (mm)
28.1 KP	Hinuma Riv.	190	Paddy field, upland field	12–13 Sep. 2002 30 Sep.–2 Oct. 2002	21 80
Kataniwa-A	Kataniwa Riv.	1.1	Forest	10–11 July 2002 30 Sep.–2 Oct. 2002	105 80
Kataniwa-B	Kataniwa Riv.	3.4	Forest, quarry	10–11 July 2002 30 Sep.–2 Oct. 2002	105 80
Inadasawa	Inadasawa Riv.	4.5	Paddy field, quarry	30 Sep.–2 Oct. 2002	80
Maekawa	Hinuma Mae Riv.	80	Paddy field, upland field	30 Sep.–2 Oct. 2002	80
Yotsugi	Shira Riv.	480	Paddy field, upland field, forest	24–25 June 2002	127

where L_k expresses the load of the chemical component of sampling number k, Q_k expresses the flow rate of the river at the sample k, and n means the total number of samples + 1 (Figure 2).

The value of the absolute H means the magnitude of the hysteresis, which is the fraction of S_1 to $(S_1 + S_2)$ in Figure 3. The maximum value of the H is one. Positive H means that the increasing rate of the load exceeds that of flow rate of the river, and the decreasing rate of the load exceeds that of flow rate, and clockwise rotation as shown in Figure 3. On the other hand, negative H means that the decreasing rate of the load exceeds that of flow rate of the river, and decreasing rate of the load exceeds that of flow rate, and anticlockwise rotation as shown in Figure 3. If absolute H is nearly equal to zero, it means that there is less hysteresis about flow rate and load. To calculate H, the curve needs to be closed. Therefore, L_1 and Q_1 are given as L_n and Q_n for the sake of simplicity.

Results and discussion

Concentration of chemical components in the storm period

Flow weighted averaged concentrations of chemical components in river water are shown in Table 2. SS concentration was up to 1,000 mg/L at the river whose basins have quarries. Over 80% of TOC, 90% of TP, and 30% of TN were particulate. Therefore, accurate estimation of concentration of particulate components will lead to accurate estimation of total discharged load.

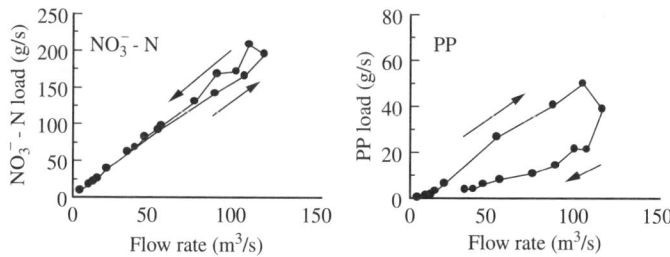

Figure 1 Relationships between nutrient loads L and flow rate Q, observed in the storm period, 1–2 Oct., 2002 at 28.1 KP in the Hinuma River, NO_3^--N (left), particulate phosphorus (right)

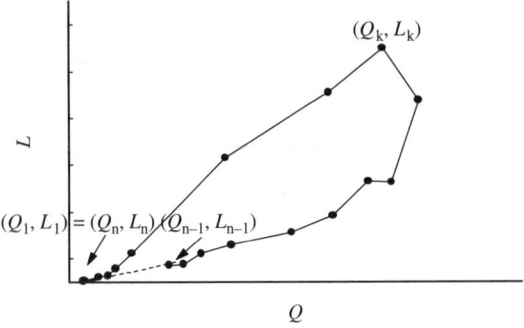

Figure 2 Example of the relationships between flow rate Q and nutrient load L; for the sake of simplicity, Q_1 and L_1 are given as Q_n and L_n

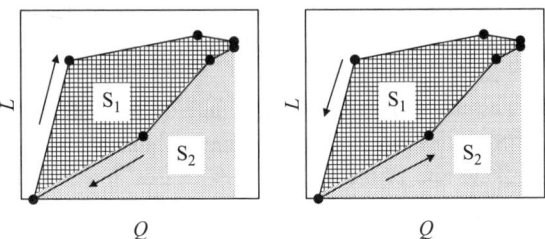

Figure 3 Two patterns of the fluctuation of nutrient load L with flow rate Q; clockwise rotation type (left) and anticlockwise rotation type (right)

Classification of characteristics of discharged nutrient loads in the storm period

Modeling of the discharged load L by the power function of flow rate Q is widely used, as below

$$L = CQ^n \tag{5}$$

where C and n are constants. It is recognized that if $n > 1$, concentration rises with the flow rate (Increasing type; I), and if $n < 1$, concentration drops with flow rate (Decreasing type; D). If value of n is nearly equal to 1, it means that the concentration is constant irrespective of flow rate (Constant type; C). Moreover, using the NLHC, we could classify the patterns of nutrient discharged load into 15 patterns as shown in Table 3. The results of the classification of the characteristics of fluctuation of the discharged load are shown in Table 4. Turbidity, SS, PN, PP, TP, TOC and POC loads have large value of n and H, and are classified into I^{++}. It is clear that particulate components have quite strong hysteresis and their concentrations rise with increasing flow rate. DP load also has strong hysteresis despite the concentration tending to be constant. The groups of + or ++ are the groups of the components which are supplied from the watershed immediately with increasing flow rate. Though DP has a strong hysteresis, the proportion of it in TP is negligible in the storm period. On the other hand, inorganic nitrogen compounds and DOC loads have less hysteresis and can be modeled with the function of the flow rate. It is clear that the load of the components classified into I^{++} group can hardly be modeled with flow rate.

Estimation of particulate components by turbidity

Effect of the particle size distribution on output of the sensor of the turbidity meter. The type of the turbidity meter we used is measuring the strength of the backscatter of light

Table 2 Flow weighted averaged concentration of chemical components in river water in the storm period in Table 1

St. Unit	Period	n	Turb. mg/L	SS mg/L	PN mg/L	DN mg/L	NO_3^--N mg/L	NO_2^--N μg/L	NH_4^+-N μg/L	PP μg/L	DP μg/L	DRP μg/L	POC mg/L	DOC mg/L
28.1 KP	Sep.02	12	390	419	1.24	2.07	1.85	1	11	491	19	16	11.6	1.4
	Oct. 02	19	273	258	0.68	1.93	1.76	2	2	247	19	18	7.9	2.2
Kataniwa-A	Jul. 02	8	359	392	1.11	1.17	0.90	1	1	179	1	1	33.3	1.6
	Oct. 02	16	98	90	0.47	1.17	1.08	1	1	52	1	1	6.6	1.8
Kataniwa-B	Jul. 02	17	813	1,498	1.05	0.92	0.81	1	1	479	1	1	19.5	0.6
	Oct. 02	15	640	1,188	0.62	1.00	0.96	1	1	287	3	1	9.9	1.0
Inadasawa	Oct. 02	12	1,614	2,444	1.18	0.54	0.43	3	3	311	2	1	26.5	2.0
Maekawa	Oct. 02	15	299	377	1.43	4.69	4.28	5	5	549	43	39	16.1	6.7
Yotsugi	Jun. 02	27	156	202	0.74	1.43	1.27	2	2	444	45	37	7.2	0.7

Table 3 Classification of runoff load of nutrients L by modeling with $L = CQ^n$ and Nutrient Load Hysteresis Coefficient H

	$n < 0.9$	$0.9 \leq n \leq 1.1$	$1.1 < n$
$0.25 < H$	D^{++}	C^{++}	I^{++}
$0.1 < H \leq 0.25$	D^{+}	C^{+}	I^{+}
$-0.1 \leq H \leq 0.1$	D	C	I
$-0.25 < H \leq -0.1$	D^{-}	C^{-}	I^{-}
$H < -0.25$	D^{-}	C^{-}	I^{-}

(wavelength = 880 nm in air). Therefore, the output of the turbidity meter is affected by the particle size distribution of the SS in water. Turbidity (*Turb.*) and SS concentration can be expressed as

$$Turb = a \cdot SS^b \tag{6}$$

where a and b are constants. The output of the turbidity is mainly controlled by the concentration of the fine fraction ($d < 20\,\mu m$) of the SS (Yokoyama, 2002). The fraction of coarse sand does not affect the output of the turbidity meter. If we use the same particle size of SS, b in Eq. (6) will be close to 1 if the fine particle dominates in SS.

Relationship between turbidity and particulate nutrient concentration. In general, the correlation coefficient of SS concentration and turbidity is high and SS concentration can be estimated by turbidity meter with proper calibration. We found that the correlation coefficients of particulate nutrients and POC were also high. An example of the relationships between particulate nutrient concentrations and turbidity is shown in Figure 4. The relationships between particulate nutrient concentrations and SS concentration are also shown in Figure 4. PN concentration and PP concentration could be expressed as the power functions of the SS concentration and turbidity as follows:

$$C = \alpha \cdot Turb^m \tag{7}$$

$$C = \beta \cdot SSC^n \tag{8}$$

Table 4 Results of classification of runoff load of nutrients L by modeling with power number n in the $L = CQ^n$ and Nutrient Load Hysteresis Coefficient H of the time series in Table 1

Components	n				H				Classification
	Max.	Av.	Min.	S.D.	Max.	Av.	Min.	S.D.	
Turbidity	2.13	1.67	1.18	0.25	0.86	0.46	0.00	0.29	I^{++}
SS	2.29	1.83	1.05	0.34	0.79	0.51	0.11	0.27	I^{++}
NH_4^+-N	1.33	1.05	1.00	0.10	0.61	−0.03	−0.38	0.28	C
NO_2^--N	1.61	1.16	1.00	0.22	0.27	−0.02	−0.66	0.28	I
NO_3^--N	1.37	1.04	0.88	0.21	0.13	−0.12	−0.30	0.15	C^{-}
DN	1.33	1.04	0.87	0.18	0.14	−0.10	−0.28	0.14	C
TN	1.40	1.20	1.06	0.11	0.31	0.17	0.02	0.10	I^{+}
PN	2.07	1.57	1.04	0.28	0.70	0.43	−0.01	0.27	I^{++}
DRP	1.19	0.89	0.21	0.28	0.73	0.16	−0.05	0.25	D^{+}
DP	1.38	0.91	0.35	0.30	0.96	0.33	−0.16	0.38	C^{++}
TP	1.91	1.51	0.91	0.26	0.77	0.51	0.07	0.26	I^{++}
PP	1.94	1.59	0.92	0.30	0.77	0.50	−0.04	0.29	I^{++}
TOC	2.04	1.62	1.04	0.29	0.81	0.34	−0.40	0.42	I^{++}
DOC	1.67	1.14	0.95	0.21	0.66	−0.01	−0.80	0.54	I
POC	2.24	1.75	1.06	0.32	0.84	0.39	−0.40	0.44	I^{++}

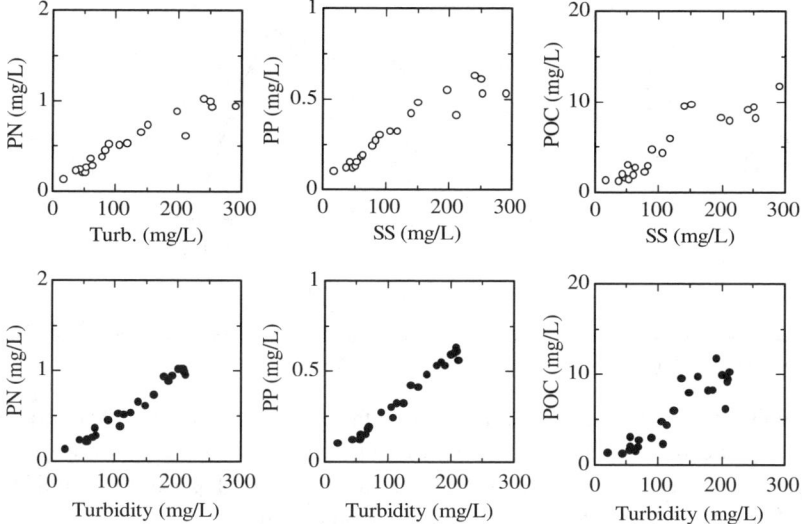

Figure 4 The relationships between turbidity, SS and particulate components on 24–25 June, 2002, at the Yotsugi Bridge of the Shira River

where C is concentration of particulate components, *SSC* is SS concentration, and *Turb* is turbidity. α, β, m, and n are constants. Averaged n and m for SS, PN, PP, POC are shown in Table 5. The values of the power coefficients of the n are lower than the values of m and 1. This means that higher concentration of SS leads to lower values of PN/SS, PP/SS, POC/SS. On the other hand, the values of m are more close to 1. Particularly, PP concentration can be expressed by the linear function of turbidity. The correlation coefficient of PN and POC is lower than that of PP.

Particulate distribution and the fraction of particulate nutrients. Particulate phosphorus exists as a component of the body of organisms and organic debris, or it is adsorbed on the surface of minerals or hydroxides. In the case of being adsorbed on the surface of minerals, it is thought that the amount of the adsorbed phosphorus increases with the specific surface area. Therefore, it is assumed that warse materials contained lower concentration. Typical particle size distribution of SS in the river water, which was collected from the surface water of the river in the storm period, is shown in Figure 5. More than 90% of the total volume of SS did not exceed the particle size of 100 μm.

It was reported that the concentration of Particulate Reactive Phosphorus (PRP) in the fraction of SS from 8 μm to 44 μm was much higher than that of SS over 44 μm (Kawabe et al., 1997). To clarify the accumulation of nutrients in the fine fractions, multiple linear regression analysis was applied. We set SS concentration in each fraction and POC concentration as explaining variables, and PN and PP concentrations as objective variables. The results of the analysis showed that the fine fraction of SS ($0 \mu m < d < 23 \mu m$) was correlated with PP at 1% significant level and partial correlation coefficient of 0.580 (Table 6). Concentration of PP was dependent on the concentration of the fine particle of SS. On the other hand, PN concentration was affected by POC concentration because PN

Table 5 Averaged power coefficients in Eqs. (7) and (8) for nine time series of the floods shown in Table 1

	Turb	SS	PN	PP	POC
m	–	1.21 (R = 0.963)	0.812 (R = 0.925)	0.994 (R = 0.966)	1.11 (R = 0.914)
n	0.798 (R = 0.963)	–	0.689 (R = 0.890)	0.867 (R = 0.946)	0.911 (R = 0.906)

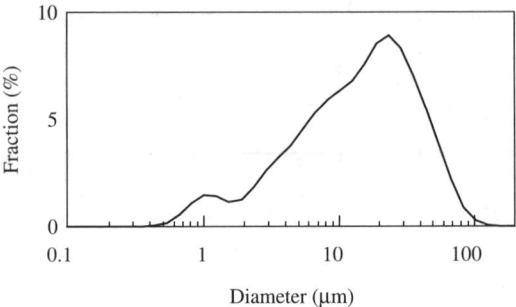

Figure 5 Volumetric particle size distribution of SS in river water in the storm period, Yotsugi Station in the Shira River, 25 June 2002

was found in the organic substances or debris of the organisms. We found that the accumulation of particulate nutrient concentration on fine particles and higher sensitivity of the turbidity meter for fine particles led to the linear relationship between turbidity and particulate nutrient concentration, particularly for phosphorus.

Estimation of runoff load of nutrients using turbidity meter (L-Q-Turb Method)

We proposed an advanced method of estimating runoff load of nutrients using turbidity in the L-Q-Turb method. Loads of dissolved nitrogen L_{DN} and phosphorus L_{DP} are modeled by an ordinary L-Q regression curve,

$$L_{DN} = C_{DN} Q^n \tag{9}$$

$$L_{DP} = C_{DP} Q^p \tag{10}$$

where Q is flow rate, and C_{DN}, C_{DP}, n and p are constants. Particulate nitrogen concentration, C_{PN} and particulate phosphorus concentration, C_{PP} are modeled by

$$C_{PN} = \alpha_{PN} Turb^\beta \tag{11}$$

$$C_{PP} = \alpha_{PP} Turb^\gamma \tag{12}$$

where $Turb$ is turbidity, α_{PN}, α_{PP}, β and γ are constants.

Table 6 Results of the multiple regression analysis of the river water of 28.1 KP station on 12–13 Sep., 2002 ($n = 12$) and 1–2 Oct., 2002 ($n = 17$)

	Explaining variable	PN	PP
Correlation coefficient	SS (0 μm < d < 23 μm)	0.935*	0.952**
	SS (23 μm < d < 54 μm)	0.842	0.851
	SS (54 μm < d < 103 μm)	0.464	0.461
	SSC (103 μm < d)	0.195	0.187
	POC	0.811*	0.735
Partial correlation coefficient	SSC (0 μm < d < 23 μm)	0.418	0.580**
	SSC (23 μm < d < 54 μm)	0.240	0.151
	SS (54 μm < d < 103 μm)	−0.297	−0.157
	SS (103 μm < d)	0.076	−0.029
	POC	0.498*	0.105
Standard partial regression coefficient	SS (0 μm < d < 23 μm)	0.524*	0.795**
	SS (23 μm < d < 54 μm)	0.370	0.225
	SS (54 μm < d < 103 μm)	−0.284	−0.157
	SS (103 μm < d)	0.033	−0.012
	POC	0.286	0.052
Coefficient of determination		0.891	0.887

**1% of significant level; *5% of significant level

Table 7 Comparison of the calculated cumulative discharged load of Total Nitrogen (upper) and Total Phosphorus (lower)

Station	Storm period	Measured load (t)	L-Q Regression		L-Q-Turb Method	
			Load (t)	Error (%)	Load (t)	Error (%)
(a) Total Nitrogen						
28.1 KP	12–13 Sep.02	2.39	2.12	11.2	2.28	4.7
	1–2 Oct.02	15.1	14.9	1.1	15.1	0.0
Kataniwa-A	10–11 Jul.–02	0.126	0.105	16.9	0.134	5.9
	1–2 Oct.–02	0.166	0.179	8.0	0.158	4.7
Kataniwa-B	10–11 Jul.–02	0.311	0.297	4.6	0.284	9.4
	1–2 Oct.–02	0.269	0.286	6.2	0.259	3.9
Maekawa	1–2 Oct.–02	9.65	9.65	0.0	9.49	1.7
Yotsugi	24–25 Jun.–02	17.2	17.4	1.5	17.2	0.1
Averaged				6.2		3.8
(b) Total Phosphorus						
28.1 KP	12–13 Sep.02	375	154	58.9	330	13.9
	1–2 Oct.02	1,580	1,700	7.7	1,710	7.7
Kataniwa-A	10–11 Jul.–02	9.74	3.40	65.2	9.02	8.0
	1–2 Oct.–02	5.37	5.53	2.9	5.07	5.9
Kataniwa-B	10–11 Jul.–02	69.8	35.8	48.7	64.0	9.1
	1–2 Oct.–02	46.2	33.6	27.2	57.7	19.9
Maekawa	1–2 Oct.–02	845	792	6.3	760	11.2
Yotsugi	24–25 Jun.–02	3,500	3,400	2.8	3,100	12.9
Averaged				27.5		11.1

Load: cumulative discharged load. Error: root mean square of error which was calculated by $\sqrt{(L_M - L_E)^2/L_M^2}$, where L_M: measured cumulative load, and L_E: estimated cumulative load

As a consequence, time series of the total nitrogen load, $L_{TN}(t)$ and total phosphorus load, $L_{TP}(t)$ are calculated as follows:

$$L_{TN}(t) = C_{DN}Q(t)^n + \alpha_{PN}Turb(t)^\beta Q(t) \tag{13}$$

$$L_{TN}(t) = C_{DP}Q(t)^p + \alpha_{PP}Turb(t)^\gamma Q(t) \tag{14}$$

where $Turb(t)$ is the time series of turbidity and $Q(t)$ is the time series of flow rate.

Evaluation of estimation method of runoff load of nutrients

Total discharged nutrient loads were calculated by the following three methods to evaluate the accuracy of the estimation of the load of nutrients: (1) actual discharged nutrient loads from the measured flow rate and nutrient concentrations; (2) runoff load modeled by L-Q regression curve and calculated total discharged load from the time series of flow rate; (3) calculation by L-Q-Turb method. The results of the comparison are shown in Table 7. Root mean square of errors by the L-Q-Turb method was half of that by L-Q regression curve. L-Q-Turb method could improve the accuracy of the estimation of the runoff load of nutrients in flood periods.

Conclusions

To establish a new estimation method of the runoff loads of nutrients, the relationship of particulate nutrient concentrations and turbidity was studied. The main conclusions are as follows. (1) The "Nutrient Load Hysteresis Coefficient (NLHC)" was proposed to evaluate the hysteresis of the nutrient loads to flow rate quantitatively. As a result of the NLHC analysis combined with the power number of the L-Q regression curve, we could classify the runoff patterns into 15. It was clarified that the fluctuation of the particulate nutrient loads has strong hysteresis; therefore they can hardly be modeled with flow rate.

(2) Linear relationships between the concentrations of particulate nutrients and turbidity were observed. It was clarified that the linearity was caused by the influence of the particle size on turbidity output and accumulation of nutrients on the smaller particles (diameter <23 μm). (3) The *L-Q-Tb* method, which is the new method for the estimation of the nutrient loads using regression curve between the turbidity and the concentrations of particulate nutrients, was developed. This method could raise the precision of estimation of nutrient loads even if they had strong hysteresis to flow rate.

References

Kawabe, H., Aita, T., Tachibana, H., Wang, B. and Yoshida, K. (1997). Composition of particulate phosphorus and AGP. In *Proceedings of Annual Conference of the Japan Society of Civil Engineering*, **52**, pp. 172–173 (in Japanese).

Tachibana, H., Yamamoto, K., Yoshizawa, K. and Magara, Y. (2001). Non Point Pollution of Ishikari River, Hokkaido, Japan. *Water Science & Tech.*, **44**(7), 1–8.

Yokoyama, K. (2002). The influence of particle size on turbidity output and the instructions for using turbidity sensor in the field. *Journal of Hydraulic, Costal and Environmental Engineering*, **698**(58), 93–98 (in Japanese).

Application of monitored natural attenuation to remediate a petroleum-hydrocarbon spill site

C.M. Kao*, W.Y. Huang*, L.J. Chang*, T.Y. Chen*, H.Y. Chien* and F. Hou**

*Institute of Environmental Engineering, National Sun Yat-Sen University, Kaohsiung, Taiwan
**Department of Occupational Safety and Health, China Medical University, Tai-Chung, Taiwan

Abstract Contamination of groundwater by petroleum-hydrocarbons is a serious environmental problem. The Monitored Natural Attenuation (MNA) approach is a passive remediation to degrade and dissipate groundwater contaminants *in situ*. In this study, a full-scale natural bioremediation investigation was conducted at a gasoline spill site. Results show that concentrations of major contaminants (benzene, toluene, ethylbenzene, and xylenes) dropped to below detection limit before they reached the downgradient monitor well located 280 m from the spill location. The results also reveal that natural biodegradation was the major cause of the observed contaminant reduction. The calculated natural first-order attenuation rates for BTEX and 1,2,4-trimethylbenzene (1,2,4-TMB) ranged from 0.051 (benzene) to 0.189 1/day (1,2,4-TMB). Evidence for the occurrence of natural attenuation includes the following: (1) depletion of dissolved oxygen, nitrate, and sulfate; (2) production of dissolved ferrous iron, sulfide, and CO_2; (3) decreased BTEX concentrations and BTEX as carbon to TOC ratio along the transport path; (4) increased alkalinity and microbial populations; (5) limited spreading of the BTEX plume; and (6) preferential removal of certain BTEX components along the transport path. Additionally, the biodegradation capacity (44.73 mg/L) for BTEX and 1,2,4-TMB was much higher than other detected contaminants within the plume. Hence, natural attenuation can effectively contain the plume, and biodegradation processes played an important role in contaminant removal.

Keywords BTEX; groundwater; methanogenesis; MNA; natural attenuation; petroleum-hydrocarbon

Introduction

Accidental release of petroleum products from underground storage tanks (USTs) is one of the most common causes of groundwater contamination. There are more than three million USTs storing petroleum products in the US, and as many as 500,000 may be leaking petroleum into the ground (AFCEE, 1994; US EPA, 2004). Current attention is focused on human and environmental safety concerning the release of hydrocarbons to the environment. Petroleum hydrocarbons contain benzene, toluene, ethylbenzene, and xylene isomers (BTEX), the major components of fuel oils (especially gasoline); they are hazardous substances regulated by many nations. In addition to BTEX, other gasoline constituents such as methyl-*t*-butyl ether (MTBE), naphthalene, 1,3,5-trimethylbenzene (1,3,5-TMB), and 1,2,4-trimethylbenzene (1,2,4-TMB) are also toxic to humans. Results of field investigation suggest that many of these spills have been naturally biodegraded before the contaminants reach a drinking water receptor (Borden *et al.*, 1995; Kao and Wang, 2000; Kao and Prosser, 2001). Because the petroleum-hydrocarbon caused plumes that could be quite diffuse and widespread, some more economic approaches are desirable for *in situ* remediation to provide for a long-term control of the contaminated groundwater. The Monitored Natural Attenuation (MNA) approach is a passive remedial method that depends upon natural processes to degrade and dissipate contaminants in soil and groundwater. This natural process includes physical, chemical, and biological transformations, e.g. aerobic/anaerobic biodegradation, cometabolism, dispersion, volatilization,

doi: 10.2166/wst.2006.066

oxidation, reduction, and adsorption (Rifai et al., 1995; Bedient et al., 1999; Surampalli and Banerji, 2002). Aerobic and anaerobic biodegradations are believed to be the major mechanisms that account for both containment of the petroleum-hydrocarbon plume and reduction of the contaminant concentrations. The aerobic biodegradation relies on dissolved oxygen (DO) as the electron acceptor used by the subsurface microorganisms while the anaerobic processes depend on a variety of biodegradation mechanisms to use nitrate, ferric iron [Fe(III)], sulfate, and carbon dioxide (CO_2) as terminal electron acceptors (Hunt et al., 1997; Kota, 1998; Kao and Wang, 2000).

Environmental conditions and microbial competition will ultimately determine which anaerobic biodegradation processes would dominate. When oxygen is depleted and nitrate is present, the latter can be used as an electron acceptor by facultative denitrifiers to mineralize the fuel hydrocarbons in denitrification (Kao and Wang, 2000; Johnson et al., 2003). Once the available DO and nitrate in the aquifer are depleted, ferric iron can be used as an electron acceptor. A large quantity of ferric iron is present in the sediments of most aquifers, which could potentially provide a large reservoir of electron acceptor for hydrocarbon biodegradation (Kota, 1998; Kao and Wang, 2000). The available evidence suggests that the iron-reducing process can significantly influence the fate and transport of hydrocarbons in the subsurface (Kota, 1998; Kao et al., 2001). After the depletion of DO, nitrate, and ferric iron, sulfate-reducing bacteria could then degrade petroleum hydrocarbons using sulfate as the electron acceptor in the sulfate reduction process. After sulfur is depleted, methanogenic consortia could potentially biodegrade fuel hydrocarbons, and CO_2 could be used as the electron acceptor. The presence of elevated methane levels in the groundwater relative to background methane concentrations is a good indicator of methane formation (Brown et al., 1997; Bedient et al., 1999; Seagren and Becker, 2002).

At present, many research efforts have been directed toward applying the natural attenuation to control the dissolved contaminant migration. However, results of full-scale demonstrations of the natural attenuation process and its mechanisms, especially at the demonstration site where mixed biodegradation patterns (e.g. aerobic degradation, denitrification, iron reduction, methanogenesis) are occurring within the plume have not been well documented in the literature. The objectives of this study were to: (1) characterize a hydrocarbon contaminated aquifer at a petroleum-hydrocarbon spill site, (2) characterize the variations in aqueous phase geochemistry, (3) assess the occurrence and mechanisms of natural attenuation, (4) calculate the field-scale natural attenuation rates of the contaminants, and (5) evaluate the effects of using MNA as a remedial option.

Study site description

A government owned tank farm facility site located in southern Taiwan (Kaohsiung County) was selected for this MNA study. In 1998, leakage from a fuel-oil pipeline resulted in groundwater contamination by petroleum hydrocarbons (mainly BTEX). During the following six-year investigation period (from 1998 to 2004), more than 100 soil gas and soil samples were collected; meanwhile 15 monitor wells and two recovery wells were installed for site characterization and contaminated-groundwater extraction. On-site borings encountered up to 25 m of mostly brownish to graying, fine to medium sand to silty sands. The average groundwater elevation within the shallow aquifer is approximately 1.5 to 3 m below the ground. Groundwater in the unconfined aquifer, according to the groundwater elevation in on-site monitor wells, flows to the southwest. The measured effective porosity is 0.3, and the average hydraulic conductivity for the surficial, unconfined aquifer is 2.9×10^{-5} m/sec (2.5 m/day). The calculated site groundwater flow velocity is 3.7×10^{-5} m/sec (3.2 m/day). The measured groundwater temperature in the surficial aquifer varies from 17 to 29 °C. Since the year 2001, MNA has been applied to remediate the contaminated groundwater. Figure 1 presents the site map showing the estimated plume boundary, locations of monitor wells and soil sampling locations, and groundwater flow direction.

Natural attenuation investigation

In the spring of 2001, a MNA study at the Kaohsiung site was conducted with the funding provided by the Chinese Petroleum Corp. and Taiwan's National Science Council. During the two-year investigation period, a monitoring network system was installed to delineate the dissolved contamination (Figure 1). The most probable number (MPN) enumeration studies were performed to define the distribution of microorganisms; natural attenuation rates were calculated to evaluate the effectiveness of MNA on plume containment.

Monitoring network installation

A total of 15 monitor wells (5.1-cm I.D.; 0.025-cm slot screen) were installed to delineate the longitudinal distribution of the dissolved BTEX plume (Figure 1). For simplicity and ease of identification, the plume was divided into four different areas, i.e. Area 1 to Area 4, to represent the source area, mid-plume area, downgradient area, and background area, respectively. The representative monitor wells in these four areas were designated as

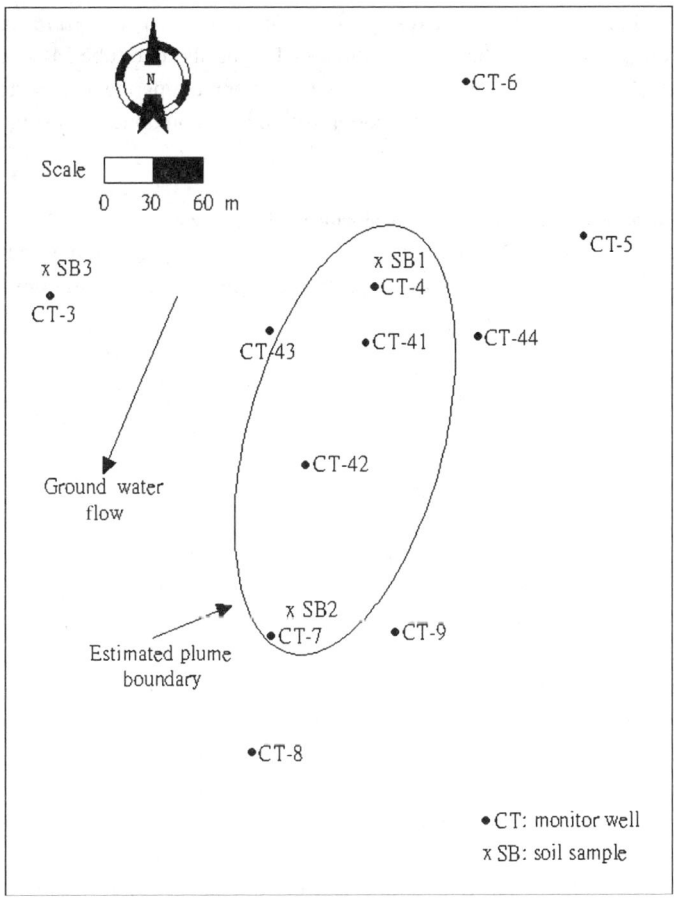

Figure 1 Site map showing the groundwater flow direction, estimated plume boundary, and soil and groundwater sampling locations

CT4, CT42, CT7, and CT3 respectively. Groundwater samples collected from these four representative wells were analyzed and compared in this study. All four wells were screened from 2 to 5.1 m below land surface (bls).

Groundwater monitoring and microbial enumeration

Groundwater samples from the monitor wells were collected and analyzed for organic compounds and geochemical indicators including BTEX, 1,3,5-TMB, 1,2,4-TMB, CH_4, CO_2, inorganic nutrients, anions, pH, redox potential (Eh), and DO. Organic compound analyses were performed in accordance with US EPA Method 502.2, using a Varian 3800 Gas Chromatograph (GC). Methane was analyzed using a Shimadzu GC-9A GC using headspace techniques. Ion chromatography (Dionex) was used for analyzing inorganic nutrients and anions (NO_3^-, NO_2^-, SO_4^{-2}, PO_4^{-3}). On-site measurements of Eh and pH were done with two MP120 pH/Eh meters (Mettler-Toledo). A WTW DO meter (Oxi 330) was used for on-site DO and temperature measurements, and a Hach digital titrator cartridge was used for on-site CO_2 measurements. Aquifer sediments were collected from the soil borings SB1, SB2, and SB3, which were adjacent to CT4, CT7, and CT3, respectively. The collected aquifer sediments were used for microbial enumeration to determine the number of total heterotrophs, total heterotrophic anaerobes (total anaerobes), and methanogens. Total plate counts were conducted using plate count agar (Difco) to assess the approximate number of the total heterotrophic bacterial growth using the spread plate method (APHA, 1995). Prepared plates were incubated at 30 °C for 48 hours, then counted for colony forming units (CFU). Total anaerobes were enumerated using the five-tube MPN assay (Kota, 1998; Kao and Wang, 2000). The total anaerobe tubes containing media described by Kota (1998) and Kao and Wang (2000) were score positive based on optical density.

Calculation of natural attenuation rates and biodegradation capacity

The first-order decay model (Equation 1) was applied to estimate the natural attenuation rate between CT4 and CT42 using the measured contaminant concentrations in CT4 and CT42 in the following equation:

$$C_{42} = C_4 \exp(-K_t) \tag{1}$$

where
- C_{42} = measured contaminant concentration (μg/L) at CT42
- C_4 = measured contaminant concentration (μg/L) at CT4
- K = first-order decay rate (natural attenuation rate) (day^{-1})
- t = distance between two cross-sections/groundwater flow velocity

Based on the calculated natural attenuation rates, the order of preferential removal of certain BTEX components can be determined.

The biodegradation capacity was calculated using the total amount of electron acceptor available for biological reactions. The procedure includes: (1) calculating the difference between the upgradient wells and source zone wells for oxygen, nitrate, and sulfate; and (2) measuring the production of by-products ferrous iron and methane in the source zone (US EPA, 1998). Using stoichiometry, a utilization factor can be developed to convert the mass of oxygen, nitrate, and sulfate consumed to the mass of dissolved hydrocarbons that are used in the biodegradation reactions. Similarly, the utilization factor can be developed to convert the mass of metabolic by-products that are produced to the mass of dissolved hydrocarbon to be used in the subsequent biodegradation reactions. Thus, the biodegradation capacity can be calculated using the following equation:

biodegradation capacity (mg/L) = (upgradient DO concentration

− DO concentration in the source zone)/3.15

+ (upgradient nitrate concentration

− nitrate concentration in the source zone)/4.86

+ (upgradient sulfate concentration

− sulfate concentration in the source zone)/4.71

+ (ferrous iron concentration in the source zone)/21.89

+ (methane concentration in the source zone)/0.78

Results and discussion

Groundwater samples were collected from monitor wells CT4, CT42, CT7, and CT3, which were located near the spill location, mid-plume area, downgradient area, and background area of the plume, respectively. Table 1 shows the averaged results of seven groundwater sampling events during the two-year investigation period. The aquifer sediment samples collected from soil borings SB1, SB2, and SB3 were used for microbial enumeration. Results of the bacterial population assessment are also shown in Table 1. Figure 2 presents variations in BTEX and 1,2,4-TMB concentrations along the transport path from source (monitor well CT4) to the most downgradient monitor well (CT8).

The decline in Eh and DO near the source area reflects the change from oxidizing to reducing conditions. High CO_2 concentrations and alkalinity were observed in CT4. This indicates that significant microbial activity and natural bioremediation occurred in this area. The lower nitrate and sulfate concentrations within the plume reveal that both nitrate and sulfate were used as the electron acceptors after the depletion of oxygen. The production of sulfide in CT4 also confirmed the occurrence of the sulfate reduction process. High ferrous concentrations were detected in CT4 indicating that ferric irons might have also been used as the electron acceptor around the source area. Moreover, a relatively higher concentration of methane was also detected in CT4. This indicates that mixed anaerobic biodegradation processes occurred within the most contaminated zone. The decrease in BTEX concentrations from CT4 to CT7 suggests the occurrence of natural attenuation of BTEX. The declined BTEX expressed as carbon to TOC ratio (BTEX as C/TOC) along the transport path reveals that total BTEX caused the high TOC measurements near the source area and increasing BTEX degradation by-products along the transport path. The results show that a significant amount of total heterotrophs and total anaerobes ($>10^6$ cells per g of soil) were detected in the SB1 soil sample collected from the most contaminated area. This points out that higher petroleum-hydrocarbon concentrations caused a more measured bacterial population.

The calculated first-order decay rates for contaminants between CT4 and CT42 are presented in Table 1. Results show that 1,2,4-TMB had the highest first-order decay rate ($0.189\,day^{-1}$), followed by toluene ($0.166\,day^{-1}$), $m + p$-xylene ($0.149\,day^{-1}$), o-xylene ($0.124\,day^{-1}$), ethylbenzene ($0.098\,day^{-1}$), and benzene ($0.051\,day^{-1}$). Benzene was the least biodegradable compound under anaerobic conditions and toluene was the most biodegradable compound among BTEX. The observed biodegradation trend for BTEX

Table 1 Average concentrations of petroleum-hydrocarbons and indicator parameters in a profile along the plume centerline

Monitor well	CT-3	CT-4	CT-42	CT-7	Decay rate[4]
Distance to CT-4 (m)	–[1]	0	100	210	–
Location	Background	Source	Mid-plume	Downgradient	–
Benzene (μg/L)	BDL[2]	170.1	34.4	0.8	0.051
Toluene (μg/L)	BDL	88.8	0.5	BDL	0.166
Ethylbenzene (μg/L)	BDL	182.6	8.5	BDL	0.098
m,p-xylene (μg/L)	BDL	202.4	1.9	BDL	0.149
o-xylene (μg/L)	BDL	9.5	0.2	BDL	0.124
Total BTEX (μg/L)	BDL	653.4	45.5	0.8	0.085
1,2,4-TMB (μg/L)	BDL	35.2	0.1	BDL	0.189
1,3,5-TMB (μg/L)	BDL	BDL	BDL	BDL	–
DO (mg/L)	2.4	0.5	1.4	1.5	–
Nitrate (mg/L)	83.4	0.2	37.6	41.3	–
Total iron (mg/L)	1.7	19.0	1.5	2.0	–
Ferrous iron (mg/L)	0.2	11.7	0.1	0.04	–
Sulfate (mg/L)	102.2	10.1	69.4	37.8	–
Sulfide (μg/L)	10	17	11	6	–
Carbon dioxide (mg/L)	167	250	177	179	–
Methane (mg/L)	0.002	2.1	0.056	0.010	–
pH	6.8	6.7	7.1	6.5	–
Redox potential (mV)	168.2	–44.0	158.8	178.4	–
Alkalinity (mg/L as $CaCO_3$)	164.9	374.2	232.4	154.9	–
Ammonia nitrogen (mg/L)	0.13	0.95	0.13	0.12	–
TOC (mg/L)	7.0	18.9	5.7	17.7	–
BTEX as C/TOC	–	0.033	0.008	4×10^{-5}	–
Total heterotrophs (cell/g)	5×10^5 (SB3)[3]	4.6×10^7 (SB1)	–	3.0×10^6 (SB2)	–
Total anaerobes (cell/g)	2×10^4 (SB3)	1.3×10^6 (SB1)	–	7.4×10^4 (SB2)	–

[1] –: not available
[2] BDL: below detection limit
[3] SB: soil sample
[4] First order decay rate (1/day)

matched with the results presented by other investigations (Hunt et al., 1997; Kota, 1998; Kao and Wang, 2000).

The observed decay rate for 1,2,4-TMB has drawn more attention. Most researchers indicate that 1,2,4-TMB might not be biodegradable, and thus, it has been suggested as a tracer to evaluate the natural biodegradation rate of BTEX (Wiedemeier et al., 1996; Kao and Prossor, 2001). However, results from this study indicate that the 1,2,4-TMB has a much higher decay rate than BTEX. This reveals that 1,2,4-TMB might be biodegradable under certain conditions, especially under anaerobic conditions. Thus, preliminary studies must be done prior to using 1,2,4-TMB as a tracer during the MNA study. Moreover, the observed decay rates for BTEX are higher than those reported by other researchers (Borden et al., 1995; Hunt et al., 1997; Kota, 1998; Kao and Wang, 2000; Kao and Prossor, 2001) demonstrating that significant natural attenuation processes occurred at this site to cause the removal of contaminants in groundwater.

If dilution was the only cause for the observed BTEX reduction, the plume width should increase inversely proportional to the decreasing BTEX concentrations along the transport path. However, the plume width was not observed to increase over the distance

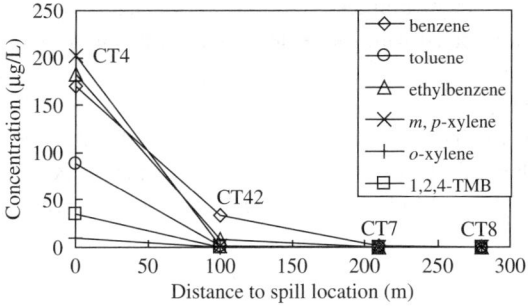

Figure 2 Variations in BTEX and 1,2,4-TMB concentrations along the transport path

while significant drops of the BTEX concentrations were detected. As the plume spread slightly in width due to dispersion, oxygen in the uncontaminated groundwater mixed with BTEX at the plume sides causing enhanced biodegradation. Thus, the limited spreading of the plume was attributed to some mixed natural attenuation processes including natural biodegradation, dilution, and dispersion, with the natural biodegradation playing an important role in plume containment. Based on the measured concentrations of DO, nitrate, ferrous iron, sulfate, and methane, the calculated biodegradation capacities for BTEX and 1,2,4-TMB aerobic biodegradation were 0.55 mg/L for denitrification, 20.96 mg/L for iron reduction, 0.48 mg/L for sulfate reduction, and 3.6 mg/L for methanogenesis. These values are much higher than the measured contaminant concentrations in groundwater thus indicating that the total amount of electron acceptors available for natural biodegradation would be sufficient for removing the total contaminant mass (BTEX and 1,2,4-TMB) found in the contaminated groundwater.

Conclusions

A detailed full-scale investigation on natural attenuation has been completed at a petroleum-hydrocarbon contaminated site located in Kaohsiung County, Taiwan. This study was conducted to assess if the MNA processes can be applied as a remedial option to reduce the contaminant concentrations to below regulatory standards before the contaminants are transported to downgradient potential receptors. The two-year investigation results show that natural attenuation mechanisms occurred at this site to cause the removal of the groundwater contaminants through mixed physical, chemical, and biological processes. Evidence for natural attenuation includes:

(1) depletion of DO, nitrate, and sulfate within the plume;
(2) production of dissolved ferrous iron, sulfide, and CO_2 within the plume;
(3) decreased BTEX concentrations and BTEX measured as the carbon to TOC ratio along the transport path;
(4) increased alkalinity and microbial populations within the plume;
(5) limited spreading of the BTEX plume; and
(6) preferential removal of certain BTEX components along the transport path.

The monitoring results, which show high contaminant decay rates and high biodegradation capacity, demonstrate that the contaminants could be attenuated naturally before they reach any farther downgradient receptors. The results also point out that the BTEX plume is not growing but reaching a steady-state condition when it is subject to natural bioremediation under mixed aerobic and anaerobic conditions. Results from this study indicate that the MNA approach is an acceptable remedial option; it is effective in remediating and managing this petroleum hydrocarbon spill site. With increasing knowledge on the mechanism of natural attenuation less conservative regulatory decisions may be possible such that the costs associated with site remediation can be minimized.

Acknowledgements

This study was funded by National Science Council in Taiwan and Chinese Petroleum Corp., Taiwan. Additional thanks to Mr. C.Y. Yu of Chinese Petroleum Corp., Taiwan, and Mr. J.K. Fu and C.M. Tang of National Sun Yat-Sen University for their assistance throughout this project.

References

AFCEE (Air Force Center for Environmental Excellence), Armstrong Laboratory, and Air Force Institute of Technology (1994). Use of risk-based standards for cleanup of petroleum contaminated soil. Brooks Air Force Base, TX.

APHA/AWWA/WEF (1995). *Standard Methods for the Examination of Water and Wastewater*, 11th edn., Washington, DC, USA.

Bedient, P.B., Rifai, H.S. and Newell, C.J. (1999). *Ground Water Contamination – Transport and Remediation*, PTR Prentice-Hall, Inc., New Jersey.

Borden, R.C., Gomez, C.A. and Becker, M.T. (1995). Geochemical indicators of natural bioremediation. *Ground Water*, **33**, 180–189.

Brown, K., Sekerka, P., Thomas, M., Perina, T., Tyner, L. and Sommer, B. (1997). Natural attenuation of jet fuel-impacted groundwater. In Situ *and On-site Bioremediation*, Alleman, B.C. and Leeson, A. (eds), Battelle Press, Columbus, Ohio, Vol. 1, pp. 83–88.

Hunt, M.J., Shafer, M.B., Barlaz, M.A. and Borden, R.C. (1997). Anaerobic biodegradation of alkylbenzenes in aquifer material under methanogenic and iron-reducing conditions. *Bioremediation J.*, **1**, 53–64.

Johnson, S.J., Woolhouse, K.J., Prommer, H., Barry, D.A. and Christofi, N. (2003). Contribution of anaerobic microbial activity to natural attenuation in groundwater. *Engineering Geology*, **70**, 343–349.

Kao, C.M., Kota, S., Ress, B., Barlaz, M.A. and Borden, R.C. (2001). Effects of subsurface heterogeneity on natural bioremediation at a gasoline spill site. *Wat. Sci. Tech.*, **43**(5), 341–348.

Kao, C.M. and Prosser, J. (2001). Evaluation of natural attenuation rate at a gasoline spill site. *J. of Hazardous Materials*, B**82**, 275–289.

Kao, C.M. and Wang, C.C. (2000). Control of BTEX migration by intrinsic bioremediation at a gasoline spill site. *Wat. Res.*, **34**, 3413–3423.

Kota, S. (1998). Biodegradation in contaminated aquifers: influence of microbial ecology and iron bioavailability. Ph.D. Dissertation, North Carolina State University, Raleigh, NC.

Rifai, H.S., Borden, R.C., Wilson, J.T. and Ward, C.H. (1995). Intrinsic bioattenuation for subsurface restoration. In *Intrinsic Bioremediation*, Hinchee, R.E., Wilson, J.T. and Downey, D.C. (eds), CRC Press, Boca Raton, FL, pp. 1–30.

Seagren, E. and Becker, J. (2002). Review of natural attenuation of BTEX and MTBE in groundwater. *Practice periodical of hazardous, toxic, and radioactive waste management*, **6**, 156–172.

Surampalli, R. and Banerji, S. (2002). Long-term performance monitoring at natural attenuation site. *Practice periodical of hazardous, toxic, and radioactive waste management*, **6**, 173–176.

Wiedemeier, T.H., Wilson, J.T., Kampbell, D., Jansen, J.E. and Haas, P. (1996). Technical protocol for evaluating the natural attenuation of chlorinated ethenes in groundwater. In *Proceedings of the 1996 Petroleum Hydrocarbons and Organic Chemicals in Ground Water: Prevention, Detection, and Remediation Conference*, Nat. Water Well Asso., Houston, TX, pp. 425–444.

US EPA (1998). *BIOPLUME III: Natural Attenuation Decision Support System, User's Manual*, Ver. 1.0. EPA/600/R-98/010.

US EPA (2004). How to Evaluate Alternative Cleanup Technologies for Underground Storage Tank Sites: a Guide for Corrective Action Plan Reviewers. EPA/510/R-04-002.

Comparison of several methods for BAP measurement

J. Nakajima, Y. Murata and M. Sakamoto

Department of Environmental Systems Engineering, Faculty of Science and Engineering, Ritsumeikan Univ., 1-1-1 Nojihigashi, Kusatsu, Shiga, 525-8577, Japan (E-mail: *jnakajim@nisiq.net*)

Abstract It has been more important for management of water quality to estimate the amount of bioavailable phosphorus (BAP) in suspended solids (SS) entering lakes and estuaries. AGP test or extraction by 0.1 mol l^{-1} NaOH (C-BOD) is widely used. Recently, highly bioavailable phosphorus (HBAP) was introduced to indicate a more easily soluble and bioavailable fraction using successive extraction by 0.1 mol l^{-1} HCl and 0.1 mol l^{-1} NaOH. New biologically measured BAP (B-BAP) using bacterial respiration activity was introduced in this paper. B-BAP was estimated from oxygen uptake rate (OUR), which was measured by a respiratory meter for BOD measurement using a pressure sensor. B-BAP is useful for a rapid and direct measurement of phosphorus bioavailability. B-BAP, HBAP and C-BAP in river SS were measured and compared with each other. The percentages of HBAP and B-BAP to PP were large in the urban river, while the percentage of NaOH-P or C-BAP was large in the rivers flowing in agricultural areas. By comparison with phosphorus fractions in paddy soil and activated sludge it was suggested that SS in the rivers flowing in agricultural areas mainly consisted of clay, silt or sand, while the SS in the urban river consisted of a large percentage of organic particles as well. Phosphorus in SS was suggested to be more easily bioavailable in the urban river than the rivers in agricultural areas. The ratio of C-BAP/B-BAP was large in the rivers in agricultural areas and small in the urban river. As HBAP contents were almost similar to B-BAP contents in the river SS, HBAP can be a suitable index of phosphorus indicating easily and rapidly the bioavailable fraction in SS.

Keywords Bacterial respiration; bioavailable phosphorus; eutrophication; lake water management; oxygen uptake rate; phosphate analysis

Introduction

It has become more important to estimate bioavailable phosphorus (BAP) amount entering lakes or estuaries from non point sources in their basin for management of their water quality and environment. In order to measure BAP, chemical and biological methods have been applied (Okubo, 1996). In chemical methods 0.1 mol l^{-1} NaOH extraction is the most common to measure BAP in agricultural runoff (Dorich *et al.*, 1985; Butkus *et al.*, 1988; Parker, 1991; Sharpley *et al.*, 1991, 1992). Nakajima and Okubo (2003) improved the sensitivity of the BAP measurement by collecting SS using filtration procedures before 0.1 mol l^{-1} NaOH extraction. They also introduced 0.1 mol l^{-1} HCl extraction before 0.1 mol l^{-1} NaOH extraction in order to estimate the more easily soluble and bioavailable fraction (HBAP; highly bioavailable phosphorus) by applying the successive extraction method of soil phosphorus fractions (Chang and Jackson, 1957; Williams *et al.*, 1967; Sekiya, 1973). By measuring HBAP as well as BAP the seasonal changes of the phosphorus in SS and sediments were shown more clearly (Nakajima and Okubo, 2003).

In the biological method green algae such as *Selenastrum capricornutum* has been widely used as an algal growth potential (AGP) test to measure BAP (Cowen and Lee, 1976; Huettl *et al.*, 1979; Dorich *et al.*, 1980; Williams *et al.*, 1980; Ekholm and Krogerus, 1998). The chemical methods estimate the bioavailability of phosphorus by microorganisms indirectly, while the biological methods measure directly the algal growth potentials. However, the latter takes several weeks to get to the maximum growth of the algae. Bacteria have been also used for measurement of phosphorus in water using growth of *Pseudomonas fluorescens* (Lehtola et al., 1999) and DO consumption by respiration (Nakamoto, 1977) as well as phosphorus in soil using CO_2 production by respiration (Nordgren, 1992; Demetz and Insam, 1999). The periods of the bioassay using bacteria are within several days and rather shorter than the case of using algae. Good correlations were shown as for the relationship between chemical and biological methods for BAP measurement (Williams et al., 1980; Dorich et al., 1980, 1985; Butkus et al., 1988; Sharpley et al., 1991).

Nakamoto (1977) introduced MBOD-P (modified biochemical oxygen demand) showing AGP by DO consumption in BOD assay using media added organics and nutrients without phosphorus. We applied MBOD-P to BAP measurement in this study because AGP without phosphorus addition seemed to be strongly related to BAP in the sample. Firstly the Winkler method using a DO bottle was examined in the measurement but the range of detection was small. Hence respiratory meters that were used in biodegradation tests (Larson and Perry, 1981; OECD, 1993; Reuschenbach et al., 2003) were used and good results were obtained. The purpose of this study is to examine this new biological measurement method of BAP using bacterial respiration by applying it to river suspended solids (SS) and by comparing it with BAP measurement using the chemical method we proposed previously.

Materials and methods

Samples

Surface water samples were taken from five rivers in Shiga Prefecture entering Akanoi Bay in the southern part of Lake Biwa from January 2003 to May 2004 (96 samples), the Nakae River in Gifu Prefecture from April 2003 to June 2004 (41 samples) and seven points in the Yamato River in the southern part of Osaka from July to November 2003 (14 samples). The rivers entering Akanoi Bay and the Nakae River are located in paddy field and their water quality is influenced largely by agricultural cultivation, while Yamato River flows in an urban basin and the water quality was strongly influenced by sewerage and industrial wastewater as well as urban diffuse pollution. Samples were brought back to the laboratory and filtered through a glass fiber filter GS-25 (Advantec) to collect the suspended solids. In order to examine the characteristics of BAP in river SS, BAP in soil and sludge samples were also measured. Paddy soil samples were taken in the basin of the rivers entering Akanoi Bay (15 samples). Activated sludge samples were taken from experimental treatment plants in our laboratory (seven samples).

Measurement of B-BAP by bacterial respiration

Measurement of B-BAP (biologically measured BAP) fundamentally followed Nakamoto (1977) while the DO bottle was exchanged to a commercial respiratory meter for the BOD test in order to measure a precise DO change and to obtain its wide dynamic range. The filter paper collecting SS was put in a 125 mL bottle and 80 mL of modified BOD seeding media without phosphorus (Table 1) were added in the bottle. After putting a seal cup filled with 0.8 g of soda lime, in order to adsorb CO_2, the bottle was capped and connected to a pressure sensor part of BOD Track (HACH). The bottle was stored in a dark place for 36 hours at 25 °C with mixing by a magnetic stirrer. The DO concentration in the water layer in the bottle was estimated by measuring the pressure in the upper gas layer in the bottle and the data were stored in a memory. The DO consumption data were plotted according to the time and oxygen uptake rate (OUR: $mgO_2/L/hr$) was determined

Table 1 Media composition used for B-BAP measurement

Compound	Concentration (mg/L)
NH_4Cl	170
$MgSO_4 \cdot 7H_2O$	22.5
$CaCl_2$	27.5
$FeCl_3 \cdot 6H_2O$	0.25
Glucose	400
$NaHCO_3$	168
Seeding bacteria*	1 capsule/2×10^4 L

*Polyseed-US (Console)

by the change of the DO consumption. The B-BAP concentration of a sample was estimated by the calibration curve using the OUR of the standard PO_4-P solutions. B-BAP content (mgP/g) was then calculated using SS of the sample.

Measurement of HBAP and C-BAP by acid and alkali extraction

HBAP and C-BAP (chemically measured BAP) were extracted from the samples fundamentally according to Pearson (1940) as shown in Figure 1 (Nakajima and Okubo, 2003). The filter paper collecting SS (usually after drying and measuring its weight to estimate the SS value of the sample) was put in a 140 mL PTFE bottle. 50 mL of 0.1 mol l^{-1} HCl was added to the bottle and after shaking for 3 minutes the acid extract was obtained by filtration (Advantek No.5C). HBAP was determined as PO_4-P in the acid extract after neutralization. The residue of the filtration with the filter paper was put into the bottle again and 100 mL of 0.1 mol l^{-1} NaOH was added to the bottle. After shaking for 17 hrs the alkali extract was obtained by filtration. NaOH-P was determined as PO_4-P in the alkali extract after neutralization. C-BAP was obtained as the sum of HBAP and NaOH-P. HBAP and C-BAP contents (mgP/g) were then calculated using SS of the sample. Total phosphorus and dissolved phosphorus of the sample water were also measured and particulate phosphorus content (PP: total phosphorus in SS) was determined by their difference as shown in Figure 1.

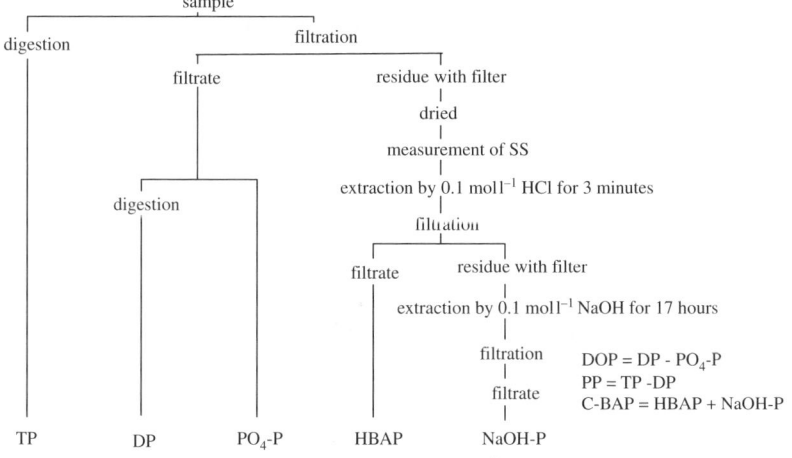

Figure 1 Measurement procedure of HBAP, C-BAP and PP

Results and discussion

Performance of B-BAP measurement method

The optimum experimental conditions such as incubation temperature, effect of equipment sterilization, concentration of nitrogen and glucose, kind of buffer solution and seeding as well as treatment method of filter paper were obtained by results of preliminary experiments. An example of the change in DO consumption using PO_4-P standard solution is shown in Figure 2. Usually a lag phase of 12 h was observed. DO consumption increased after that and then OUR was determined by its slope. An example of the calibration curve is shown in Figure 3. In this case, OUR was saturated at a concentration of 300 μgP/L. The range of determination was from 5 μgP/L to 250 μgP/L as the concentration of the solution in the bottle. The range was developed by increasing the concentrations of NH_4Cl and glucose in the media. The coefficient of variation of this method applied to river SS was 10% showing that the repeatability of the determination was good enough as a bioassay.

This method is carried out in less than 2 days, which is rather short compared to BAP bioassay using the growth of *Selenastrum capricornutum*. Moreover it seems reasonable that the phosphorus fraction that bacteria can utilize in a short period must also be easily utilized by algae directly or after bacterial utilization. Therefore B-BAP measured by bacterial respiration in this method will be useful as a rapid and direct measurement of phosphorus bioavailability.

BAP contents in river SS

The average values and the standard deviations of PP, HBAP, C-BAP and B-BAP contents of river SS in the three areas are shown in Table 2. PP was high in the Nakae River because of high C-BAP (or high NaOH-P) while HBAP and B-BAP were high in the Yamato River. The average percentages of the phosphorus fractions to PP are shown in Figures 4 and 5. The percentages of HBAP and B-BAP were small in the rivers entering Akanoi Bay and the Nakae River and large in the Yamato River. On the contrary the percentage of NaOH-P or C-BAP was large in the rivers entering Akanoi Bay and the Nakae River and small in the Yamato River.

River SS mainly consists of clay, silt or sand as well as organic particles such as microorganisms or organic detritus. The paddy soil was measured as a typical example of clay, silt or sand and the activated sludge was measured as a typical example of microorganisms or organic detritus. The average percentages of phosphorus fractions in paddy soil and activated sludge samples are also shown in Figures 4 and 5. The paddy soil had a large percentage in C-BAP and a small percentage in B-BAP, while the activated

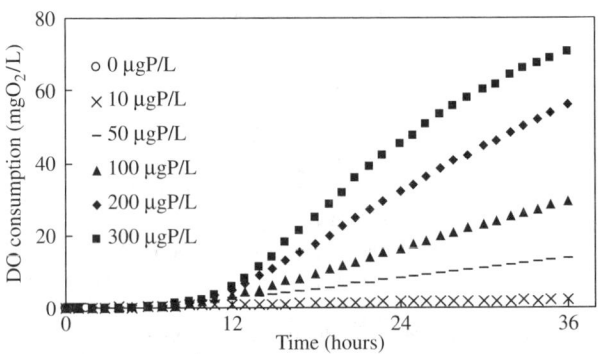

Figure 2 Change of DO consumption after blank correction

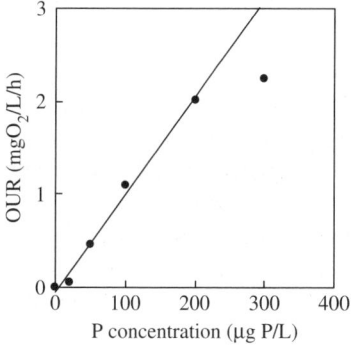

Figure 3 Calibration curve

Table 2 BAP contents in SS of three river groups

	n	Average ± standard deviation (mgP/g)			
		PP	HBAP	C-BAP	B-BAP
Rs. Akanoi	96	8.65 ± 3.29	1.37 ± 0.86	5.89 ± 2.75	1.29 ± 0.59
R. Nakae	41	20.8 ± 12.4	1.13 ± 0.79	18.7 ± 12.4	2.29 ± 1.33
R. Yamato	14	8.19 ± 1.71	2.14 ± 0.59	3.59 ± 0.80	2.78 ± 1.01

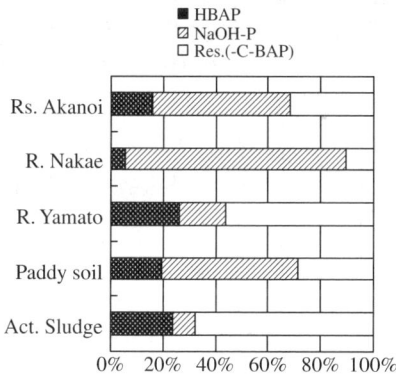

Figure 4 Percentages of HBAP and NaOH-P to PP

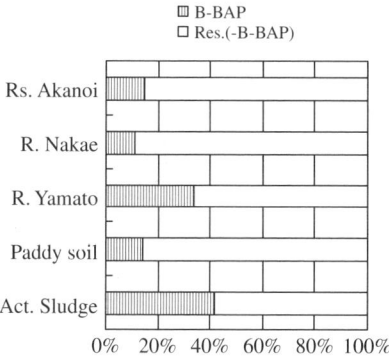

Figure 5 Percentages of B-BAP to PP

sludge had large percentages in B-BAP as well as HBAP and a small percentage in NaOH-P.

The percentages in the rivers entering Akanoi Bay and the Nakae River were similar to the percentages in the paddy soil, while the percentages in the Yamato River were similar to the percentages in the activated sludge. This suggests that the SS in the rivers entering Akanoi Bay and the Nakae River mainly consists of clay, silt or sand, while the SS in the Yamato River consists of a large percentage of organic particles as well. This seems reasonable because the former rivers flow in an agricultural area and the latter river flows in an urban area. Moreover the results suggest that phosphorus in SS is more easily bioavailable in urban rivers than rivers in agricultural areas.

Correlations between B-BAP and C-BAP

The relationships between B-BAP and C-BAP as well as HBAP are shown in Figure 6. HBAP and C-BAP increased with the increase of B-BAP in all cases, although their ratio was different in each river area. In the rivers entering Akanoi Bay and the Nakae River the difference between HBAP and C-BAP was large, while the difference was small in the Yamato River. The ratio of C-BAP/B-BAP was large in the rivers entering Akanoi Bay and the Nakae River and small in the Yamato River. The ratio of HBAP/B-BAP was around 1 in all cases. This means HBAP contents were almost similar to B-BAP contents in the river SS.

The relationships between B-BAP and C-BAP in the paddy soil and activated sludge samples are shown in Figure 7. The correlations were clear in the paddy soils between B-BAP and C-BAP as well as HBAP. The ratio of C-BAD/B-BAP was large like SS in the rivers entering Akanoi Bay and the Nakae River. On the contrary the ratio is small in the activated sludge like SS in the Yamato River. These results also suggest SS in the

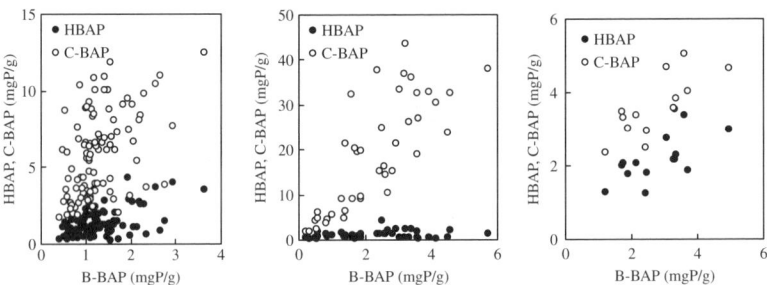

Figure 6 Relationships between B-BAP and C-BAP in the rivers entering Akanoi Bay (left), the Nakae River (center) and the Yamato River (right)

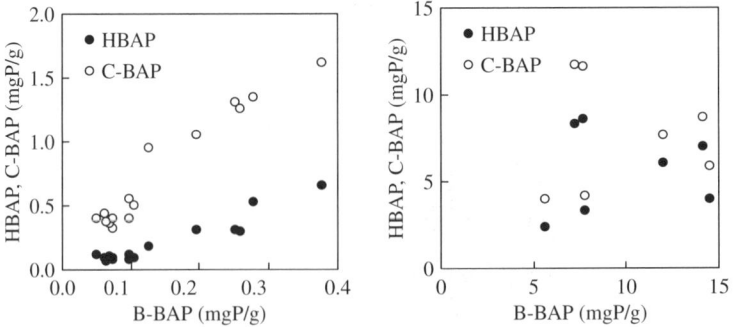

Figure 7 Relationships between B-BAP and C-BAP in paddy soil (left) and activated sludge (right)

agricultural area and urban area largely consists of soil and organic particulate, respectively.

The result that HBAP contents were similar to B-BAP contents supports the proposal by Nakajima and Okubo (2003) that HBAP is a suitable index of phosphorus indicating easily and rapidly bioavailable fraction. HBAP is able to be measured very simply and the sensitivity of the detection can be improved by filtration of a large volume of sample water. HBAP in river SS is also a part of reactive phosphorus (R-P) in water that can be dissolved from SS by acid reagent addition.

Conclusions

(1) A new biological BAP measurement method using bacteria respiration was introduced to rapid measurement of easily bioavailable phosphorus in river SS.
(2) The percentages of HBAP and B-BAP were large in the urban river, while the percentage of NaOH-P or C-BAP was large in the rivers flowing in the agricultural area.
(3) SS in the rivers flowing in agricultural areas mainly consists of clay, silt or sand, while the SS in the urban river consists of a large percentage of organic particles as well.
(4) Phosphorus in SS was suggested to be more easily bioavailable in the urban river than the rivers in the agricultural area.
(5) The ratio of C-BAP/B-BAP was large in the rivers in the agricultural areas and small in the urban river.
(6) HBAP contents were almost similar to B-BAP contents in the river SS. Therefore HBAP is a suitable index of phosphorus indicating easily and rapidly the bioavailable fraction in SS.

Acknowledgements

The authors thank the Living Environment Section in Wanouchi Town Office for their cooperation.

References

Butkus, S.R., Welch, E.B., Horner, R.R. and Spyridakis, (1988). Lake response modeling using biologically available phosphorus. *J. Water Pollut. Control. Fed.*, **60**, 1663.
Chang, S.C. and Jackson, M.L. (1957). Fractionation of soil phosphorus. *Soil Sci.*, **84**, 133–144.
Cowen, W.F. and Lee, G.F. (1976). Phosphorus availability in particulate materials transported by urban runoff. *J. WPCF*, **48**(3), 580–591.
Demetz, M. and Insam, H. (1999). Phosphorus availability in a forest soil determined with respirometory compared to chemical methods. *Geoderma*, **89**, 259–271.
Dorich, R.A., Nelson, D.W. and Sommers, L.E. (1980). Algal availability of sediment phosphorus in drainage water of Black Creek Watershed. *J. Environ. Qual.*, **9**(4), 557–563.
Dorich, R.A., Nelson, D.W. and Sommers, L.E. (1985). Estimating Algal available phosphorus in suspended sediments by chemical extraction. *J. Environ. Qual.*, **14**(3), 400–405.
Ekholm, P. and Krogerus, K. (1998). Bioavailability of phosphorus in purified municipal wastewaters. *Water Research*, **32**(2), 343–351.
Huettl, P.J., Wendt, R.C. and Corey, R.B. (1979). Prediction of algal-available phosphorus in runoff suspensions. *J. Environ. Qual.*, **8**(1), 130–132.
Larson, R.J. and Perry, R.L. (1981). Use of the electrolytic respirometer to measure biodegradation in natural waters. *Wat. Res.*, **15**, 697–702.
Lehtola, M.J., Miettinen, I.T., Vartianen, T. and Martikainen, P.J. (1999). A new sensitive bioassay for determination of microbially available phosphorus in water. *Appl. Environ. Microbiol.*, **65**, 2032–2034.

Nakajima, J. and Okubo, T. (2003). BAP measurement using successive extraction by acid and alkali. In *Proceedings of 7th International Specialised Conference on Diffuse Pollution and Basin Management*, pp. 13, 16–22.

Nakamoto, T. (1977). Use of heterotrophic activity as a measure of algal growth potential of waters. *Journal of Water and Waste*, **19**(6), 87–94.

Nordgren, A. (1992). A method for determining microbially available N and P in an organic soil. *Biol. Fertil. Soils*, **13**, 195–199.

OECD (1993). OECD 301 A: DOC Die-Away Test; OECD 301 B: CO_2 Evolution Test; OECD 301 C: Modified MITI Test(|); OECD 301 D: Closed Bottle Test; OECD 301 E: Modified OECD Screening Test; OECD 301 F: Manometric Respirometry Test, *OECD Guidelines for testing of chemicals*.

Okubo, T. (1996). Bioavailability of particle phosphorus. *Journal of Water and Waste*, **38**(3), 228–240.

Parker, M. (1991). Relations among NaOH-extractable phosphorus, suspended solids, and ortho-phosphorus in streams of Wyoming. *J. Environ. Qual.*, **20**, 271–278.

Pearson, R.W. (1940). Determination of organic phosphorus in soils. *Industrial and Engineering Chemistry Analytical Ed.*, **12**, 198–200.

Reuschenbach, P., Pagga, U. and Strotmann, U. (2003). A critical comparison of respirometric biodegradation tests based on OECD 301 and related test methods. *Wat. Res.*, **37**, 1571–1582.

Sekiya, K. (1973). Fractionation of inorganic phosphate. In *Analysis Method of Soil Nutrient*, Yokendo, Tokyo, pp. 235–238.

Sharpley, A.N., Troeger, W.W. and Smith, S.J. (1991). The measurement of bioavailable phosphorus in agricultural runoff. *J. Environ. Qual.*, **20**, 235–238.

Sharpley, A.N., Smith, S.J., Jones, O.R., Berg, W.A. and Coleman, G.A. (1992). The transport of bioavailable phosphorus in agricultural runoff. *J. Environ. Qual.*, **21**, 30–35.

Williams, J.D.H., Syers, J.K. and Walker, T.W. (1967). Fractionation of soil inorganic phosphate by a modification of Chang and Jackson's procedure. *Soil Sci. Soc. Amer.Proc.*, **31**, 736–739.

Williams, J.D.H., Shear, H. and Thomas, R.L. (1980). Availability to Scenedesmus quadricauda of different forms of phosphorus in sedimentary materials from the Great Lakes. *Limnol. Oceanogr.*, **25**(1), 1–11.

Evaluation of atmospheric deposition of nitrogen to the Feitsui Reservoir in Taipei

S.L. Lo and H.A. Chu

Graduate Institute of Environmental Engineering, National Taiwan University 71 Chou-Shan Rd., Taipei, 106, Chinese Taiwan (E-mail: sllo@ccms.ntu.edu.tw)

Abstract This research studied how the air pollutants of urban areas affect a neighboring reservoir and its water quality. Through the atmospheric dispersion process, air pollutants move from the Taipei metropolitan to the Feitsui reservoir and enter the water body through dry and wet depositions. ISCST3 (Industrial Source Complex Short Term Model), an air quality model, was used to simulate dispersion, dry deposition and wet deposition of the air pollutants. Then the nitrogen loadings to the Feitsui Reservoir were evaluated. The results indicate that wet deposition places a greater burden than dry deposition does on the water body. Wet and dry deposition of NH_4^+ together make up a rather large proportion of the total pollution. The ratio ranged from 21.9 to 25.2%. Those of nitrate make up a smaller proportion, ranged from 2.0 to 2.3%. If we take indirect deposition into account and calculate the NO_3^- and NH_4^+ together, the proportion is 15.9–17.6%.
Keywords Atmospheric deposition; dry deposition; nitrogen loading; wet deposition

Introduction

Although the average annual rainfall in Taiwan (2,500 mm/yr) is approximately 2.5 times that of the global average, the precipitation index per capita per year is one-sixth lower than that of the world average because of a heavy population density. Natural factors that make water management in Taiwan difficult include steep river slopes, fragile topsoil in watersheds, and significantly uneven temporal and spatial distribution of precipitation. According to hydrological statistics, the rainfall from May to October accounts for 78% of the total annual precipitation. The precipitation from November to April, which is the so-called dry season, accounts for 22% only. In southern Taiwan, the rainfall during the wet season is as high as 90% of the total annual precipitation. Such an uneven rainfall distribution pattern creates difficulties in efforts to utilize water resources.

Since population and economic activity are mostly concentrated in the northern and southern ends of western Taiwan, water shortages occur in these areas when the supply cannot meet the demand. During the dry season, nearly all of the water in the rivers is utilized, and the rivers in the southwest are almost dry except during the typhoon season. Thus, the construction of reservoirs along river systems has long been considered an important approach to fulfill the demands of fresh water supply. Many reservoirs in Taiwan have been found either to have eutrophication problems or are in danger of becoming eutrophic. Extensive highland farming, road construction, and community development destroy the vegetation cover of watershed areas. Excessive use of pesticides and fertilizers and the seasonal use of river banks or beds as cultivating grounds bring nutrients and other pollutants to the streams and reservoirs. In addition, air pollutants can be transported to the water body of the reservoir through the atmospheric dispersion process, and dry and wet depositions (Wu *et al.*, 1992; Leeuw *et al.*, 2003). Most of the reservoirs in Taiwan are public water supply sources. Protecting the quality of these water bodies is very important to maintain a sufficient drinking water supply as well as to ensure continuing social and economic development.

Trophic status of the Feitsui Reservoir

Lush vegetation, a high concentration of salts, and high turbidity are the defining characteristics of eutrophication. When reservoirs are polluted with large amounts of nitrogen and phosphorus, however, rapid growth of algae and eutrophication occur, thereby causing the water quality to deteriorate and increasing the cost of water treatment. The trophic state index is often based on total phosphorus (TP) concentration, total nitrogen (TN) concentration, chlorophyll a (chl a) concentration, and Secchi disk depth (SD). Since eutrophication involves complex changes in the water, the results obtained from using only one parameter may easily mislead or bias the user. For this reason, the multivariable trophic state indexing methods were developed. The most commonly used multivariable indices are the Carlson (1977) and Morihiro et al. (1981) indices. The Environmental Protection Administration (EPA, Chinese Taiwan) uses the Carlson trophic state index (CTSI) to conduct an overall assessment of water quality in all reservoirs, but also emphasizes that both algae cell density and TN are two other important parameters which must be monitored.

The Feitsui Reservoir (Figure 1) is the main water supply source in Taipei. Its watershed area covers a total area of 30,300 hectares. Based on a survey carried out by the EPA recently, land use areas that have a larger environmental impact within this watershed include farmland, market space, and exposed ground, comprising a total of 2,018 hectares. According to the Taipei Feitsui Reservoir Administration, monitoring data from 1987 to 2002 show that CTSI values rose from 40.8 in 1993 to 46.3 in 1998 (Table 1). Thereafter, from 1999 to 2002, CTSI remained around 46.0. A CTSI of 50 or more indicates that the water body has reached a state of eutrophication. Based on monthly measurements of water quality over the last five years (Figure 2), CTSI has already exceeded 50 on nine occasions.

Recently, Taipei County's Pinglin Township residents held a local referendum vote in favor of building a highway interchange in Pinglin. The issue has raised people's concerns about the water quality within this water quality protection area. The area of Pinglin Township is over 17,000 hectares, or 59% of the total watershed area of the Feitsui reservoir. The EPA therefore warns against further development and construction within

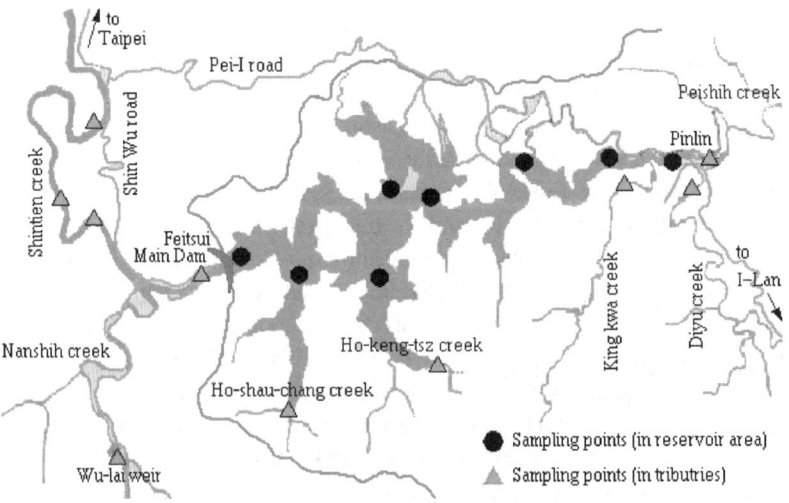

Figure 1 Layout of water quality sampling points of the Feitsui Reservoir

Table 1 Water quality and Carlson trophic state index of the Feitsui Reservoir (by average)

Year	Total P (µg/L)	Chl a (µg/L)	SD (m)	CTSI	Algae No. (cells/mL)	TN (mg/L)
1987	20.57	2.07	3.0	42.84	1,687	1.05
1988	20.16	7.28	3.2	45.00	15,189	1.12
1989	12.12	4.99	3.5	41.63	2,171	0.94
1990	10.53	4.85	2.9	42.92	3,303	0.88
1991	9.07	5.00	3.1	41.63	4,525	0.80
1992	8.88	7.00	3.3	41.97	7,887	0.86
1993	8.03	4.40	3.1	40.83	8,155	0.90
1994	9.93	4.80	3.4	41.70	4,889	0.98
1995	11.67	4.10	3.8	41.31	39,705	0.78
1996	16.84	5.16	2.8	44.41	37,218	0.87
1997	13.88	3.67	2.5	43.40	20,082	1.18
1998	14.89	6.18	2.2	46.32	28,520	1.31
1999	20.13	4.53	2.3	45.99	26,977	1.19
2000	21.91	3.70	2.9	45.53	55,085	1.00
2001	27.97	2.16	2.4	46.37	42,844	1.01
2002	33.61	1.27	2.0	46.11	51,680	1.08

the vicinity of the Feitsui Reservoir watershed area in order to prevent further pollution damage to the water source quality.

There are two main types of pollution that affect the water quality of reservoir watersheds. As for point source pollution, sewage systems are available to only 40% of residential and recreational areas due to the rural and mountainous characteristics of this region, and therefore such systems are unable to effectively solve point source pollution. As for non-point source pollution from farmland and forestland, if easy access was provided for motor vehicles, it can be foreseen that the level of traffic to Pinglin would greatly increase, followed by an increasing demand for recreational orchards, villas, and campgrounds. If such a situation is not appropriately managed, human activities would lead to increased levels of nitrogen and phosphorus and cause even more severe eutrophication. At the same time, alterations to the surrounding terrain and lay of the land would result in more exposed ground and overuse of the land. Heavy rainfalls would wash away soil and in severe instances would lead to severe soil erosion. The EPA therefore advises that any development should be preceded by the presentation of a complete set of

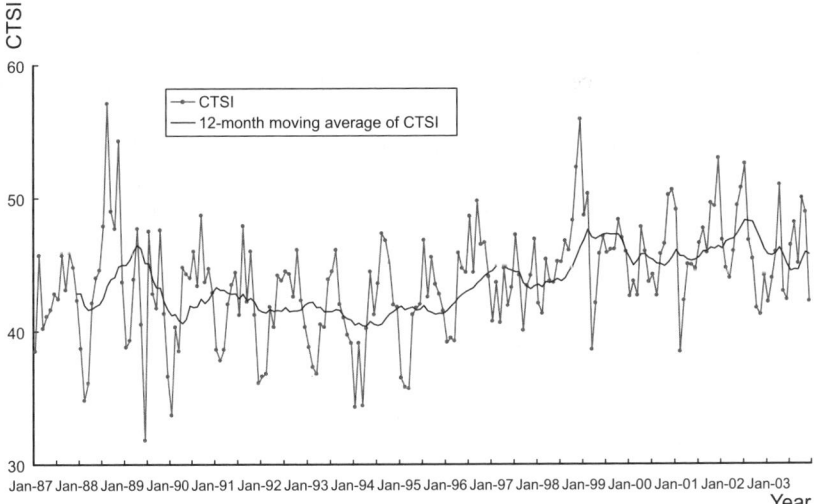

Figure 2 The CTSI of the Feitsui Reservoir from 1987 to 2003

management measures, including aspects of land use so as to ensure that water source areas have a safe quality and quantity of water.

Methods

The ISCST3 (Industrial Source Complex Short Term) model was used in this study. The effects of air pollutants of the Taipei metropolitan (including Taipei City, Taipei County and Keelung City) on the water quality of the Feitsui Reservoir were estimated through the atmospheric dispersion process, dry and wet depositions. Basic data needed in the ISCST3 model include: pollutant sources data (point, area and volume sources of SO_x, NO_x, Pb and TSP), meteorological data (wind velocity, wind direction, temperature, stability, friction velocity and Monin–Obukov length), receptor data (2 km × 2 km receptor grid points, 28 sensitive receptors), and topographical data (grid size 1 km × 1 km). The geocoding function of Arc View was used to input the topographical data.

The dry deposition of particulates was calculated by the Acid Deposition and Oxidant Model (ADOM, Pleim *et al.*, 1984).

$$F(x,y)_d = C(x,y) \times V_d \tag{1}$$

where $F(x, y)_d$ is the dry deposition flux (g/cm^2.s), $C(x, y)$ is the pollutant concentration (g/cm^3), and V_d is the dry deposition velocity of particulates (cm/s). The dry deposition velocity can be calculated from the gravitational settling velocity (V_g), aerodynamic resistance (R_a) and deposition-layer resistance (R_d).

$$V_d = V_g + 1/(R_a + R_d + R_a R_d V_g) \tag{2}$$

The gravitational settling velocity (V_g) is the function of the diameter of particulate (μm), shape of particulate and its density (g/cm^3). The aerodynamic resistance (R_a) is the function of wind velocity, stability and surface roughness length (cm). All influence factors mentioned above must be input into the ISCST3 model to calculate the dry deposition.

The wet deposition of gases and particulates can also be estimated by the ISCST3 model. The scavenging coefficient approach is used in the ISCST3 model to estimate the wet deposition.

$$F(x,y)_w = \int_0^\infty \Lambda C(x,y,z)dz \tag{3}$$

where $F(x, y)_w$ is the wet deposition flux (g/cm^2.s), Λ is the scavenging coefficient, and $C(x, y, z)$ is the pollutant concentration (g/cm^3). The scavenging coefficients of particulates with different diameters must be input into the ISCST3 model. The scavenging coefficient is the function of the particulate radius (Jindal and Heinold, 1991).

There are five major input parts of model parameters in ISCST3: CO (job control), SO (source), RE (receptor), ME (meteorology) and OU (output). The options of the CO input file includes default values, pollutants (SO_2, NO_x, PM10, TSP), urban or rural dispersion model, calculating concentrations, dry deposition (DDEP) or wet deposition (WDEP), dry depletion, wet depletion, flat or elevated topography. The concentrations and dry and wet depositions of particulates can be calculated in the ISCST3 model. But for gases, only concentrations and wet deposition can be obtained. SO is the input file pollution source, including location, elevation, altitude and emission rate of pollution source. The units of point and area pollution sources are g/s and g/s.m^2, respectively. The scavenging coefficient of particulate during rainfall (PARTLIQ) was selected from Jindal and Heinold's results (1991), while it was set as $0.0\,s^{-1}$ during snowfall (PARTICE). For gases, the scavenging coefficients of SO_2 and NO_x were selected as $1.22 \times 10^{-5}\,s^{-1}$ (the average value of in-clouds and below-clouds) and $7.95 \times 10^{-11}\,s^{-1}$ (the average value of NOs and NO_2s) (Hertel et al., 1995). Schwede and Paumier (1996) studied sensitivity of the ISCST3 model to input deposition parameters. Their results are quite useful to choose the input parameters.

Results and discussion

The mass fraction (%) of suspended particulates with different diameters in northern Taiwan are shown in Table 2 (Chiang, 1997). The average mass fraction data were used to input the MASSFRAC values. Figures 3, 4 and 5 show the comparison results between monitored values and simulated results for total suspended particulate (TSP), SO_2 and NO_2, respectively, in 1994 and 1996. Excluding the extreme values of the ratio of monitored value to simulated results for each group, the average ratios for TSP, SO_2 and NO_x are 1.6, 1.2 and 0.56 in 1994, and 1.58, 0.86 and 1.02 in 1996, respectively. The modeling errors are acceptable. Because Xindian (station 4) is the nearby station to the watershed of the Feitsui Reservoir, its ratios of TSP, SO_2 and NO_2 (Table 3) were used as the correcting coefficients for all modeling concentration results of receptors to calculate dry and wet depositions.

Some extreme ratios in Figures 3–5, such as TSP at station 3 (15.47 and 10.59), SO_2 at station 2 (2.40 and 2.69) and NO_2 at station 1 (2.61 and 2.77), result from their locations too near the northern seashore. Lacking in pollution source data from sea area results in the modeling results underestimated at these three monitoring stations.

Dry deposition comes from the settling of TSP. The direct dry deposition means atmospheric pollutants precipitating directly onto the water surface of the Feitsui Reservoir. The precipitation of particulates onto the watershed of reservoir causes the indirect deposition scoured by the surface runoff (assume 20% of pollutants will enter the water body). The average dry depositions of TSP upon the water surface of the Feitsui Reservoir are 8.64 and 6.27 g/m². yr (Table 4) calculated by ISCST3 model in 1994 and 1996, respectively.

For the watershed of the Feitsui Reservoir, the average dry depositions of TSP are 6.45 and 5.05 g/m².yr. The mass fractions of SO_4^{2-}, NO_3^- and NH_4^+ within particulates are 8.0%, 5.3% and 3.0% by chemical constituent analysis. The total surface area of the Feitsui Reservoir is 10.24 km², and the total watershed area is 303 km². Thus, the direct and indirect dry depositions of SO_4^{2-}, NO_3^- and NH_4^+ can be obtained by the TSP deposition values, mass fractions, M/S ratios and surface areas (as shown in Table 5).

Wet deposition derives from the scavenging of TSP, SO_2 and NO_x by rainfall. The average wet depositions of TSP, SO_2 and NO_x directly upon the water surface and upon watershed are shown in Table 4. Through a similar calculating procedure, the direct and indirect wet depositions of SO_4^{2-}, NO_3^- and NH_4^+ can be obtained, as shown in Table 5.

Table 2 Mass fraction of suspended particulates in northern Taiwan (%)

Monitoring station	Period	Particulate diameter (μm)		
		≥10	2.5–10	<2.5
Fushing	1994.9–1995.4	69.10	17.24	13.66
	1995.8–1996.3	41.00	31.86	27.14
	1996.9–1997.3	50.00	23.00	27.00
Sanchong	1994.9–1995.4	64.60	18.76	16.64
	1995.8–1996.3	41.00	29.50	29.50
	1996.9–1997.3	64.00	7.92	28.08
Average		54.95	21.38	23.67

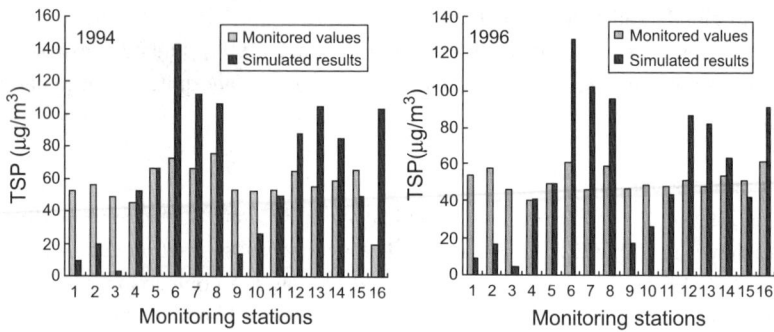

Figure 3 The comparison of monitored values and simulated results for TSP

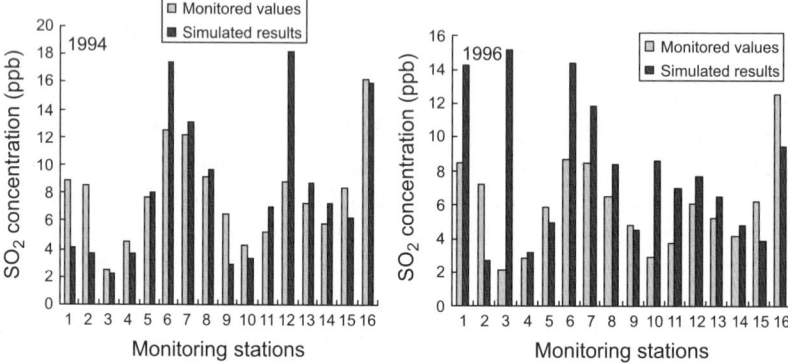

Figure 4 The comparison of monitored values and simulated results for SO_2

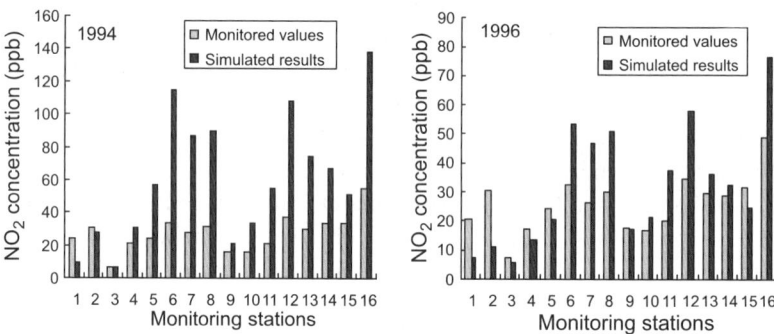

Figure 5 The comparison of monitored values and simulated results for NO_2

Table 3 Monitored and simulated results for Xindian station

Pollutant	Year	Monitored value	Simulated result	M/S ratio
TSP ($\mu g/m^3$)	1994	44.84	52.47	0.85
	1996	40.20	41.27	0.97
SO_2 (ppb)	1994	4.46	3.64	1.23
	1996	2.80	3.18	0.88
NO_2 (ppb)	1994	21.66	31.11	0.70
	1996	17.40	13.89	1.25

Table 4 The average dry deposition and wet deposition calculated by ISCST3

Type	Year	Upon water surface (Direct) (g/m².yr)	Upon watershed (Indirect) (g/m².yr)
Dry deposition			
TSP	1994	8.64	6.45
	1996	6.27	5.05
Wet deposition			
TSP	1994	12.91	8.83
	1996	10.15	7.12
SO_2	1994	0.15	0.13
	1996	0.17	0.22
NO_x	1994	0.00	0.00
	1996	0.00	0.00

Table 5 Dry and wet depositions onto the Feitsui Reservoir

Pollutant	Deposition	Year	Direct deposition (ton/yr)	Indirect deposition (ton/yr)	D/I ratio
SO_4^{2-}	Dry	1994	6.0	26.6	0.23
		1996	5.0	23.7	0.21
	Wet	1994	9.0	36.4	0.25
		1996	8.1	33.5	0.24
	Wet (from SO_2)	1994	2.8	14.8	0.19
		1996	2.3	17.6	0.13
NO_3^-	Dry	1994	4.0	17.6	0.23
		1996	3.3	15.7	0.21
	Wet	1994	6.0	24.1	0.25
		1996	5.3	22.1	0.24
NH_4^+	Dry	1994	2.3	10.0	0.23
		1996	1.9	8.9	0.21
	wet	1994	3.4	13.6	0.25
		1996	3.0	12.6	0.24

The ratios of direct depositions to indirect depositions are near 22%. The wet deposition places a greater burden than dry deposition does on the water body.

According to the report of Taipei Feitsui Reservoir Administration, the annual loadings of SO_4^{2-}, NO_3^- and NH_4^+ are 8,942.4, 436.4 and 22.2 ton/yr, respectively. The results of direct depositions indicate that wet and dry depositions of NH_4^+ together make up a rather large proportion of the total pollution loadings. The ratio is 25.2% in 1994 and 21.9% in 1996. Those of nitrate make up a smaller proportion, 2.3% in 1994 and 2.0% in 1996. Those of sulfate make up a rather low proportion (0.2%). If we take indirect deposition into account and calculate the NO_3^- and NH_4^+ together, the proportion is 15.9–17.6%. Beddig et al. (1997) reported that the atmospheric contribution to the nitrogen load of the German Bight during 1989–1992 was estimated to be about 30%.

Conclusions

The water quality of the Feitsui Reservoir is gradually deteriorating. Up to now, only 40% of point source pollution can be controlled by the sewage systems. Consequently, non-point source pollution control within its watershed is urgently needed. This article shows that 15.9–17.6% of NO_3^- and NH_4^+ of the Feitsui Reservoir derives from atmospheric deposition. Efforts should begin as soon as possible for basic data collection and formulating an institutional framework for implementation of control strategies.

References

Beddig, S., Brockmann, U., Dannecker, W., Korner, D., Pohlman, T., Plus, W., Radach, G., Robers, A., Rick, H.J., Schatzmann, M., Schluenzen, H. and Schultz, M. (1997). Nitrogen fluxes in the German Bight. *Marine Pollution Bulletin.*, **34**(6), 294–382.

Carlson, R.E. (1977). A trophic state index for lakes. *Limnol. Oceanog.*, **22**(2), 361–369.

Chiang, P.C. (1997). *Air Pollution Problems and Control Strategies of Suspended Particulates in Taiwan.* National Taiwan University, Chinese Taiwan, EPA-86-FA42-09-19.

Hertel, O., Christensen, J. and Runge, E.H. (1995). Development and testing of a new variable scale air pollution model–ACDEP. *Atmospheric Environment*, **29**(11), 1267–1290.

Jindal, M. and Heinold, D. (1991). Development of particulate scavenging coefficients to model wet deposition from industrial combustion sources. In *84th Annual Meeting Exhibition of AWMA*, Vancouver, BC, June 16–21.

Leeuw, G., Skjoth, C.A., Hertel, O., Jickells, T., Spokes, L., Vignati, E., Frohn, L., Frydendall, J., Schulz, M., Tamm, S., Sorensen, L.L. and Kunz, G.J. (2003). Deposition of nitrogen into the North Sea. *Atmospheric Environment*, **37**(S1), S145–S165.

Morihiro, A., Outoski, A., Kawai, T., Hosome, M. and Muraoka, K. (1981). Application of modified Carlson's trophic state index to Japanese lakes and its relationship to other parameters related to trophic state. *Research Report Natl. Inst. Environ. Stud.*, **23**, 12–30.

Pleim, J., Venkatram, A. and Yamartino, R. (1984). *ADOM/TADAP Model Development Program*, Vol. 4. The Dry Deposition Module, ERT Document P-B980-520, USA.

Schwede, D.B. and Paumier, J.O. (1996). Sensitivity of the industrial source complex model to input deposition parameters. *Journal of Applied Meteorology*, **36**, 1096–1106.

Wu, Y.L., Davidson, C.I., Dolske, D.A. and Sherwood, S.I. (1992). Dry deposition of atmospheric contaminants: The relative importance of aerodynamic, boundary layer, and surface resistances. *Aerosol Sci. Tech.*, **16**, 65–81.

Keyword Index

agricultural nonpoint source pollution 1
agriculture 289, 303
airborne particulate matter 185
AnnAGNPS 263
annual loading 33
atmospheric concentration 215
atmospheric deposition 53, 79, 337

bacterial respiration 329
best management practice 225, 263
best management practices 175
bioavailable phosphorus 329
biomarker 93
bog 63
BTEX 321

Ca/Mg ratio 111
Caltrans 225
CALUX assay 11
challenge 1
China 1
chitosan 155
coagulation 155
compost 111
correlation analysis 235

delivery coefficient 271
diffuse pollution 271
dioxins 11
dry deposition 337

effective 147
electric conductivity 203
eutrophication 119, 329

farmland 111, 119
first flush 225, 245
first flush runoff 193
fish toxicity 155
forest 79

GIS 23
GPC-TC analysis 193
groundwater 321

half-life 131
heavy metal accumulation 111
herbicide 131
herbicide concentrations 139

highway runoff 245
highway stormwater runoff 235
highways 225
hillslope 93
hydrology 263, 303

indicator 303
inductively coupled plasma mass spectrometry (ICP-MS) 185
infiltration capacity 163
interflow 93
irrigation water management 139

lagoons 131
Lake Biwa 23, 131
lake water management 329
land use change 45
land-use change 63
leaching 289
lead isotope ratio 185
litter 225
long term monitoring 53
L-Q equation 311
L-Q-Turbidity method 311

Macro Model 23
management 1
mass balance model 163
mass first flush 235
metal 203
methanogenesis 321
mineral nutrient 111
Miyun Reservoir 1
MNA 321
model development 253
MONERIS 281

N removal rate constant 101
natural attenuation 321
newly constructed reservoir 73
nitrate 289
nitrogen 53, 79, 119, 147
nitrogen loading 337
non-point pollution 281
non-point source pollution 263
nutrient 203
nutrient load 311
nutrient loads 79

organic substance 193
outflow load 147
oxygen uptake rate 329

paddy field 101, 119
paddy fields 139
paddy management 253
paddy water 139
PAH content 245
PAH profiles 245
PAHs 215
particulate matter 215
pesticide runoff 139
petroleum-hydrocarbon 321
phosphate analysis 329
phosphorus 53, 73, 79, 119, 281, 303
plant diversity 63
pollution load runoff 271
polycyclic aromatic hydrocarbons 215
prevention of eutrophication 73

quinone 93

rainfall-runoff 93
recycling irrigation system 101
recycling ratio 101
refractory 193
regional properties 33
regression model 33
replacement cost method 147
retention coefficient 271
river sediment 11
river-mouth sediment 11
road runoff 185
road surface runoff 163
run-off 119
runoff characteristics 73
runoff rate 139

sediment 73, 175, 263
sediment quality guidelines 175

sedimentation 155
shallow irrigation 253
soil 193, 289, 303
soil erosion 45
soil particle 73
soil penetration facility 163
soil purification mechanisms 163
spline technique 23
storm characteristics 235
storm events 33
storm runoff 79
storm sewer 203
stormwater 225
stormwater management 175
stormwater runoff 193
sugarcane production 45
surface soil 11
suspended solids 245
SWAT 281
system dynamic model 45

time-series model 163
total deposition 215
trace metals 175
transfer factor 111
turbidity 203, 311
turbidity removal 155

urban runoff 203

water chemistry 63
water contamination 131, 303
water level 63
water quality 33, 45, 253
water quality distribution 23
watershed modeling 281
watershed property 11
weir height 253
wet deposition 337

Zala River catchment 281